Phospholipid Metabolism in Apoptosis

Subcellular Biochemistry

Volume 36

SUBCELLULAR BIOCHEMISTRY

Recent Volumes in This Series

Volume 26	***myo*-Inositol Phosphates, Phosphoinositides, and Signal Transduction** Edited by B. B. Biswas and Susweta Biswas
Volume 27	**Biology of the Lysosome** Edited by John B. Lloyd and Robert W. Mason
Volume 28	**Cholesterol: Its Functions and Metabolism in Biology and Medicine** Edited by Robert Bittman
Volume 29	**Plant±Microbe Interactions** Edited by B. B. Biswas and H. K. Das
Volume 30	**Fat-Soluble Vitamins** Edited by Peter J. Quinn and Valerian E. Kagan
Volume 31	**Intermediate Filaments** Edited by Harald Herrmann and J. Robin Harris
Volume 32	**α-Gal and Anti-Gal: α1,3-Galactosyltransferase, α-Gal Epitopes, and the Natural Anti-Gal Antibody** Edited by Uri Galili and José Luis Avila
Volume 33	**Bacterial Invasion into Eukaryotic Cells** Edited by Tobias A. Oelschlaeger and Jörg Hacker
Volume 34	**Fusion of Biological Membranes and Related Problems** Edited by Herwig Hilderson and Stephen Fuller
Volume 35	**Enzyme-Catalyzed Electron and Radical Transfer** Edited by Andreas Holzenburg and Nigel S. Scrutton
Volume 36	**Phospholipid Metabolism in Apoptosis** Edited by Peter J. Quinn and Valerian E. Kagan

---

A Continuation Order Plan is available for this series. A continuation order will bring delivery of each new volume immediately upon publication. Volumes are billed only upon actual shipment. For further information please contact the publisher.

Phospholipid Metabolism in Apoptosis

Subcellular Biochemistry Volume 36

Edited by

Peter J. Quinn
Kings College London
London, United Kingdom

and

Valerian E. Kagan
University of Pittsburgh
Pittsburgh, Pennsylvania

Kluwer Academic / Plenum Publishers
New York, Boston, Dordrecht, London, Moscow

Library of Congress Cataloging-in-Publication Data

Phospholipid metabolism in apoptosis/edited by Peter J. Quinn and Valerian E. Kagan.
p. cm. – (Sub-cellular biochemistry, ISSN 0306-0225; v. 36)
"This series is a continuation of the journal Sub-cellular biochemistry, volumes 1±4 of which were published quarterly from 1972 to 1975."
Includes bibliographical references and index.
ISBN 0-306-46782-8
1. Apoptosis–Molecular aspects. 2. Lipid membranes. I. Kagan, Valerian E. II. Sub-cellular biochemistry. III. Series.

QH671 .P466 2002
571.6–dc21

2002023678

ISSN 0306-0225

ISBN 0-306-46782-8

This series is a continuation of the journal *Sub-Cellular Biochemistry*,
Volumes 1 to 4 of which were published quarterly from 1972 to 1975

233 Spring Street, New York, New York 10013

http://www.wkap.com

10 9 8 7 6 5 4 3 2 1

A C.I.P. record for this book is available from the Library of Congress

Printed in the United States of America

Contributors

Hervé Benoist, Inserm U466, Institut Louis Bugnard, CHU Rangueil, 31403 Toulouse Cedex 4, France

Garry R. Buettner, Department Radiology (Free Radical and Radiation Biology Graduate Program), The University of Iowa College of Medicine, and The University of Iowa Holden Comprehensive Cancer Center, Iowa City, IA 52242, USA

C. Patrick Burns, Professor of Medicine, University of Iowa, Department of Medicine, University Hospitals, Iowa City, IA 52242, USA

Raffaele D' Amelio, Seconda Facoltà di Medicina e Chirurgia, Università degli studi "La Sapienza", S. Andrea Hospital, Rome, Italy

Samer El Bawab, Department of Biochemistry and Molecular Biology; Medical University of South Carolina, 114 Doughty Street, Charleston, SC 29425, USA

Kazuo Emoto, Department of Molecular Biodynamics, The Tokyo Metropolitan Institute of Medical Science, 3-18-22 Honkomagome, Bunkyo-ku, Tokyo 113-8613, Japan

Cristiano Ferlini, Laboratory of Antineoplastic Pharmacology, Department of Obstetrics and Gynaecology, Università Cattolica Sacro Cuore, L.go A. Gemelli 1, 00168 Rome, Italy

Erich Gulbins, Department of Molecular Biology, University of Essen, Hufelandstrasse 55, 45147 Essen, Germany

Yasuf A. Hannun, Department of Biochemistry, Medical University of South Carolina, PO Box 250509, 173 Ashley Avenue, Charleston, SC 29425, USA

Martin Houweling, Department of Biochemistry and Cell Biology, Faculty of Veterinary Medicine, University of Utrecht, Yalelaan 2, 3584 CM Utrecht, The Netherlands

Jean-Pierre Jaffrézou, INSERM E9910, Institut Claudius Régaud, 20 rue du Pont St Pierre, 31052 Toulouse, France

V. E. Kagan, Department of Environmental and Occupational Health, University of Pittsburgh, Pennsylvania 15260, USA

Katarina Kågedal, Division of Pathology II, Faculty of Health Sciences, Linköping University, Linköping, Sweden

Eric E. Kelley, Departments of Medicine and Radiology (Free Radical and Radiation Biology Graduate Program), The University of Iowa College of Medicine, and The University of Iowa Holden Comprehensive Cancer Center, Iowa City, IA 52242, USA

Richard Kolesnick, Laboratory of Signal Transduction, Memorial Sloan Kettering Cancer Center, New York, NY 10021, USA

Sadaaki Komura, Institute of Applied Biochemistry, Yagi Memorial Park, Mitake, Giu, Japan

Guy Laurent, INSERM E9910, Institut Claudius Régaud, 20 rue du Pont St Pierre 31052 Toulouse France and INSERM U466, CHU Rangueil, 31403 Toulouse France

Volker Lehmann, German Cancer Research Center (DKFZ), Department of Immunochemistry (G0200), Im Neuenheimer Feld 280, 69120 Heidelberg, Germany

Thierry Levade, INSERM E9910, Institut Claudius Régaud, 20 rue du Pont St Pierre 31052 Toulouse France and INSERM U466, CHU Rangueil, 31403 Toulouse, France

S. X. Liu, Department of Environmental and Occupational Health, University of Pittsburgh, Pennsylvania 15238, USA

Cungui Mao, Department of Medicine, Medical University of South Carolina, 114 Doughty Street, Charleston, SC 29425, USA

Anne Nègre-Salvayre, Inserm U466, Institut Louis Bugnard, CHU Rangueil, 31403 Toulouse Cedex 4, France

Lina M. Obeid, Department of Medicine, Medical University of South Carolina, 114 Doughty Street, Charleston, SC 29425, USA

Nobuko Ohishi, Institute of Applied Biochemistry, Yagi Memorial Park, Mitake, Giu, Japan

Karin Öllinger, Division of Pathology II, Faculty of Health Sciences, Linköping University, Linköping 581 85, Sweden

Susan Pyne, Department of Physiology and Pharmacology, Strathclyde Institute for Biomedical Sciences, University of Strathclyde, Glasgow, G4 0NR, Scotland, UK

Peter J. Quinn, Division of Life Sciences, King's College London, 150 Stamford Street, London SE1 9NN, UK

Alessandra Rufini, Department of Experimental Medicine and Biochemical Science, University of Rome "Tor Vergata", Italy

Robert Salvayre, Inserm U466, Institut Louis Bugnard, CHU Rangueil, 31403 Toulouse Cedex 4, France

Gabriella Sarmay, Department of Immunology, Loránd Eötvös University, Pázmány Péter Sétány 1/c, H-1117 Budapest, Hungary

Giovanni Scambia, Laboratory of Antineoplastic Pharmacology, Dept. Of Obstetrics and Gynaecology, Università Cattolica Sacro Cuore, L.go A. Gemelli 1, 00168 Rome, Italy

N. F. Schor, Pediatrics and Neurology, University of Pittsburgh, Pennsylvania 15238, USA

Vladimir Shatrov, German Cancer Research Center (DKFZ), Department of Immunochemistry (G0200), Im Neuenheimer Feld 280, 69120 Heidelberg, Germany

Yoshihiro Shidoji, Cellular Biochemistry Section, Department of Nutrition and Health Sciences, Siebold University of Nagasaki, Nagayo, Nagasaki 851-2195, Japan

A. A. Shvedova, Health Effects Laboratory Division, Pathology and Physiology Research Branch, NIOSH, Morgantown, West Virginia 26505, USA

C. A. Smith, Departments of Environmental and Occupational Health, University of Pittsburgh, Pennsylvania 15238, USA

Roberto Testi, Department of Experimental Medicine and Biochemical Science, University of Rome "Tor Vergata", Italy

V. A. Tyurin, Institute of Evolutionary Physiology and Biochemistry, St. Petersburg, Russia

Y. Y. Tyurina, Institute of Evolutionary Physiology and Biochemistry, St. Petersburg, Russia

Masato Umeda, Department of Molecular Biodynamics, The Tokyo Metropolitan Institute of Medical Science, 3-18-22 Honkomagome, Bunkyo-ku, Tokyo 113-8613, Japan

Arie B. Vaandrager, Department of Biochemistry and Cell Biology, Faculty of Veterinary Medicine, University of Utrecht, P.O. Box 80176, 3508 TD Utrecht, The Netherlands

Brett A. Wagner, Department of Medicine, The University of Iowa College of Medicine, and The University of Iowa Holden Comprehensive Cancer Center, Iowa City, IA 52242, USA

Kunio Yagi, Institute of Applied Biochemistry, Yagi Memorial Park, Mitake, Giu, Japan

Preface

The last few years have witnessed an explosion of both interest and knowledge about apoptosis, the process by which a cell actively commits suicide. The number of publications on the topic has increased from nothing in the early 1980s to more than 10,000 papers annually today. It is now well recognized that apoptosis is essential in many aspects of normal development and is required for maintaining tissue homeostasis. The idea that life requires death seems somewhat paradoxical, but cell suicide is essential for an animal to survive. For example, without selective destruction of "non-self" T cells, an animal would lack immunity. Similarly, meaningful neural connections in the brain are whittled from a mass of cells. Further, developmental cell remodeling during tissue maturation involves programmed cell death as the major mechanism for functional and structural safe transition of undifferentiated cells to more specialized counterparts. Apoptosis research, with roots in biochemistry, developmental and cell biology, genetics, and immunology, embraces this long-ignored natural law.

Failure to properly regulate apoptosis can have catastrophic consequences. Cancer and many diseases (AIDS, Alzheimer's disease, Parkinson's disease, heart attack, stroke, etc.) are thought to arise from deregulation of apoptosis. As apoptosis emerges as a key biological regulatory mechanism, it has become harder and harder to keep up with new developments in this field.

The steps of apoptosis are distinct from those of necrosis which is a nonprogramed form of cell death in response to injury, in which the cell swells and bursts, causing inflammation. Apoptosing cells form a tight sphere and their membranes undulate, resulting in bulges that form blebs. The nuclear membrane breaks, and endonucleases clip chromosomes where the DNA is exposed from protective proteins. This occurs at 180-base intervals, so the DNA pieces are all the same size. Then the cell fragments, with enough membrane to sequester toxic cell contents and so prevent inflammation at the site. Finally, nearby cells consume the remains.

Little is known about the genetic control of apoptosis. Most apoptosis genes under investigation turn the process on or off. "Apoptosis on" genes include ced-3 and ced-4 in *C. elegans*, and ICE and p53 in mammals. The gene ced-4 encodes a novel protein, but ced-3 is a homologue of ICE (interleukin-1β converting enzyme). p53 encodes a transcription factor and is common in many human cancers and mediates cellular responses to some types of environmental damage. The p53 protein either temporarily halts cell division so the cell can repair altered DNA, or sends the cell to an apoptotic death. The bcl-2 gene is an "apoptosis off" control and is the mammalian equivalent of ced-9, a worm anti-death gene. The proto-oncogene is an apoptosis "brake" in the skin. The gene is expressed in the basal layers, where stem cells must divide to supply more cells. In the upper layers, lack of bcl-2 protein permits apoptosis, preventing tumor formation. The bcl-2 protein binds a protein called bax. The bcl-2/bax ratio appears to be critical to a cell's fate. If bcl-2 is in excess, all available bax is bound, apoptosis is blocked, and the cell lives. If bax is in excess, all bcl-2 is bound, the brake is released, and the cell dies. Certain viruses also have apoptosis brake genes which keeps the cell it infects from committing suicide. Such genes are found in adenovirus, Epstein-Barr virus, African swine flu virus, and vaccinia.

The molecular mechanisms of apoptosis are presently unknown and the subject of currently focused research effort. It is clear that cell membrane structure and properties play an early part in the induction process. There is increasing evidence that the arrangement of polar lipids in the membrane lipid matrix is an important factor coupled with the homeostatic mechanisms responsible for preserving membrane lipid composition and asymmetry. Changes in membrane permeability are also likely to be involved, possibly as a direct consequence of disturbances in the lipid bilayer matrix. In addition, phospholipids such as lysophospholipids, can contribute to regulation of apoptosis acting as antiapoptotic factors. Another important participation of phospholipids in apoptosis is via the association of anionic phospholipids with the plasma protein β 2-glycoprotein I (β 2-GPI) to form a complex that is recognized by many antiphospholipid autoantibodies - the process typical of patients with both primary antiphospholipid syndrome and systemic lupus erythematosus.

This volume addresses the role of membrane lipids in early events of apoptosis. In particular, the role of phospholipids in mitochondrial permeability, membrane lipid asymmetry, sphingolipid and phospholipid signaling processes in early apoptotic events are reviewed by current researchers in these fields. The organelles of central importance to apoptosis are mitochondria and the plasma membrane with supporting roles played by lysosomes and the Golgi. In the first two chapters mitochondria are considered as stress sensors that harbor apoptogenic proteins. The release of cytochrome C and initiation of the caspase cascade is considered in the light of cardiolipin metabolism and particularly the susceptibility of this polyunsaturated phospholipid to oxidation by reactive oxygen species. The role of an antioxidant mitochondrial enzyme, PHGPx, is postulated to protect cardiolipin from oxidative damage and prevent cytochrome C release into the cytoplasm.

The molecular events responsible for generating the asymmetric arrangement of phospholipids across the plasma membrane is described in one chapter which relates this to membrane lipid homeostasis and apoptosis. Another chapter focuses on the biosynthesis of phosphatidylserine and its role in apoptosis. Oxidative stress is manifest in alterations in a number of pathways involving membrane lipid oxidation and culminating in cell death. The disturbance of phosphatidylserine distribution in the plasma membrane resulting from oxidation leading to the creation of recognition sites for macrophages is the subject of one chapter and the role of •NO in oxidation of polyunsaturated membrane lipids is discussed in another. The action of oxidized lipids of low-density lipoproteins and ionizing radiation in triggering apoptosis in cells and the importance of redox-cycling quinones in maintaining an appropriate oxidant-antioxidant balance are subjects of separate chapters.

Sphingolipids and their metabolites are now recognized as important signaling molecules and their levels in cells are regulated to control a variety of physiological processes including apoptosis. The action of ceramidases in cleaving the N-acyl linkage of ceramide to form free fatty acid and sphingosine, which can be subsequently phosphorylated to form sphingosine-1-phosphate, are discussed in the context of regulation of apoptosis, cell growth, migration and proliferation. The action of ceramides in regulating glycerophospholipid biosynthesis and the generation of apoptotic signaling molecules are topics addressed in several chapters. The concluding chapters examine the role of lipid signaling in CD95-mediated apoptosis and signals originating from phosphoinosides in cell cycling.

Editing this volume has emphasized to us how much progress has been made over recent years in understanding apoptosis and the role of phospholipid metabolism in the process. Knowledge itself, however, also reveals how much is yet to be discovered before the detailed molecular mechanisms responsible for initiating and regulating apoptosis are fully understood. We are confident that this work will provide an impetus to extend our knowledge of this fundamental cellular process. We join to thank all the contributors for their invaluable assistance and cooperation in the production of this volume.

Peter Quinn
London

Valerian Kagan
Pittsburgh

Contents

Chapter 1

Lipid Metabolism and Release of Cytochrome c from Mitochondria

Volker Lehmann and Vladimir Shatrov
German Cancer Research Center, Heidelberg, Germany

1. INTRODUCTION

During the process of cell death, cytochrome c is released from mitochondria into the cytosol where it assists to activate the caspases, a family of killer caspases that trigger cell death. In this chapter evidence that transmission of cell death signals into the release of cytochrome c involves phospholipids at several stages will be presented. Thus, phospholipids target proapoptotic proteins to mitochondria, enable these or other proteins to form channels or pores and to break the mitochondrial permeability barrier. Finally, peroxidation or increased levels of Ca^{2+} disrupt the ability of cardiolipin to interact with cytochrome c, initiating a sequence of events that ultimately lead to cell death.

2. EVOLUTION OF A STABLE PARTNERSHIP

According to one favored theory, mitochondria evolved from aerobic bacteria that were endocytosed by an anaerobic ancestor of eukaryotic cells. The endocytic event is thought to have occurred when oxygen emerged in the atmosphere (about two billion years ago) and threatened most life systems. Under the selective pressure of oxygen, a stable relationship developed in which the host cell had acquired the ability to exploit the bacterial oxidative phosphorylation system for their own use (Green and

Subcellular Biochemistry, Volume 36: Phospholipid Metabolism in Apoptosis.
Edited by Quinn and Kagan. Kluwer Academic/Plenum Publishers, New York, 2002

Reed, 1998; Margulis, 1996). In the beginning, this relationship was probably a dangerous one. For instance, protomitochondria, the early descendants of endosymbiotic bacteria, generated reactive oxygen species (ROS) as a by-product of oxidative phosphorylation. Conditions that permitted extensive production of these ROS could easily destroy the host. To avoid such conflicts, many activities of the protomitochondria were placed under the control of the host genome. Essential genes of the protomitochondria were transferred to the nuclear genome (Gray *et al.*, 1999; Poyton and McEwen, 1996). Most of today's mitochondrial proteins are nuclear gene products (Figure 1). Such proteins are synthesized in the cytoplasm and then imported to mitochondria (Neupert, 1997). One of these nuclear proteins is cytochrome c, a protein with a double life: when it is attached to the inner mitochondrial membrane, it assists the production of the life sustaining ATP. However, when it detaches and escapes into the cytosol, it triggers cell death (Reed, 1997).

Today's mitochondria lack most of the genes involved in phospholipid metabolism. Therefore, mitochondria have to import most of their lipids. Phospholipids such as phosphatidylcholine, phosphatidylserine, phosphatidylglycerol, and phosphatidylinositol must be synthesized in the endoplasmatic reticulum under the control of nuclear genes and then transferred to mitochondria (Voelker, 2000) (Figure 1). Mitochondria use both nuclear and mitochondrial encoded proteins to further diversify phospholipids (Dowhan, 1997; Kent, 1995; Daum, 1985). Thus, a nuclear phosphatidylserine decarboxylase converts phosphatidylserine into phosphatidylethanolamine, or mitochondrial encoded cardiolipin synthase converts phosphatidylglycerol into cardiolipin which is incorporated exclusively into the inner mitochondrial membrane.

3. MITOCHONDRIA: STRESS SENSOR AND HOME OF APOPTOGENIC PROTEINS

Mitochondria are the center of energy production and the key regulators of cell death. Each mitochondria is bounded by two highly specialized membranes, the outer and inner membrane, which enclose the intermembrane space. The latter is home of several apoptogenic proteins (Brenner and Kroemer, 2000; Von Ahsen *et al.*, 2000; Alnemri, 1999; Green, 1998). One principal member of these proteins is cytochrome c. Also present are AIF (Susin *et al.*, 1999a), Smac (Du *et al.*, 2000) or DIABLO (Chai *et al.*, 2000) and several procaspases, including procaspases-2,-3 and 9 (Susin *et al.*, 1999b). The release of these proteins into the cytosol can be triggered by diverse stimuli which cause cell damage, metabolic perturbations, or by growth factor deprivation (Reed, 1997; Vander Heiden *et al.*, 1997). Most of the responses converge on mitochondria through the activation of pro-apoptotic members of the Bcl-2 family, including Bak, Bad, Bax and Bid (Figure 2) (Antonsson and Martinou, 2000; Tudor *et al.*, 2000; Vander Heiden and Thompson, 1999; Adams and Cory, 1998; Wang *et al.*, 1996). These molecules form a pool of inactive proteins in the cytosol.

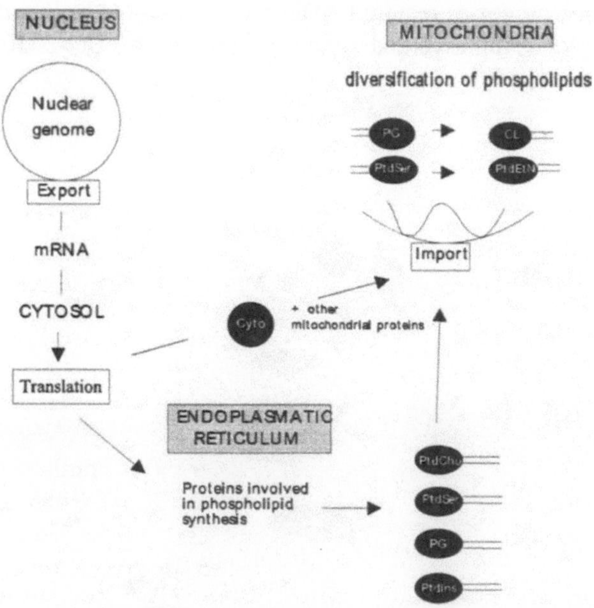

Figure 1. Control of mitochondrial biogenesis by the nuclear genome. Most mitochondrial proteins, including cytochrome c, are nuclear gene products which are subsequently imported into mitochondria. Similarly, most enzymes involved in synthesis of mitochondrial phosphoplipids are encoded in the nuclear genome. Being located in the endoplasmatic reticulum, they synthesize phosphatidylcholine (PtdCho), phosphatidylserine (PtdSer), phosphatidylglycerol (PG) and phosphatidylinositol (Ptdins). The phospholipids are transferred to the outer membrane. The imported lipids then move into the inner membrane at contact sites. Mitochondria then diversify phospholipids. They decarboxylate phosphatidylserine to phosphatidylethanolamine (PtdEtN), but the main reaction is the conversion of imported phosphatidylglycerol to cardiolipin (CL). Cardiolipins localize mainly in the outer leaflet of the inner membrane.

Proapototic signals direct these proteins to mitochondria where they compete with antiapoptotic members of the Bcl-2 family to regulate the cytochrome c release and to determine the fate of the cell: life or death (Cosulich *et al.*, 1999). Unlike proapoptotic proteins, the antiapoptotic Bcl-2 proteins reside in the outer mitochondrial membrane, anchored by a hydrophobic stretch of amino acids located at the COOH-termini,

(Antonsson and Marinou; 2000; Adams and Cory, 1998). Stress may also cause Ca^{2+} increase, nitric oxide (NO), or reactive oxygen species (ROS) production which, in turn, triggers release of apoptotic proteins from the intermembrane space (Kroemer and Reed, 2000; Vieira *et al.*, 2000). Release of these proteins from mitochondria are required for stress induced killing but are, with a few exceptions (Bergmann *et al.*, 1994, Schulze- Osthoff *et al.*, 1993), dispensible for CD95 and TNF-receptor transduced apoptosis. These other death processes require FADD and caspase-8 to be recruited into the death receptor complexes and cannot be blocked by Bcl-2 (Krammer, 2000; Scaffidi *et al.*, 1998).

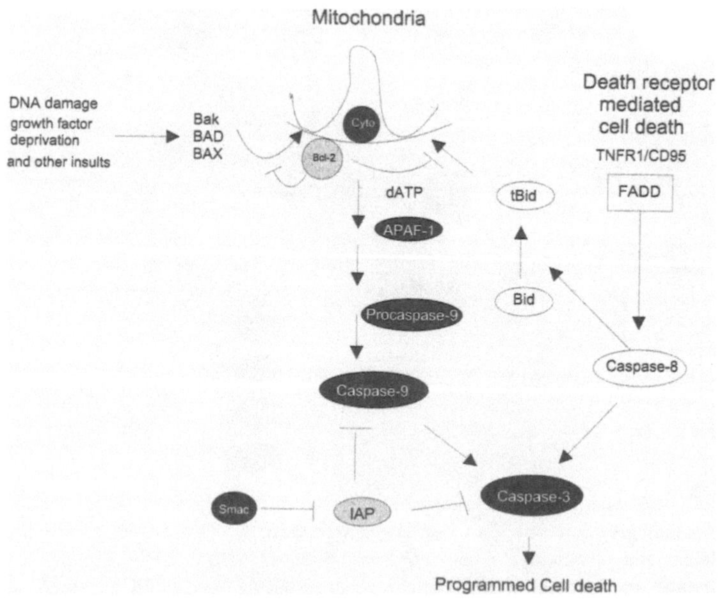

Figure 2. The routes to cell death. The signaling pathways of programmed cell death can be divided in two parts, involving either the mitochondria or death receptors. In the death receptor pathway receptors such as TNFR1 and CD95 are ligated with their cognate ligands. This event initiates the recruitment of a downstream signaling partner such as FADD and caspase-8. In the mitochondrial pathway cytochrome c (Cyto) is released from the intermembrane space of mitochondria upon activation by death signals. These signals may include proapoptotic members of the Bcl-2 family such as Bax, Bad, truncated Bid, or nonproteineous agents such as Ca^{2+} and NO. Antiapoptotic family members such as Bcl-2 inhibit the release of cytochrome c. Once released cytochrome c and dATP bind to Apaf-1 inducing it to associate with procaspase-9, thereby triggering caspase-9 and caspase-3 activation and initiating the proteolytic cascade which culminates in apoptosis. IAPs are regulators which inhibit effectors of apoptosis which, in turn, are regulated by Smac/DIABLO. Both death receptor and mitochondrial-mediated pathways can be bridged by Bid which is cleaved by caspase-8 to generate the proapoptotic tBid. It was postulated that

this cleavage involves the exposure of a hydrophobic surface that is targeted to the mitochondrial membrane. The hydrophobic surface includes a three helix domain of tBid (Lutter *et al.*, 2000). Cleavage also leads to exposure of the BH3 region, which mediates binding of tBid to the antiapoptotic Bcl-2 protein (Wang *et al.*, 1996).

A link between the pathways controlled by the TNF-receptor family members and the mitochondrial pathway has been identified. It has been shown that caspase-8 can cleave Bid, resulting in the formation of truncated Bid (tBid), a death-inducing member of the Bcl-2 family (Gross *et al.*, 1999; Li *et al.*, 1998; Luo *et al.*, 1998). Bid activation appears not to be essential in cells with high amounts of caspase-8 in the death receptor complex but may be required to amplify the cascade in cells with low amounts of caspase-8 in the death receptor complex.

4. CYTOCHROME C: LIFE SUPPORTER OR KILLER

Like the majority of mitochondrial proteins, apocytochrome c is synthesized in the cytosol and subsequently imported into mitochondria. Unlike most imported proteins, apocytochrome c does not possess a cleavable signal sequence. When it enters the intermembrane space, it is converted to holocytochrome c through addition of a heme group by the enzyme cytochrome c heme lyase (Mayer *et al.*, 1995). Once formed, holocytochrome c binds to the outer surface of the inner mitochondrial membrane where it shuttles electrons between complexes III and IV of the respiratory chain and contributes to the production of ATP. There it cannot be released into the cytosol as long the barrier function of the outer membrane remains intact. Only heme-binding holocytochrome c is competent to catalyse the activation of caspases. Once released in the cytosol, cytochrome c binds Apaf-1 (apoptotic protease-activating factor-1) (Li *et al.*, 1997) and in the presence of dATP or ATP recruits and activates procaspase 9 to form a complex designated apoptosome (Hengartner, 2000). Activated caspase-9 can, in turn, activate other caspases that ultimately destroy the cells (Thornberry and Lazebnik, 1998).

Cytochrome c is thought to be an essential component of the apoptopic pathway responsive to DNA damage and other forms of cellular stress. This notion is based on studies with cell lines from mice having the gene disrupted (Li *et al.*, 2000). These cell lines are refractory to apoptosis induced by stimuli that affect mitochondria.

Kinetic studies of cytochrome release during apoptosis in Hela cells (Goldstein *et al.*, 2000) indicate that, after an initial and variable lag period, all the cytochrome c is released within a short period of five minutes. This all-or-nothing effect suggests that the release of cytochrome c is part of a decision making process (Martinou *et al.*, 2000). However, cytochrome c release alone does not appear to be sufficient to trigger apoptosis. It has to act in concert with other pro-apoptotic proteins to ensure commitment of cells to death. Thus, Smac or DIABLO presumably follows the same exit route as cytochrome c from mitochondria into the cytosol and eliminates the inhibitory effect of IAPs (inhibitors of apoptosis.), a family of proteins which suppresses apoptosis by antagonizing caspase-3 activation (Deveraux and Reed, 1999; Miller, 1999). These results implied that loss of the barrier function of the outer membranes is the critical event responsible for the mitochondria-dependent caspase activation and cell death. There are also reports that suggest that extensive release of cytochrome c may trigger a slower nonapoptotic cell death resembling necrosis: it is independent of cytochrome c-mediated activation of downstream caspases and may be caused by a lack of ATP production, reflecting a blockade between respiratory complexes III and IV (Mootha *et al.*, 2001; Li *et al.*, 1999; Green and Reed, 1998).

5. PHOSPHOLIPIDS: PROPERTIES AND INTERACTIONS WITH CYTOCHROME C AND OTHER MITOCHONDRIAL PROTEINS

The primary role of phospholipids is to form a lipid bilayer which serves as permeability barrier and as matrix for a vast array of proteins. The properties of such bilayers is governed by the chemical structure of the lipids present. One characteristic property is the fluidity which depends on the melting transition temperature of the lipids (De Kruijff, 1997). Biomembranes contain some phospholipids, including phosphatidylethanolamine and cardiolipin which, in isolation, tend to form nonbilayer structures such as the reversed hexagonal state (H_{II} phase)(Cullis and de Kruijff, 1997). A dominant feature of these lipids is a small polar head group associated with a large hydrophobic domain. An obvious role of nonbilayer phospholipids would be to promote the formation of local, highly curved structures which affect the activities of membrane proteins or facilitate insertions of proteins into membranes (Dowhan, 1997). Interactions of membranes with proteins are often governed by the general properties of the molecules involved. For charged proteins and lipids, electrostatic interactions are usually dominant. Thus, utilizing liposome inhibition assays and reflection spectroscopy, it was shown that cytochrome c binds preferentially to monolayers of acidic phospholipids such phosphatidylserine, phosphatidylserine and cardiolipin (Demel, 1994; Demel *et al.*, 1989;

Kozarac *et al.*, 1988). A comparison between apocytochrome c and holocytochrome c revealed striking differences in phospholipid specificities. The unfolded structure of apocytochrome c showed a similar interaction for all anionic lipids. The initial electrostatic interaction is followed by penetration of the protein between the acyl chains of the lipid bilayer. Both properties are considered as prerequisite for the import of this protein into mitochondria. In contrast, the folded structure of holocytochrome c binds preferentially cardiolipin and exhibits a low potential to penetrate the lipid bilayer (Demel, 1994; Quinn and Dawson, 1969). Moreover, studies utilizing NMR and ESR spectroscopy (Heimburg *et al.*, 1991) revealed that anions at the surface of the phospholipid bilayer were able to induce conformational changes in cytochrome c. It was suggested that these changes are necessary for interaction with lipids and for opening of the heme cavity of the protein. It was also proposed that negatively charged polar head groups of deprotonated acidic phospholipids interact with Lys-72 and Lys-73 while protonated acidic phospholipids bind via hydogen bonding to Asn-52 at the site of opening of the cavity (Rytömaa and Kinnunen, 1995). The acyl chains of the lipid are pointed to opposite directions from the headgroup. One becomes accommodated within the hydrophobic cavity, while the rest of the acyl chains remain embedded in the membrane. All features of the interactions between cytochrome c and lipid are in accordance with the final localization of cytochrome c.

Cardiolipin forms also tight complexes, with the adenine nucleotide translocator (ATN) affecting its translocator activity (Beyer and Nuscher, 1996). Six cardiolipin residues are tightly bound to lysines (Beyer and Klingenberg, 1985). Removal of these lipids renders the translocator inactive, but activity can be reconstituted by adding cardiolipin. It has also a pivotal role as a boundary lipid of various proteins such as NADH: ubiquinone oxireductase (Hoch, 1992) or cytochrome c oxidase (Ushmorov *et al.*, 1999; Vik *et al.*, 1981).

6. TARGETING PROAPOPTOTIC TBID TO MITOCHONDRIA

Most of the responses which lead to cytochrome c release are triggered by activated proapoptotic members of the Bcl-2 family which are directed to mitochondria, where they insert into the outer membrane. What mediates the specific targeting of these proteins to mitochondria. In a recent work, Lutter *et al.* (2000) investigated this question using the proapoptotic Bcl-2 member (tBid) as a tool. These authors established a liposome model for cytochrome c release and recreated the lipid constitution of mitochondrial outer membrane and of contact sites between inner and outer membranes. They found that tBid bound very little to liposomes resembling the mitochondrial outer membrane but bound effectively to liposomes resembling the contact site. Binding to the contact site was comparable to that of mitochondria, indicating that binding of tBid to mitochondria can occur in the absence of proteins. In addition, liposomes containing individual phospholipids present in the contact zone were tested for their ability to bind tBid. It was found that tBid did not bind efficiently until cardiolipin reached physiological concentration (20% of the total). The binding was further augmented by the presence of phosphatidylethanolamine and matched the level of mitochondrial binding (55-60%). No binding of tBid to free cardiolipin occurred excluding a direct ligand receptor interaction. To study the mitochondrial targeting *in vivo*, the authors generated several constructs in which tBid was fused to green florescent protein (GFP). Truncation experiments using tBid as a template indicated that a three-helix domain in tBid consisting of helices H4, H5, H6 is necessary for targeting of tBid to mitochondria *in vitro* and is sufficient for targeting GFP to mitochondria *in vivo*. Finally, the authors established the importance of cardiolipin for mitochondrial targeting *in vivo*. They used a temperature-sensitive Chinese hamster overy cell line (PG cells), which is a conditionally defective phosphatidylglycerophosphate synthase. At the nonpermissive temperature at 40°C, it produced a reduced cardiolipin content, whereas other major lipids did not change. They showed that tBid is indeed unable to bind to mitochondria isolated from PGs cells kept at the nonpermissive temperature. Based on these findings the authors suggest a mechanism by which helices H4, H5 and H6 of tBid selectively bind to mitochondria. The three helices are thought to form a hydrophobic surface that is able to insert into the lipid bilayer. Since mitochondria contain cardiolipin and phosphatidylethanolamine which favor nonbilayer structures, it was postulated that unique membrane structures containing bilayer and the hexagonal H_{II} structures provide specificity for targeting tBid to mitochondria.

7. PROAPOPTOTIC BCL-2 MEMBERS, PHOSPHOLIPIDS AND THE CYTOCHROME C EXIT

The precise mechanism how proapoptotic Bcl-2 members regulate the exit of cytochrome c and how phospholipids participate in this process is still controversial. There are two prevailing models (Figure 3): In model 1 Bcl-2 members insert into the lipid bilayer of the mitochondrial outer membrane and form channels that facilitate exit of cytochrome c and other apoptogenic

proteins. This model is based on crystalography that revealed a structural similarity between members of the Bcl-2 family and the channel-forming domain of diphteria toxin and bacterial colicins. Thus, a structural comparison of the Bcl-2 member Bcl-xl and the pore-forming domain of diphtheria toxin revealed that both structures contain two central hydrophobic helices surrounded by amphipathic α helices (Fesik, 2000; Muchmore *et al.*, 1996). In addition, like these bacterial toxins, several members of the Bcl-2 family, including Bax and tBid, insert into lipid vesicles and planar lipid membranes, destabilize phospholipid bilayer.
(Basanez *et al.*, 1999), oligomerize and form channels (Kim *et al.*, 2000; Kudla *et al.*, 2000; Zhai *et al.*, 2000; Schendel *et al.*, 1999; Minn *et al.*, 1997).

These channels show multiconductance levels and are voltage and pH-dependent. Moreover, Saito *et al.* (2000) demonstrated that channels consisting of four Bax molecules are capable of transporting cytochrome c. Proapoptotic proteins may also act in concert with other proteins of the Bcl-2 family to form channels. For instance, Bax or tBid interact with Bak, a resident member of the Bcl-2 family to form channels (Korsmeyer *et al.*, 2000). TBid, following insertion and conformational change, exposes the BH3 domains that, in turn, induces an allosteric conformational activation of Bak which then oligomerize forming a channel that allows the passage of cytochrome c (Korsmeyer *et al.*, 2000; Adams and Cory, 1998). So far, there is only indirect support for this model: for instance addition of Bax, Bak and Bid to isolated mitochondria derived from liver or Hela cells results in permeabilization of the outer mitochondrial membrane, but the ultrastructure of mitochondria was preserved. Moreover, these mitochondria maintained their membrane potential and protein import capacity (von Ahsen *et al.*, 2000). In addition, Goldstein *et al.* (2000) monitored the release of cytochrome c from mitochondria in response to various stimuli employing Hela cells that stably express florescent protein (GFP)-tagged cytochrome c in mitochondria. They also found that the electron-transport chain can maintain the mitochondrial transmembrane potential even after cytochrome c has been released. In model 2 Bcl-2 members interact with other proteins resulting in rupture of the outer membrane (Martinou and Green, 2001; Constantini *et al.*, 2000; Bradham *et al.*, 1998). One candidate for such a protein is the voltage dependent anion channel (VDAC). This protein is a subunit of the mitochondrial permeabilty transition pore (PTP), a large pore which is proposed to span the inner and outer mitochondrial at sites of contact between the two membranes. Opening of the PTP, initiated by apoptotic Bcl-2 proteins such as Bax, results in changes of the mitochondrial physiology (Shimizu *et al.*, 1999; Zamzami *et al.*, 1995). These changes include an increase in permeability, a rapid loss of membrane potential and mitochondria to swelling which, in turn, increases the lateral stress within the lipid bilayer resulting ultimately in rupture of the outer mitochondrial membrane and cytochrome c exit. Involvement of the PTP is supported by observations that pharmacological inhibitors of PTP can act as potent inhibitors of cytochrome c release and hence prevent cell death (Loeffler and Kroemer, 2000).

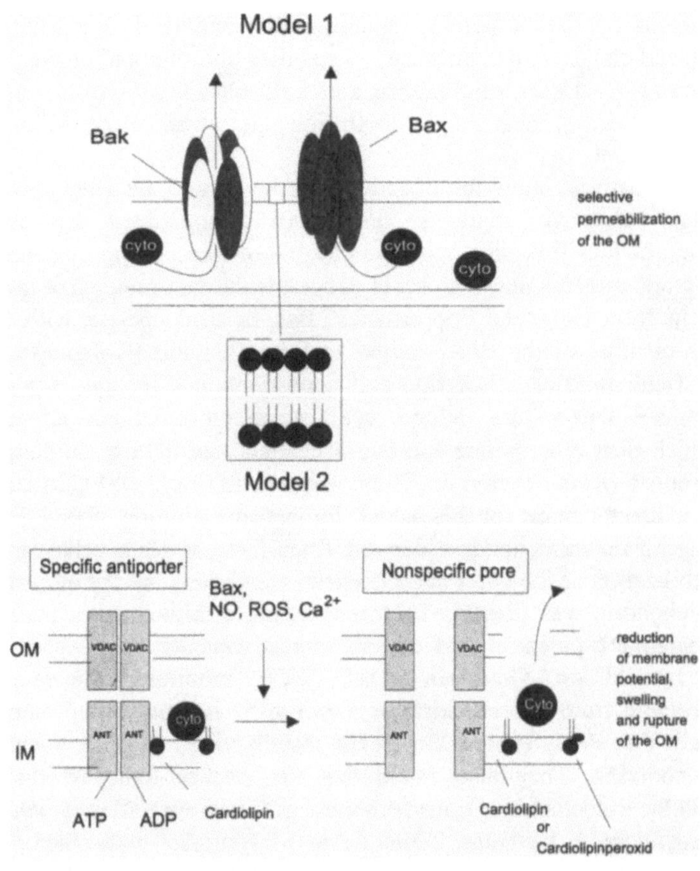

Figure 3. Possible mechanisms of actions of Bcl-2 members. Two prevailing models through which Bcl-2 members trigger cytochrome c release have been suggested. In both models phospholipids in the bilayer structure either individually and/or collectively induce a conformational change in Bcl-2 members, allowing them to insert into the outer mitochondrial membrane. In model 1 proapoptotic proteins destabilize the outer mitochondrial membrane, oligomerize and form channels through which cytochrome c and other proteins of the intermembrane space can escape.Bcl-2 proteins such as Bax or tBid act in concert with other proteins of the Bcl-2 family to form channels. In model 2 Bcl-2 members such as Bax interact with resident proteins in the outer membrane (OM) such as the voltage-dependent anion

channel (VDAC) or adenine nucleotide translocase (ANT) either to generate a pore for cytochrome c exit, or to cause a change of permeability which, in turn, leads to swelling of mitochondria, resulting in rupture of the outer membrane. Ca^{2+} has been suggested to act on ATN-associated cardiolipin.

8. DISRUPTION OF PHOSPHOLIPID/PROTEIN INTERACTIONS ELICITED BY PEROXIDATION AND CA^{2+}

Cytochrome c is well established to bind to cardiolipin which is abundant in the outer leaflet in the inner mitochondrial membrane (Kroemer and Reed, 2000; Vieira *et al.* 2000). This interaction is very tight, irreversible and stoichometrical (Rytömaa and Kinnunen, 1995), suggesting that cardiolipin may anchor cytochrome c to the inner membrane and enable it to fullfil its function in the respiratory chain. So, what molecular processes trigger the disruption of this interaction during stress-induced cell death? Shidoji *et al.* (1999) analyzed the molecular interactions between cytochrome c and cardiolipin by ^{1}H NMR-spectroscopy. They observed that unlike cardiolipin, their peroxides derivatives failed to interact with cytochrome c. This finding suggested that peroxidation of cardiolipin may induce release of cytochrome c from mitochondria. Ushmorov *et al.* (1999) demonstrated that the NO-induced cell death is indeed critically dependent on the peroxidation of cardiolipin. The authors observed that treatment of human leukemic cell lines with the NO donor glycerol trinitrate at concentrations which trigger cell death caused a decrease in the content of cardiolipin, a release of mitochondrial cytochrome c in the cytosol and activation of caspase-9 and caspase-3. These changes were accompanied by a downregulation in respiratory chain complex activities as well as by an increase of cells with a low mitochondrial membrane potential and with a high level of reactive oxygen species. The critical role of lipid degradation in NO-mediated cell death in APO-S Jurkat cells was established by experiments with the vitamin E analog trolox. It was found that this potent inhibitor of lipid peroxidation significantly inhibited NO-induced cell death and reduced the number of NO-treated Jurkat cells with low content of cardiolipin without any effect on the quantity of cells with low $\Delta\Psi m$ and high ROS production, indicating that lipid peroxidation was required for this type of cell death.

Nomura *et al.* (2000) showed that hypoglycaemia-induced apoptosis of basophile leukemia cells involves the generation of cardiolipin hydroperoxide and the subsequent release of cytochrome c from mitochondria. (Nomura *et al.*, 2000). The generation of cardiolipin hydroperoxide occurred in mitochondria of basophile leukaemia cells which had been induced to undergo apoptosis by exposure to hypoglycaemia with

2-deoxyglucose. The release was completely suppressed when the production of cardiolipin-OOH in mitochondria was inhibited by the overexpression of mitochondrial phospholipid hydroperoxide glutathione peroxidase (PHGPx). Mitochondrial PHGPx is localized in the intermembrane space, in particular at contact sites in which the PT pore locates and the detachment of cytochrome c occurs. Detachment of cytochrome c from the inner mitochondrial membrane following peroxidation of cardiolipin is an initial event in the exit of cytochrome c from mitochondria. However, the details of the mechanism by which cytochrome c leaves mitochondria remains to be resolved. There are reports that the adenine nucleotide translocator (ANT) appears to require cardiolipin for function (Beyer and Nuscher, 1996; Hoffmann *et al.*, 1994). ANT is a subunit of the mitochondrial permeability pore (PTP), and opening of PTP quickly leads to cytochrome c release and cell death. Accordingly, it was postulated that peroxidation of cardiolipin may induce a conformational change in ANT which opens the PT pores (Nomura *et al.*, 2000).

In addition, there is evidence that disruption of the interaction between ANT and cardiolipin may also operate in cell death induced by increased Ca^{2+} levels. Ca^{2+} is a classical PTP opener which triggers rupture of the outer mitochondrial membrane, allowing the release of cytochrome c and other proteins of the intermembrane space. In the presence of Ca^{2+} ANT can function as large nonspecific pores in both mitochondria and proteoliposomes containing cardiolipin and ANT. Since cardiolipin has a high affinity for Ca^{2+}, it was postulated that specific binding of Ca^{2+} to tightly associated cardiolipin disrupts its interactions with lysine residues on the ANT protein resulting in a conformational change that induces pore formation (Dowhan, 1997).

9. CONCLUSIONS

Mitochondria are now appreciated as being important in the control of cell survival and cell death. Most apoptotic signaling processes converge on the mitochondria, which releases cytochrome c, leading to the activation of effector caspases. The response of mitochondria to upstream proapoptotic signals is a critical control point for the regulation of cell death. It certainly involves phospholipids which provide specificity to target proapoptotic proteins to mitochondria and enable them to interact with Bcl-2 members or with other proteins to release cytochrome c. Despite these successes, we are mostly in the dark at the mechanism of how phospholipids regulate such interactions. The challenge in the future will be to explore how phospholipids control such interaction. This implies that more specific interactions between individual lipids and proteins associated with apoptosis have to be established and their molecular base to be identified. Given the importance of phospholipid/protein interactions for the maintenance of mitochondrial hoemostasis, it would be not surprising if they serve as useful targets for drug intervention. For instance anticancer drugs could be designed that prevent such interactions resulting in loss of mitochondrial integrity and ultimately in cell death.

REFERENCES

Adams, J.M., and Cory, S., 1998, The Bcl-2 protein family: arbiters of cell survivals, *Science* **281**: 1322-1326.

Alnemri, E.S., 1999, Hidden powers of the mitochondria, *Nature Cell Biology* **1**: E40-42.

Antonsson, B., and Martinou, J.C., 2000, The Bcl-2 protein family, *Exp. Cell Res.* **256**: 50-57.

Basanez, B., Nechushtan, A., Drozhinin, O., Chanturiya, A., Choe, E., Tutt, S., and Wood, K.A., 1999, Bax, but not Bcl-x_L decreases the lifetime of planar phospholipid bilayer membranes at subnanomolar concentrations, *Proc. Natl. Sci. USA* **96**: 5492-5497.

Bergmann, S., Shatrov, V., Ratter, F., Schiemann, S., Schulze-Osthoff, K., and Lehmann, V., 1994, Adenosine and homocysteine together enhance TNF-mediated cytotoxicity but do not alter activation of nuclear factor-κB in L929 cells, *J. Immunol.* **153**: 1736-1743.

Beyer, K., Klingenberg, M., 1985, ADP/ATP carrier protein from beef heart mitochondria has high amounts of tightly bound cardiolipin, as revealed by ^{31}P nuclear magnetic resonance, *Biochemistry* **24**: 3821-3826.

Beyer, K., and Nuscher, B., 1996, Specific cardiolipin binding interferes with labeling of sulfhydryl residues in the adenosine diphosphate/adenosine triphosphate carrier protein from beef heart mitochondria, *Biochemistry* **35**: 15784-15790.

Bradham, C.A., Qian, T., Streetz, K., Trautwein, C., Brenner, D.A., and Lemasters, J.J., 1998, The mitochondrial permeability transition is required for Tumor Necrosis Factor Alpha-mediated apoptosis and cytochrome c release, *Molecular and Cellular Biology* **18**: 6353-6364.

Brenner, C., and Kroemer, G., 2000, Mitochondria – the death signal integrators, *Science,* **289**: 1150-1151.

Chai, J., Du, C., Wu, J.W., Kyin, S., Wang, X., and Shi, Y., 2000, Structural and biochemical basis of apoptotic activation by Smac/DIABLO, *Nature* **406**: 855-862.

Constantini, P., Jacotot, E., Decaudin, D., and Kroemer, G., 2000, Mitochrondrion as a novel target of anticancer chemotherapy, *J. of the National Cancer Institute* **92**: 1042-1053.

Cosulich, S.C., Savory, P.J., and Clarke, P.R., 1999, Bcl-2 regulates amplification of caspase activation by cytochrome c, *Current Biology* **9**: 147-150.

Cullis, P.R., and de Kruiff, B., 1997, Lipid polymorphism and the functional roles of lipids in biological membranes, *Biochim. Biophys. Acta* **559**: 399-420.

Daum, G., 1985, Lipids of mitochondria, *Biochim. Biophys. Acta* **822**: 1-42

De Kruijff, B., 1997, Lipids beyond the bilayer, *Nature* **386**: 129-130.

Demel, R.A., Jordi, W., Lambrechts, H., van Damme, H., Hovius, R. and de Kruijff, B., 1989, Differential interactions of apo- and holocytochrome c with acidic membrane lipids in model systems and the implication for their import into mitochondria, *J. Biol. Chem.* **264**: 3988-3997.

Demel, R.A., 1994, Monomolecular layers in the study of biomembranes, in: Subcellular Biochemistry, Volume 23: physicochemical methods in the study of biomembranes, (H. J. Hilderson and G. B. Ralston, eds.) Plenum Press, New York, pp.83-120.

Deveraux, Q.L. and Reed, J.,C. 1999, IAP family proteins- suppressors of apoptosis, *Genes Dev.* **13**: 239-252.

Dowhan, W., 1997, Molecular basis for membrane phospholipid diversity: Why are there so many lipids?, *Annu. Rev. Biochem.* **66**: 199-232.

Du, C., Fang, M., Li, Y., Li, L., and Wang, X., 2000, Smac, a mitochondrial protein that promotes cytochrome c-dependent caspase activation by eliminating IAP inhibition, *Cell* **102**: 33-42.

Fesik, St.W., 2000, Insights into programmed cell death through structural biology, *Cell,* **103**: 273-282.

Goldstein, J.C., Waterhouse, N.J., Juin, P., Evan, G.I., and Green, D.R., 2000, The coordinate release of cytochrome c during apoptosis is rapid, complete and kinetically invariant, *Nature Cell Biology* **2**: 156-162.

Gray. M.W., Burger, G., and Lang. B.F., 1999, Mitochondrial evolution, *Science* **283**: 1476-1481.

Green, D.R., 1998, Apoptotic pathways: The roads to ruin, *Cell* **34**: 695-698.

Green, D.R., and Reed, J.C., 1998, Mitochondria and apoptosis, *Science* **281**: 1309 1316.

Gross, A., Yin, X.M., Wang, K., Wei, M.C., Jockel, J., Milliman, C., Erdjument-Bromage, H., Tempst, P., and Korsmeyer, S.J., 1999, Caspase cleaved Bid targets mitochondria and is required for cytochrome c release, while Bcl-X_L prevents this release but not tumor necrosis factor-R1/Fas death, *J. Biol. Chem.* **274**: 1156-1163.

Heimburg, T., Hildebrandt, P., and Marsh, D., 1991, Cytochrome c-lipid interactions studied by resonance Raman and ^{31}P NMR spectroscopy. Correlation between the conformational changes of the protein and the lipid bilayer *Biochemistry* **30**: 9084-9089.

Hengartner, M.W., 2000, The biochemistry of apoptosis, *Nature* **407**: 770-767.

Hoch, F.L., 1992, Cardiolipins and biomembrane function. *Biochim. Biophys. Acta* **1113**: 71-133.

Hoffmann, B., Stöckl, A., Schlame, M., Beyer, K.,and Klingenberg, M., 1994, The reconstituted ADP/ATPcarrier activity has an absolute requirement for cardiolipin as shown in cysteine mutants, *J. Biol. Chem.* **269**: 1940-1944.

Kent, C., 1995, Eukaryotic phospholipid biosynthesis, *Annu. Rev. Biochem.* **64**: 315-343.

Kim, T.H., Zhao, Y., Barber, M.J., Kuharsky, D.K., and Yin, X.M., 2000, Bid-induced cytochrome c release is mediated by a pathway independent of mitochondrial permeability transition pore and Bax, *J. Biol. Chem.* **275**: 39474-39481.

Korsmeyer, S.J., Wei, M.C., Saito, M., Weiler, S., Oh, K.J., and Schlesinger, P.H., 2000,Proapoptotic cascade activates Bid, which oligomerizes Bak or Bax into pores that result in the release of cytochrome c, *Cell Death and Differentiation* **7**: 1166-1173.

Kozarac, Z., Dhathathreyan, A., Möbius, D., 1988, Adsorption of cytochrome c to phospholipd monolayers studied by reflection spectroscopy, *FEBS Lett.* **229**: 372-376.

Krammer, P.H., 2000, CD95's deadly mission in the immune system, *Nature* 407: 789-795.

Kroemer, G., and Reed, J.C., 2000, Mitochondrial control of cell death, *Nature Medicine* **6**: 513-519.

Kudla, G., Montessuit, S., Eskes, R., Berrier, C., Martinou, J.C., Ghazi, A., and Antonsson, B., 2000, The destabilization of lipid membranes induced by the C-terminal fragment of caspase 8-cleaved Bid is inhibited by the N-terminal fragment, *J. Biol. Chem.* **275**: 22713-22718.

Li, K., Li, Y., Shelton, J.M., Richardson, J.A., Spencer, E., Chen, Z.J., Wang, X., and Williams, R.S., 2000, Cytochrome c deficiency causes embryonic lethality and attenuates stress-induced apoptosis, *Cell* **101**: 389-399.

Li, H., Zhu, H., Xu, C.J., and Yuan, J., 1998, Cleavage of Bid by caspase 8 mediates the mitochondrial damage in the fas pathway of apoptosis, *Cell* **94**: 491-501.

Li, P., Nijhawan, D., Budihardjo, I., Srinivasula, S.M., Ahmad, M., Alnemri, E.S., and Wang, X., 1997, Cytochrome c and dATP-dependent formation of Apaf-1/caspase-9 complex initiates an apoptotic protease cascade, *Cell* **91**: 479-489.

Li, P.F., Dietz, R., and von Harsdorf, R., 1999, P53 regulates mitochondrial membrane potential through reactive oxygen species and induces cytochrome c-independent apoptosis blocked by Bcl-2, *EMBO J.* **18**: 66027-6036.

Luo, X., Budihardjo, I., Zou, H., Slaughter, C., and Wang, X., 1998, Bid, a Bcl-2 interacting protein, mediates cythochrome c release from mitochondria in response to activation of cell surface death receptors, *Cell* **94**: 481-490.

Loeffler, M., and Kroemer, G., 2000, The mitochondrion in cell death control: certainties and incognita, *Exp. Cell Res.* **256**: 19-26.

Lutter, M., Fang, M., Luo, X., Nishijima, M., Xie, X.S., and Wang, X., 2000, Cardiolipin provides specificity for targeting of tBid to mitochondria, *Nature Cell Biology* **2**: 754-756.

Margulis, L., 1996, Archaeal-eubacterial mergers in the origin of Eukarya: Phylogenetic classification of life, *Proc. Natl. Acad. Sci. USA* **93**: 1071-1076.

Martinou, J.C., Desagher, S., and Antonsson, B., 2000, Cytochrome c release from mitochondria: all or nothing, *Nature Cell Biology* **2**: E41-43.

Martinou, J-C., and Green, D.R., 2001, Breaking the mitochondrial barrier, *Molecular Cell Biology*, **2**: 63-67.

Mayer, A., Neupert, W., and Lill, R., 1995, Translocation of apocytochrome c across the outer membrane of mitochondria, *J. Biol. Chem.* **270**: 12390-12397

Miller, L.K., 1999, An exegesis of IAPs: salvation and surprises from BIR motifs. *Trends Cell Biol.* **9**: 323-328.

Minn, A.J., Vélez, P., Schendel, S.L., Liang, H., Muchmore, S.W., Fesik, S.W., Fill, M., and Thompson, C.B., 1997, Bcl-x_L forms an ion channel in synthetic lipid membranes, *Nature* **385**: 353-357.

Mootha, V.K., Wie, M.C., Buttle, K.F., Scorrano, L., Panoutsakopoulou, V., Mannellla, C.A., and Korsmeyer, S.J., 2001, A reversible component of mitochondrial respiratory dysfunction in apoptosis can be rescued by exogenous cytochrome c, *EMBO J.* **20**: 661-671.

Muchmore, S.W., Sattler, M., Liang, H., Meadows, R.P., Harlan, J.E., Yoon, H.S., Nettesheim, D., Chang, B.S., Thompson, C.B., Wong, S.L., Ng, S.C., and Fesik, S.W., 1996, X-ray and NMR structure of human Bcl-x_L, an inhibitor of programmed cell death, *Nature* **381**: 335-341.

Neupert, W., 1997, Protein import into mitochondria, *Annu. Rev. Biochem.* **66**: 863-917.

Nomura, K., Imai; H., Koumura; T., Kobayashi, T., and Nakagawa, Y., 2000, Mitochondrial phospholipid hydroperoxide glutathione peroxidase inhibits the release of cytochrome c from mitochondria by suppressing the peroxidation of cardiolipin in hypoglycaemia-induced apoptosis, *Biochem. J.* **351**: 183-193.

Poyton, R.O., McEwen, J.E., 1996, Crosstalk between nuclear and mitochondrial genomes, *Annu. Rev. Biochem.* **65**: 563-607.

Quinn, P.J., and Dawson, R.M.C., 1969, Interactions of cytochrome c and [^{14}C] carboxylated cytochrome c with monolayers of phosphatidylcholine, phosphatidic acid and cardiolipin, *Biochem. J.* **115**: 65-75.

Reed, J.C., 1997, Cytochrome c: Can't live with it – can't live without it, *Cell* **91**: 559-562.

Rytömaa, M., and Kinnunen, P.K.J., 1995, Reversibility of the binding of cytochrome c to liposomes: implications for lipid-protein interactions, *J. Biol. Chem.* **270**: 3197-3202.

Saito, M., Korsmeyer, S.J., and Schlesinger, P.H., 2000, BAX-dependent transport of cytochrome c reconstituted in pure liposomes, *Nature Cell Biology* **2**: 553-555.

Scaffidi, C., Fulda, S., Scrinivasan, A., Friesen, C., Li, F., Tomaselli, K.J., Debatin, K.M., Krammer, P.H., and Peter, M.E., 1998, Two CD95 (APO-1/Fas) signaling pathways. *EMBO J.* **17**: 1675-1687.

Schendel, S.L. Azimov, R., Pawlowski, K., Godzik, A., Kagan, B.L., and Reed, J.C., 1999, Ion channel activity of the BH3 only Bcl-2 family member, Bid, *J. Biol. Chem.* **274**: 21932-21936.

Schulze-Osthoff, K., Beyaer, R., Vandevoorde, V., Haegeman, G.,and Fiers, W., 1993, Depletion of the mitochondrial electron transport abrogates the cytotoxic and gene inductive effects of TNF, *EMBO J.* **12**: 3095-

Shidoji, Y., Hayashi, K., Komura, S., Ohishi, N., and Yagi, K., 1999, Loss of molecular interaction between cytochrome c and cardiolipin due to lipid peroxidation, *Biochem. Biophys. Res. Commun.* **264**: 343-347.

Shimizu S., Narita, M., and Tsujimoto, Y., 1999, Bcl-2 family proteins regulate the release of apoptogenic cytochrome c by the mitochondrial channel VDAC, *Nature* **399**: 483-487.

Susin, S.A., Lorenzo, H.K., Zamzami, N., Marzo, I., Snow, B.E., Brothers, G.M., Mangion, J., Jacotot, E., Costantini, P., Loeffler, M., Larochette, N., Goodlett, D.R., Aebersold, R., Siderovski, D.P., Penninger, J.M., and Kroemer, G., 1999a, Molecular characterization of mitochondrial apaptosis-inducing factor, *Nature* **397**: 441-446.

Susin, S.A., Lorenzo, H.K., Zamzami, N., Marzo, I., Brenner, C., Larochette, N., Prevost, M.C., Alzari, P.M. and Kroemer, G., 1999b Mitochondrial release of caspase-2 and –9 during the apoptotic process, *J. Exp. Med.* **189**: 381-393.

Thornberry, N.A., and Lazebnik, Y., 1998, Caspases: enemies within, *Science* **281**: 1312-1316.

Tudor, G., Aguilera, A., Halverson, D.O., Laing, N.D., and Sausville, E.A., 2000, Susceptibility to drug-induced apoptosis correlates with differential modulation of Bad, Bcl-2 and Bcl-x_L protein levels, *Cell Death and Differentiation* **7**: 574-586.

Ushmorov, A., Ratter, F., Lehmann, V., Dröge, W., Schirrmacher, V., and Umansky, V., 1999, Nitric oxide-induced apoptosis in human leukemic lines requires mitochondrial lipid degradation and cytochrome c release, *Blood* **93**: 2342-2352.

Vander Heiden, M.G., Chandel, N.S., Williamson, E.K., Schumacker, P.T., and Thompson, C.B., 1997, Bcl-xL regulates the membrane potential and volume homeostasis of mitochondria, *Cell* **91**: 627-637

Vander Heiden, M.G., and Thompson, C.B., 1999, Bcl-2 proteins: Regulators of apoptosis or of mitochrondrial homeostasis?, *Nature Cell Biology* **1**: E209-216.

Vieira, H.L.A., Haouzi, D., Hamel, C.E., Jacotot, E., Belzacq, A.S., Brenner, C., and Kroemer, G., 2000, Permeabilization of the mitochrondrial inner membrane during apoptosis: impact of the adenine nucleotide translocator, *Cell Death and Differentiation* **7**: 1146-1154.

Vik, S.B., Georgevich, G., and Capaldi, R.A., 1981, Diphosphatidylglycerol is required for optimal activity of beef heart cytochrome c oxidase, *Proc. Natl. Acad. Sci. USA* **78**: 1456-1460.

Voelker, D.R., 2000, Interorganelle transport of aminoglycerophospholipids, *Biochim. et Biophys. Acta* **1486**: 97-107.

Von Ahsen, O., Waterhouse, N.J., Kuwana, T., Newmeyer, D.D., and Green, D.R., 2000, The harmless release of cytochrome c, *Cell Death and Differentiation* **7**: 1192-1199.

Wang, K., Yin, X.M., Chao, D.T., Milliman, C.L., and Korsmeyer, S.J., 1996, Bid: a novel BH3 domain-only death agonist, *Genes dev.* **10**: 2859-2869.

Zamzami, N., Marchetti, Ph., Castedo, M., Decaudin, D., Macho, A., Hirsch, T., Susin, S.A., Petit, P.X., Mignotte, B., and Kroemer, G., 1995, Sequential reduction of mitochondrial transmembrane potential and generation of reactive oxygen species in early programmed cell death, *J. Exp. Med.* **182**: 367-377.

Zhai, D., Huang, X, Han, X., and Yang, F., 2000, Characterization of tBid-induced cytochrome c release from mitochondria and liposomes, *FEBS Lett.* **472**: 293-296.

Chapter 2

Interaction Between Cytochrome c and Oxidized Mitochondrial Lipids

Yoshihiro Shidoji[*], Sadaaki Komura[#], Nobuko Ohishi[#], and Kunio Yagi[#]
[*]Cellular Biochemistry Section, Department of Nutrition and Health Sciences, Siebold University of Nagasaki, Nagayo, Japan; [#]Institute of Applied Biochemistry, Yagi Memorial Park, Mitake, Giu, Japan

1. INTRODUCTION

Cardiolipin or diphosphatidyl glycerol is one of the most ancient membrane phospholipids from phylogenic aspects. It is surprising for such a complex molecule as cardiolipin to have evolved as one of the major membrane lipids in prokaryotics, when steroids such as cholesterol and phytosterols did not. In eukaryotic cells, cardiolipin is exclusively localized within the mitochondria where it is particularly enriched in the outer leaflet of the inner membrane. Even though a molecular structure of cardiolipin has been conserved in entire organisms, its biological significance has escaped attention except in the case of anti-cardiolipin auto-antibodies which are clinically associated with the Wasserman reaction.

In this chapter, we describe a molecular interaction between cytochrome *c* and cardiolipin which prevents peroxidation of the lipid, implying that cardiolipin may play an important role in determination of the fate of the cell. In other words, a peroxidation of cardiolipin may well allow cytochrome *c* to discharge from the mitochondrial inner membrane into the cytosolic space during apoptotic cell death.

Subcellular Biochemistry, Volume 36: Phospholipid Metabolism in Apoptosis.
Edited by Quinn and Kagan. Kluwer Academic/Plenum Publishers, New York, 2002

2. ENHANCED PEROXIDATION REACTIONS DURING APOPTOTIC CELL DEATH

Studies of cell death have been one of the most active areas of the recent biological research. Prior to the recent scrutiny of apoptotic cell death, oxidative stress has long been recognized as a causal factor in necrotic cell death or cell injury. Reactive oxygen species such as superoxide anion radical, hydrogen peroxide, and hydroxyl radical may cause a degradation of genomic DNA, membrane lipids and cellular proteins. Besides necrotic cell death, convincing evidence has now accumulated to show that intracellular peroxidation reactions are enhanced and may be involved in conveying death-signals from mitochondria to nuclei during apoptosis.

In fact, we found that α-tocopherol, an antioxidant lipid-soluble vitamin, prevented geranylgeranoic acid-induced apoptotic cell death of human hepatoma cells. These results are presented in Figure 1 (Shidoji *et al.*, 1997).

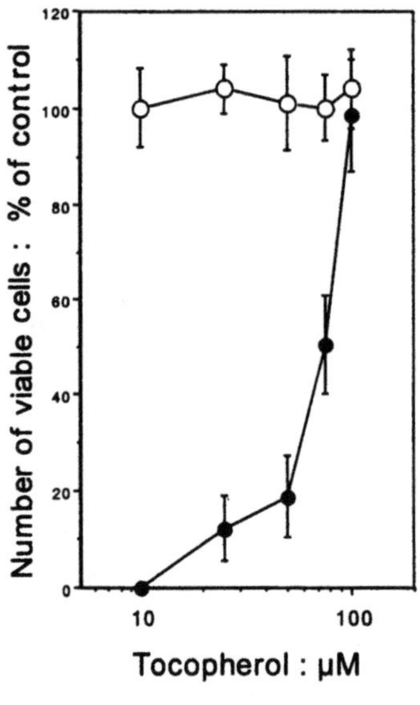

Figure 1. Inhibition of geranylgeranoic acid-induced apoptotic cell death by α-tocophrol. Increasing concentrations (10 ~ 100 μM) of α-tocopherol were added to HuH-7 cell cultures with (closed circle) or without (open circle) 10 μM geranylgeranoic acid. After overnight incubation, the number of the viable cells was counted using a Trypan blue dye exclusion method. Means ± S.E. (n=3) are shown.

Geranylgeranoic acid is a synthetic acyclic diterpenoid acid having 4 ene-bonds, which itself is unlikely to be an oxidant or prooxidant in cells. It is now known, however, that geranylgeranoic acid induces a hyper-production of superoxide anion 15 min after its addition to cell growth medium and a subsequent accumulation of hydrogen peroxide is observed in cells (Shidoji *et al.*, unpublished observations). Simultaneous addition of α-tocopherol reduced these reactive oxygen species and also prevented apoptotic cell death induced by geranylgeranoic acid treatment. We concluded, therefore, that the enhanced peroxidation reactions are responsible for geranylgeranoic acid-induced apoptosis.

In the literature, most drug-induced apoptosis is well known to require intracellular hyper-production of reactive oxygen species and several antioxidants inhibit apoptotic cell death. While it is clear that peroxidation reactions induce apoptosis, the precise molecular mechanism of how reactive oxygen species convey death-signals is unknown.

2.1 Peroxidation of cardiolipin

We directed our initial attention to the mitochondrion-specific phospholipid, cardiolipin because it is particularly rich in polyunsaturated fatty acids which are vulnerable to oxidative attack. It seemed reasonable to speculate that mitochondrial cardiolipin may degrade by the enhanced peroxidation reactions during apoptotic cell death,

As shown in Table 1, the acyl moiety of cardiolipin is comprised almost entirely of unsaturated fatty acids. Other membrane phospholipids such as phosphatidyl choline and phosphatidyl ethanolamine contain 10~40 mol of saturated fatty acids such as palmitic acid (C16:0) and stearic acid (C18:0) per 100 mol of total fatty acids. In particular, linoleic acid (C18:2) is the most abundant polyunsaturated fatty acid consisting of 80 mol%, linolenic acid (C18:3) 8 mol%, and oleic acid (C18:1) 6 mol%. Therefore, by using a commercially available cardiolipin purified from bovine heart mitochondria, we characterized auto-oxidation products by reverse phase HPTLC and reverse phase HPLC.

Table 1. Fatty acy molecular species of cardiolipin from bovine heart mitochondria (slightly modified from Schlame *et al.* (1993))

Molecular species (fatty acid combination)	Composition
	mol/100mol
(18:2-18:2)-(18:2-18:2)	47.7
(18:3-18:2)-(18:2-18:2)	21.0
(18:1-18:2)-(18:2-18:2)	14.6
(18:3-18:2)-(18:3-18:2)	5.2
(18:1-18:2)-(18:1-18:2)	4.6
(18:1-18:1)-(18:1-18:2)	0.7
not identified	6.0

As a result, we obtained three more polar spots than the intact cardiolipin on an RP-8 HPTLC plate. These 3 spots corresponded to 3 peaks monitored by UV absorption at 234 nm on an RP-8 HPLC column; and these spots or peaks contained cardiolipin mono-hydroperoxides, di-hydroperoxides and tri-hydroperoxides in increasing order of Rf (relative to the solvent front) values on the plate and in decreasing order of tR (retention time) values from the column as shown in Table 2. We failed to isolate tetra-hydroperoxides of cardiolipin, which may be very unstable.

Exposure of cardiolipin to oxygen gas resulted in a substantial loss of the lipid and most of the degradation products were hydroperoxide derivatives. Even though we have not done the comparative experiment, our experience tells us that cardiolipin is more sensitive to oxidative stress than free linoleic acid or trilinolein. Taking into account that mitochondria is the site where reactive oxygen species are often produced, we propose that peroxidation of cardiolipin may easily take place once the intracellular oxidative stress occurs.

Table 2. Peroxidized molecular species of cardiolipin from bovine heart mitochondria (from Shidoji *et al.* (1999))

Molecular species (hydroperoxide combination)	Molar ratio (weight*: conjugated diene: hydroperoxide: phosphate)
Intact cardiolipin	1:0:0:2 (theoretical)
Cardiolipin monohydroperoxide	1:0.98:1.11:2.11
Cardiolipin dihydroperoxide	1:2.10:1.97:2.15
Cardiolipin trihydroperoxide	1:3.49:3.31:2.23

*Bovine heart cardiolipin was assumed as all tetra-linoleoyl.

2.2 Decrease in cellular content of cardiolipin during apoptosis

Recently, Vogelstein's group (Polyak *et al.,* 1997) reported an important finding that a dramatic decrease in intracellular content of mitochondrial cardiolipin occurred during p53-induced apoptosis. They monitored the cardiolipin content using a probe, nonylacridine orange (NAO; cardiolipin-specific fluorogenic reagent). We tested interaction between hydroperoxide derivatives of cardiolipin and NAO and observed no increase in fluorescence intensity at 520 nm to indicate that NAO probes intact cardiolipin. Polyak *et al.* (1997) also reported that the cardiolipin level decreased soon after p53-induced reactive oxygen species were detected. Although they refer to the decrease in cardiolipin content as mitochondrial damage, oxidative degradation of cardiolipin has not been discussed in terms of a molecular involvement of cardiolipin in a death-signal transduction system.

In addition, there is experimental evidence showing that mitochondrial cardiolipin content markedly decreases following ischemia and reperfusion injury due to cardiolipin peroxidation (Soussi *et al.*, 1990) and that a decrease in the mitochondrial phospholipid cardiolipin occurred in aged rat hearts (Pepe, 2000). These decreases may be attributable to the hydroperoxide-formation of cardiolipin after exposure to intense or repeated oxidative stress during disease state or normal aging, respectively.

3. MOLECULAR INTERACTION BETWEEN CYTOCHROME *C* AND CARDIOLIPIN

It is now well established that activation of the caspase cascade is an indispensable and sufficient process in the execution phase of apoptosis (Nuñez *et al.*, 1998). As for mitochondria-mediated apoptosis, cytochrome *c* released from the mitochondrial inner membrane is well known to play an important role in the activation of caspase 9, one of the upstream proteases in the cascade (Zou *et al.*, 1997). For activation of caspase 9, cytochrome *c* or apoptotic protease activating factor 2 (Apaf 2) induces the formation of the complex between Apaf 1 and caspase 9. The resultant activated caspase 9 then activates caspase 3, which in turn leads to the genomic DNA fragmentation and apoptotic cell death.

Mitochondrial cytochrome *c* is a water-soluble basic protein, residing in the mitochondrial inner membrane where complexes III and IV of the respiratory chain can reversibly interact with the protein. When the basic protein shuttles between complexes III and IV, cardiolipin facilitates the binding of cytochrome *c* to cytochrome *c* oxidase-containing membranes (Salamon and Tollin, 1996). In other words, cytochrome *c* oxidase in complex IV efficiently receives electrons from cytochrome *c* only in the presence of cardiolipin. It is well documented that cytochrome *c* oxidase binds one cardiolipin molecule as an activator and 2 others for anchoring itself to the inner membrane (Awasthi *et al.*, 1971; Suter *et al.*, 2000).

3.1 A model of cytochrome *c* / cardiolipin complex

The small basic protein, cytochrome *c,* has often been reported to bind strongly to the acidic phospholipid, cardiolipin, rich in outer leaflet of the mitochondrial inner membrane (Gallet *et al.*, 1997). The binding of cytochrome *c* to cardiolipin is very tight, apparently irreversible, and stoichiometric (Rytömaa and Kinnunen, 1995), suggesting that this unique phospholipid, cardiolipin, may anchor the small heme protein to the inner membrane outer leaflet and keep the protein functional in the respiratory chain.

Therefore, we investigated a molecular interaction of cardiolipin with cytochrome *c*. As shown in Figure 2, cytochrome *c* consists of 5 helices and inter-helical loops which harbor a heme *c* prosthetic group by covalent thioether-bonds through cysteine-14 and -17 residues. Ferric ion, centered in the pyrrole ring, is axially liganded by histidine-18 and methionine-80 residue. The lower half of the protein consisting of flexible random coils is a rather soft structure and has a space between the heme *c* plate and β-loops inside the small basic protein.

The propionate-side (lower part in Figure 2) of the heme *c* group is exposed to the solvent so that the underside of cytochrome *c* in Figure 2 makes an entrance to the belly of the protein.

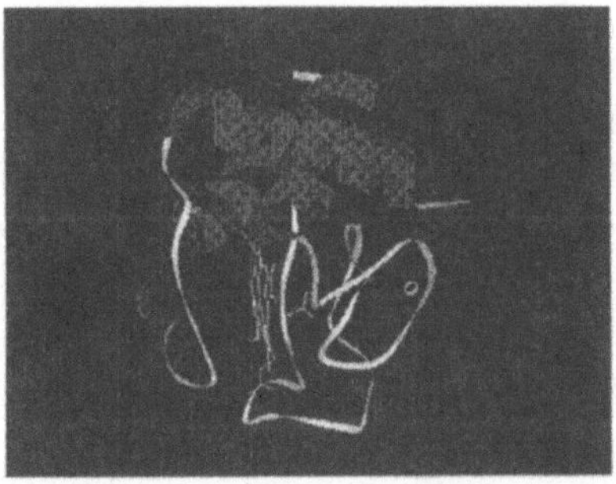

Figure 2. Three-dimensional structure of human cytochrome *c* created by Protein Adviser, ver 3.0 (FQS, Hakata, Japan) with PDB file of human cytochrome *c* down-loaded from protein structure database of NCBI. α-Helices are shown as purple ribbons, random coils as white strands, and β-turns are blue (see separate colour tip). Heme *c* is depicted in white straight lines inside the protein.

We have performed a computer-aided 3-D modeling of cardiolipin tetralinoleoyl with molecular mechanics software of MM2 in Chem 3D Pro, ver 4.0 (CambridgeSoft Co. MA). After 342 iterations, a stable conformation shown in Figure 3 was obtained by minimization of total steric energy, which included electrostatic energy as well as energy of bend, stretch, torsion, and van der Waals. As seen in the figure, 2 linoleic acids on 1 glycerol are maximally apart from each other and these 2 pairs of the acids face each other through 3 glycerol moieties in a rectangularly twisted state. As a membrane lipid, this model of cardiolipin does not fit in a lipid bilayer, because the hydrophilic moiety is located between 2 separate hydrophobic stretches. A quite different conformation has been proposed for cardiolipin

Figure 2. Three-dimensional structure of human cytochrome *c* created by Protein Adviser, ver 3.0 (FQS, Hakata, Japan) with PDB file of human cytochrome *c* down-loaded from protein structure database of NCBI. α-Helices are shown as purple ribbons, random coils as white strands, and β-turns are blue. Heme *c* is depicted in white straight lines inside the protein.

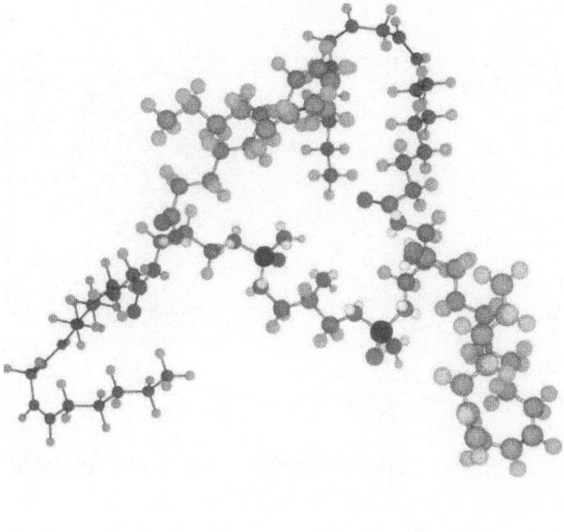

Figure 3. Three-dimensional modeling of cardiolipin tetralinoleoyl created by Chem3D Pro, ver 4.0 with running of MM2, a software of molecular mechanics, to minimize steric energy *in vacuo*. Each atom is displayed in colour, H (blue), C (black), O (red), and P (purple).

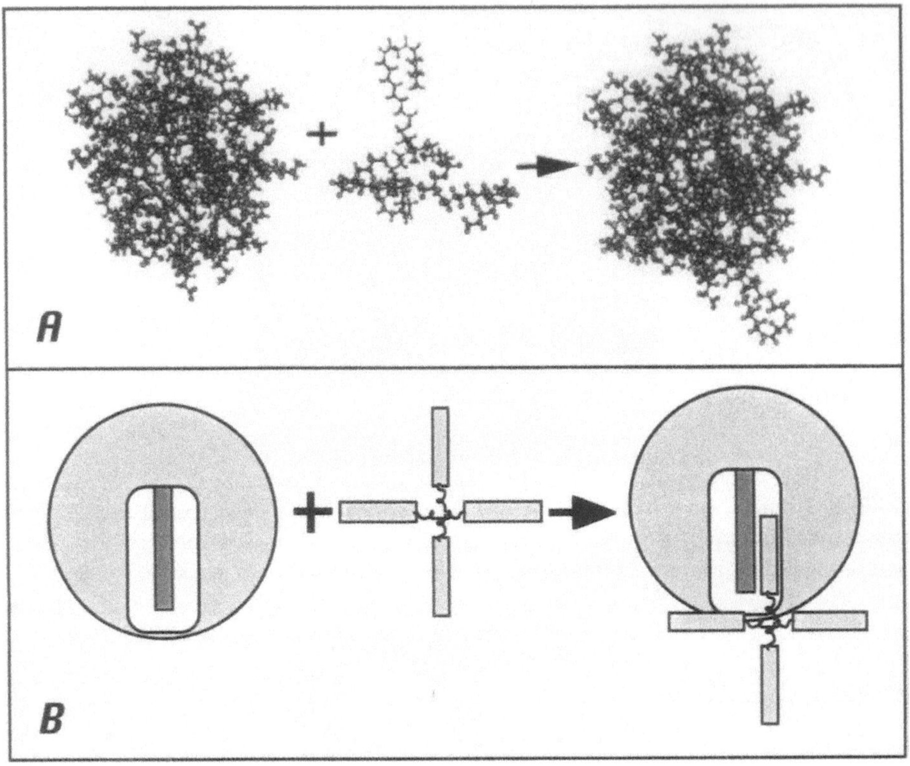

Figure 4. Computor simulation of the interaction between human cytochrome *c* and cardiolipin tetralinoleoyl (*A*) and a schematic drawing ("kendama" model) of the interaction (*B*).

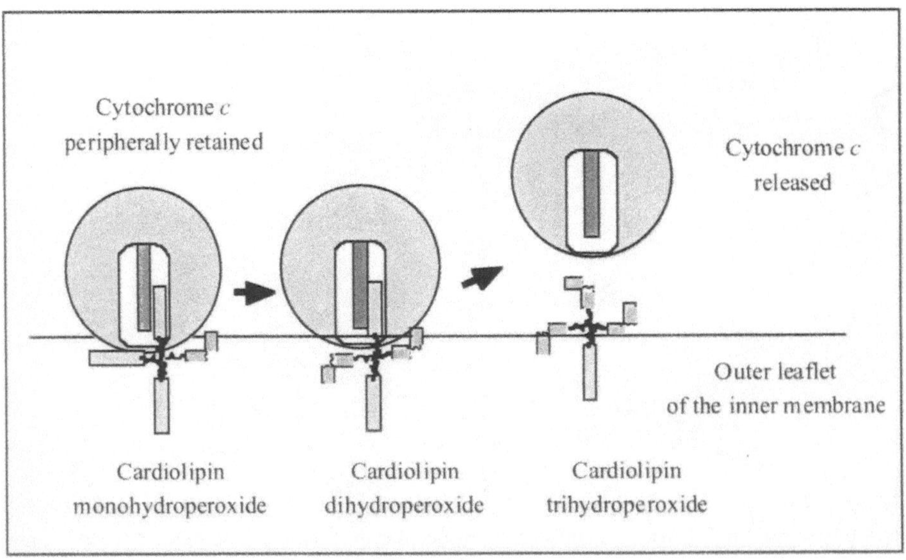

Figure 6. Proposed mechanism underlying a cytosolic release of cytochrome c from mitochondrial inner membrane during apoptosis induced by reactive oxygen species.

in the membrane plane (Schlame *et al.*, 1993). As has already been noted, since no crystal structure data of cardiolipin is available, any models have to be considered as preliminary. Our model, described below, is tentatively applied to molecular simulation analysis to examine the interaction between cardiolipin and cytochrome *c*.

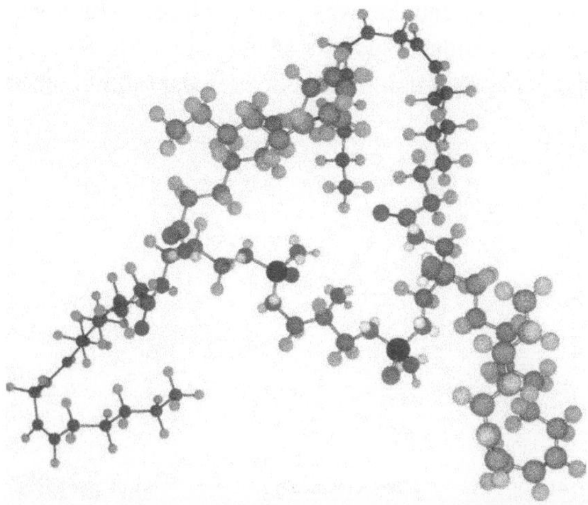

Figure 3. Three-dimensional modeling of cardiolipin tetralinoleoyl created by Chem3D Pro, ver 4.0 with running of MM2, a software of molecular mechanics, to minimize steric energy *in vacuo*. Each atom is displayed in colour, H (blue), C (black), O (red), and P (purple) (see separate colour tip).

Prior to conducting *in silico* chemistry, we experimentally analyzed the interaction between cardiolipin and cytochrome *c*. Using proton-NMR, we were able to confirm the previous finding that cardiolipin causes a conformational change at the lower part of the solvent-exposed heme edge (Soussi *et al.*, 1990). The spectral features of cytochrome *c* are well established with signal assignments for paramagnetically resolved resonances. Soussi *et al.* (1990) found that the excess linewidth of the downfield signals at 31.5 and 34.5 ppm was a function of increasing amounts of cardiolipin in the presence of cytochrome *c* in phosphatidylcholine vesicles. These resonances have been assigned to the heme methyl groups-8 and -3, both residing on the pyrrole ring in the belly of the protein in Figure 2. Addition of cardiolipin induced a dramatic broadening of the linewidth of the downfield resonances, indicating that a microenvironment of the heme was perturbed (Shidoji *et al.*, 1999; Soussi *et al.*, 1990).

It is noteworthy that in our NMR experiment 0.1 M sodium phosphate buffer remarkably attenuated the interaction between cardiolipin and cytochrome *c*, indicating that electrostatic interactions between anionic phosphate moieties on cardiolipin and basic amino acid residues of cytochrome *c* may be involved in the interaction between the acidic lipid and the basic protein. This finding provides important information for computor simulation of the molecular interaction. In other words, when we started the *in silico* docking experiments using Chem 3D software, each phosphate moiety on the cardiolipin moleculae was located close to the lower entrance where basic amino acid residues such as Lys-53, Lys-55, and Lys-79 were clustered. A subsequent dock command conducted a minimization of steric

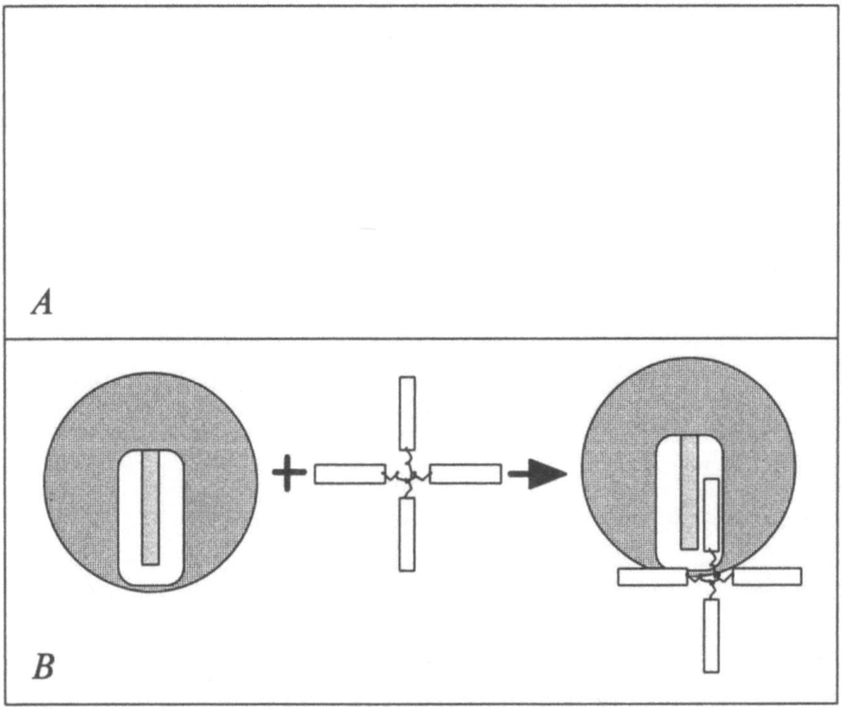

Figure 4. Computer simulation of the interaction between human cytochrome *c* and cardiolipin tetralinoleoyl (*A*) and a schematic drawing ("kendama" model) of the interaction (*B*).

energy to create a model of cytochrome *c* and cardiolipin complex as shown in Figure 4A. The cardiolipin was arranged as an energetically stable complex with the protein so that one acyl moiety was inserted into the cavity of the protein, two acyl chains were adjacent to the lower surface of the protein, and remaining other acyl moiety projected perpendicularly out of the protein. A rough idea of the complex formation is shown in the cartoon in

Figure 4B. Although this model was assembled *in vacuo*, one protruding acyl moiety may be accomodated within the outer leaflet of the mitochondrial inner membrane to retain cytochrome *c* in a peripheral location.

This "kendama" (Japanese traditional wooden toy consisting of cross-shaped stick and ball) model of the molecular interaction is consistent with a previous model for the binding of cytochrome *c* to acidic phospholipids (Rytömaa and Kinnunen, 1995).

The molecular interaction of cytochrome *c* and cardiolipin has been extensively studied. A mode of the interaction has been confirmed to be both electrostatic and hydrophobic, by using infrared spectroscopy (Choi and Swanson, 1995), fluorescence resonance energy transfer method (Rytömaa and Kinnunen, 1994), protease digestion (de Jongh *et al.*, 1995), cyclic voltammetry (Salamon and Tollin, 1997), deuterium and phosphorus NMR measurements (Spooner *et al.*, 1993), and surface plasmon resonance spectroscopy (Salamon and Tollin, 1996).

Such studies have revealed that the binding of cardiolipin induces partial unfolding of cytochrome *c*. However, we did not observe partial unfolding of cytochrome *c* in the presence of equimolar cardiolipin, because the iron-sulfur charge transfer band at 695 nm did not change significantly by addition of cardiolipin and no band appeared around 620 nm, at which the high-spin heme group of the unfolded heme protein shows absorption (de Jongh, *et al.*, 1995).

3.2 Models of cytochrome *c* discharge from the putative complex

As mentioned above, mammalian cardiolipin contains more than 80% of acyl moieties as linoleic acid (Schlame *et al.*, 1993) and is highly vulnerable to oxidative damage, prompting us to hypothesize that peroxidation of cardiolipin may cause a loss of binding between this phospholipid and cytochrome *c* and may release the protein into the cytosol, leading the cells to apoptosis. Figure 5 shows that cardiolipin trihydroperoxides barely interacted with cytochrome *c* in contrast to a tight binding of intact cardiolipin to the heme protein, as judged by proton-NMR spectroscopy of the heme methyl groups (Shidoji *et al.*, 1999).

A dissipation of mitochondrial membrane potential ($\Delta\Psi m$) is often detectable during apoptosis (Green, 1998; Shidoji *et al.*, 1997). A loss of $\Delta\Psi m$ is thought to be mediated by the opening of the mitochondrial permeability transition pore which is proposed to be involved in mitochondrial efflux of cytochrome *c* into the cytosol (Scarlett and Murphy, 1997). However, a recent report by Bossy-Wetzel *et al.* (1998) provided unequivocal evidence that the translocation of cytochrome *c* occurred prior to the dissipation of $\Delta\Psi m$. The underlying molecular mechanism for the translocation of cytochrome *c* has been virtually ignored at least that operating earlier than the opening of the mitochondrial permeability transition pore.

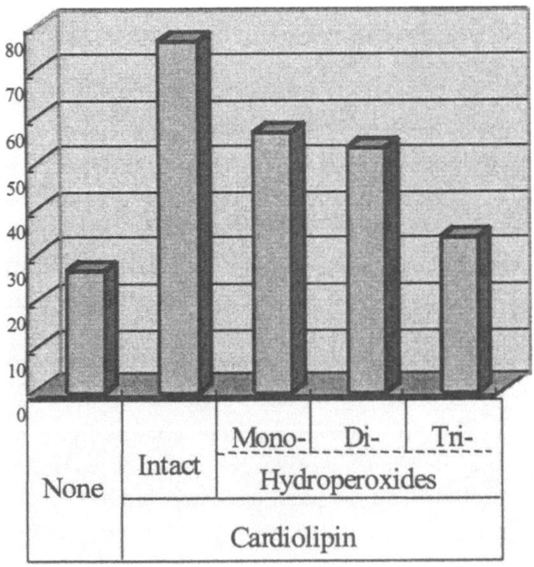

Figure 5. Effect of degree of cardiolipin peroxidation on the signal in proton NMR spectroscopy of cytochrome *c* (Shidoji *et al.*, 1999). A half-width of the downfield signal at 34.2 ppm were plotted.

Hence, in the present chapter, we addressed the important questionas to how cytochrome *c* is released from mitochondrial inner membrane by apoptotic signals independently of $\Delta\Psi m$ dissipation. We have proposed that peroxidation of cardiolipin may result in a discharge of cytochrome *c* due to a weakening of the interaction between the lipid and the protein.

Our proposed mechanism is outlined in Figure 6, with a tentative assumption that once the acyl chains of cardiolipin are peroxidized, their hydrophobicity will diminish and a resulting conformational change cannot be accommodated within the cavity between the heme plane and the inner surface of cytochrome *c*.

Our model of cytochrome c release during apoptosis is not an alternative mechanism to the Bid/Bax-regulated release of cytochrome *c*. The voltage-dependent anion channel (VDAC) is known to be converted to a cytochrome *c*-permeant conduit by Bax. Furthermore, Bax protein can oligomerize and form a non-specific cytochrome *c*-permeant pore with aid of Bid (Kroemer and Reed, 2000). Both channels may afford a selective permeabilization of mitochondrial "outer" membrane. However, we are addressing how cytochrome *c* can be discharged from the periphery of the mitochondrial "inner" membrane during apoptosis.

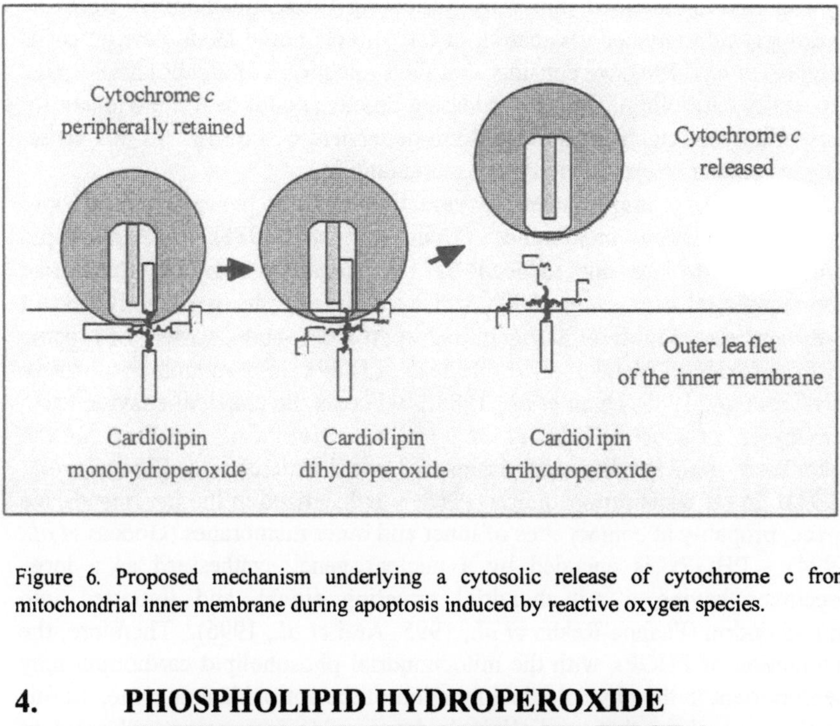

Figure 6. Proposed mechanism underlying a cytosolic release of cytochrome c from mitochondrial inner membrane during apoptosis induced by reactive oxygen species.

4. PHOSPHOLIPID HYDROPEROXIDE GLUTATHIONE PEROXIDASE (PHGPx): A PROTECTOR FOR CARDIOLIPIN

As described in section 3 of this chapter, we have postulated that cytochrome *c* would be released from mitochondrial inner membrane to the cytosol due to peroxidation of cardiolipin. In other words, peroxidation of cardiolipin might be involved in apoptotic cell death through the discharge of cytochrome *c* from the mitochondrial inner membrane. There is now increasing evidence showing a significant decrease in the concentration of cardiolipin during cell death, for example, in p53-induced apoptotsis (Polyak *et al.*, 1997) and in nitric oxide-induced cell death of Jurkat leukemic cells (Ushmorov *et al.*, 2000). A recent study revealed that the generation of cardiolipin hydroperoxide in mitochondria occurred before the release of cytochrome *c* in basophilic leukemia cells and that the release of cytochrome *c* was suppressed when the production of cardiolipin hydroperoxide in mitochondria was inhibited by the overexpression of phospholipid hydroperoxide glutathione peroxidase (PHGPx) (Nomura *et al.*, 2000). From these findings together with our previous observation that expression of PHGPx protected the host cells from exogenous phospholipid hydroperoxide-mediated injury by protecting their mitochondria from the hydroperoxide-induced dissipation of the mitochondrial membrane potential (Shidoji *et al.* 1998), we consider that the protection of mitochondrial lipids, especially cardiolipin, from peroxidation damage could be a prerequisite for protecting cells against mitochondrion-dependent cell death. In this sense, the antioxidant enzyme PHGPx has a crucial role.

PHGPx is a unique selenoenzyme that directly reduces phospholipid hydroperoxides in membranes (Ursini *et al.*, 1985). Phospholipid hydroperoxides are not reduced by the so-called classical glutathione peroxidase (Gibson *et al.*, 1980), which catalyzes reductive inactivation of hydrogen peroxide, free fatty acid hydroperoxides, and a variety of organic hydroperoxides (Wendel *et al.*, 1981). PHGPx is a monomeric protein (Ursini *et al.*, 1985, Duan *et al.*, 1988,), whereas the classical enzyme has a tetrameric structure (Flohé *et al.*, 1971). PHGPx is localized in the cytoplasm, mitochondria, and plasma and nuclear membranes (Roveri *et al.*, 1994). In rat testis mitochondria, PHGPx is localized in the intermembrane space, probably at contact sites of inner and outer membranes (Godeas *et al.*, 1994). PHGPx is encoded by a nuclear gene, synthesized as a large precursor having a mitochondrial targeting signal, and imported into mitochondria (Pushpa-Rekha *et al.*, 1995, Arai *et al.*, 1996). Therefore, the interaction of PHGPx with the mitochondrial phospholipid cardiolipin may be important in the physiological function of the antioxidant enzyme. In this section, we show that cardiolipin hydroperoxide is a potent substrate of PHGPx and that cardiolipin itself potentially activates PHGPx. On the basis of these findings, we discuss the physiological role of PHGPx in protecting mitochondrial function from oxidative disruption.

4.1 Cardiolipin monohydroperoxide is a potent substrate of PHGPx

Using pure cardiolipin monohydroperoxide described in section 2.1, we examined the activity of PHGPx toward this lipid hydroperoxide as a substrate. PHGPx was purified from the cytosol of transfectant guinea pig cells that are highly expressed human PHGPx (Yagi *et al.*, 1996). Although an early study briefly reported that PHGPx from pig heart was active toward hydroperoxide of cardiolipin and has almost the same reactivity to phosphatidylcholine hydroperoxide (Ursini *et al.*, 1985), we observed that monohydroperoxide of cardiolipin was a better substrate of PHGPx than dilinoleoyl phosphatidylcholine monohydroperoxide. When PHGPx activity was determined with increasing concentrations of cardiolipin monohydroperoxide, the reaction velocity reached its maximum at 60 μM substrate. The activity toward cardiolipin monohydroperoxide was 3.5 times higher than that measured with dilinoleoyl phosphatidylcholine monohydroperoxide under the optimum conditions.

The reaction of PHGPx toward each substrate was analyzed by a Hill plot (Figure 7). The Hill coefficient was calculated to be 2.33 for cardiolipin monohydroperoxide and 1.37 for dilinoleoyl phosphatidylcholine monohydroperoxide, indicating a higher positive cooperativity with the former substrate than that of the latter. According to the Hill coefficient values obtained, we can assume that the PHGPx protein possesses two binding sites for cardiolipin monohydroperoxide or cardiolipin, probably, one catalytic site and another effector site.

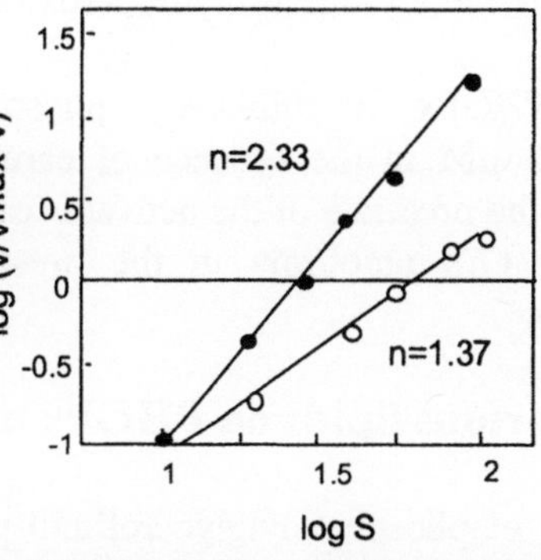

Figure 7. Hill plot analysis of PHGPx reaction. PHGPx activity was determined with either dilinoleoyl phosphatidylcholine monohydroperoxide (○) or cardiolipin monohydroperoxide (●) as a substrate. The amount of PHGPx protein in a reaction mixture was about 0.4 μg.

4.2 Cardiolipin is an endogenous activator of PHGPx

It is well recognized that activity of many mitochondrial enzymes is modulated by the mitochondrial acidic phospholipid cardiolipin. Adenine nucleotide translocase (ANT) of bovine heart mitochondria is known to be tightly associated with six molecules of cardiolipin (Beyer and Klingenberg, 1985), and cardiolipin is absolutely required for the ANT activity (Hoffmann *et al.*, 1994). ADP control of the carrier transition was gradually lost after phospholipase treatment and the ADP control could be fully restored through the addition of cardiolipin (Beyer and Nuscher, 1996). Cardiolipin-mediated activation of cytochrome *c* oxidase (Abramovitch *et al.*, 1990) and cytochrome P450SCC (Kisselev *et al.*, 1999) has also been reported. We describe here results showing a marked activation of PHGPx by cardiolipin. PHGPx used in this study was purified from the cytosol fraction of the transfectant 104C1/O4C cells, which had been transfected with the human PHGPx gene containing a code for the mitochondrial targeting signal. The enzyme activity of the purified PHGPx toward dilinoleoyl phosphatidylcholine monohydroperoxide increased with an increase in cardiolipin concentration in the reaction mixture. The maximum activity was found in the presence of 35 μM cardiolipin and was almost 2.5 times higher than that determined without the addition of cardiolipin. When the reaction was carried out in the absence of the substrate dilinoleoyl phosphatidylcholine monohydroperoxide, but with the addition of intact cardiolipin, no NADPH was consumed, supporting the notion that the preparation of cardiolipin used in the experiment was free from any phospholipid hydroperoxides as contaminant or degradation products. Accordingly, we can exclude the possibility that the stimulation of PHGPx activity by cardiolipin is merely due to hydroperoxide contaminants in the cardiolipin preparation.

The K_m value of PHGPx for dilinoleoyl phosphatidylcholine monohydroperoxide was 163 μM in the absence of cardiolipin. However, it decreased to 70 μM in the presence of the activator cardiolipin. The values of V_{max} are 11.9 and 17.8 nmol/min in the presence and absence of cardiolipin, respectively.

4.3 Effect of various lipids on PHGPx activity

Cardiolipin consists of phosphatidylglycerol and phosphatidic acid. In order to see structural specificity of lipids as an activator, we examined the effect of various cardiolipin-related lipids in the place of cardiolipin on PHGPx activity. Figure 8 shows that cardiolipin from *Eschericia coli* was as potent an activator as cardiolipin from bovine heart. The content of unsaturated fatty acids is quite different between the acidic lipids obtained from *E. coli* and bovine heart. Cardiolipin from mammalian sources is abundant in linoleic acid. On the contrary, the acidic phospholipids from bacteria have a lower content of unsaturated fatty acids. Therefore, the type of acyl moieties in cardiolipin is not a fundamental requirement for the ability of cardiolipin to activate PHGPx. Phosphatidylglycerol and phosphatidic acid were also effective in activating PHGPx, though the extent of activation by these lipids was less than that by cardiolipin. Since the activation by L-α-lysophosphatidylglycerol palmitoyl was less compared with that by phosphatidylglycerol, the presence of 2 acyl moieties appears to be necessary for the functional interaction of the lipid with PHGPx.

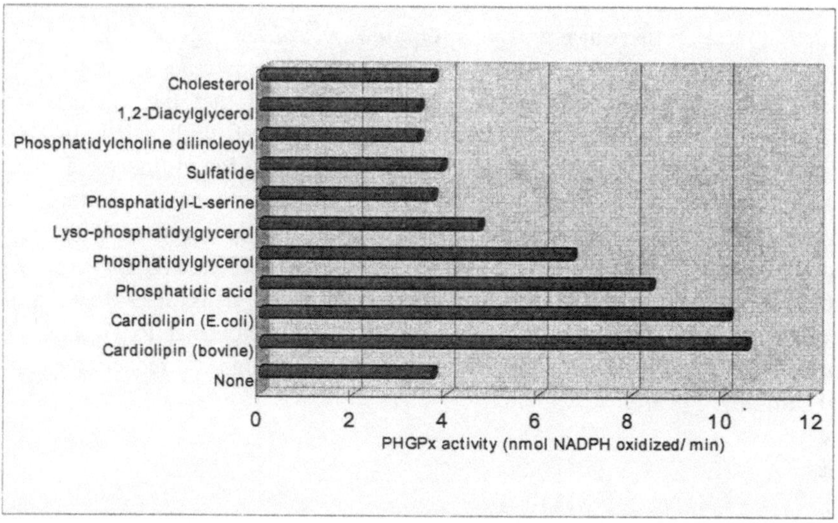

Figure 8. Effect of cardiolipin and related lipids on PHGPx activity. Each lipid was added to the reaction mixture to give a final concentration of 35 μM

Neither 1,2-diacylglycerol, phosphatidyl-L-serine, phosphatidylcholine dilinoleoyl, sulfatide, or cholesterol activated PHGPx. In the 104C1/O4C cells mentioned above, markedly increased levels of PHGPx activity and its immunoreactive protein were also found in the crude mitochondrial fraction (Yagi *et al.*, 1998). When PHGPx activity in the mitochondrial fraction was measured, as expected, an activating effect of exogenously added cardiolipin was observed. It is reasonable that PHGPx prepared from cytosol fractions has not yet become associated with endogenous cardiolipin and that once PHGPx protein has been translocated into the mitochondrial membrane, it exhibits optimum function through the specific interaction with cardiolipin.

Protein-lipid interactions are physiologically important for the function of mitochondrial proteins. As is well known, ANT is an essential component of the mitochondrial permeability transition pore. The pore is formed from a complex of VDAC (outer membrane), ANT (inner membrane), and cyclophilin-D (matrix) at contact sites between the mitochondrial outer and inner membranes. The activity of ANT is susceptible to oxidative damage both through direct modification of the protein as well as peroxidation of surrounding lipids. A recent report indicates that the oxidation of a critical cysteine residue of ANT leads to mitochondrial membrane permeabilization and cell death (Costantini *et al.*, 2000). An analgesic drug with antioxidant activity, flupirtine, protected lipids and also protein from attack by reactive oxygen species in rat liver mitochondria and ANT is thought to be one of the membrane proteins particularly protected by the antioxidant (Schluter *et al.*,

2000). In relation to aging, it was reported by Yan and Sohal (1998), that ANT was the only protein in the mitochondrial membranes exhibiting a detectable age-associated increase in carbonyls and that the age-related elevation in the ANT carbonyl content was correlated with a loss of the functional activity of this pore component. They also found that ANT was the only protein exhibiting adducts of the lipid peroxidation product 4-hydroxynonenal. Taking into account these results together with the topography, *i.e.*, that PHGPx is located close to ANT at contact sites of mitochondrial inner and outer membranes (Kroemer *et al.*, 1997), we are convinced that a major molecular function of mitochondrial PHGPx is to protect ANT from irreversible oxidative damage through a direct protection of cardiolipin *in situ*.

5. CONCLUSIONS

In this chapter, we have postulated that cardiolipin may participate in determination of cell fate by allowing cytochrome *c* release from mitochondrial inner membrane through its peroxidative damage during apoptotic cell death. And anti-oxidant mitochondrial enzyme, PHGPx may protect cardiolipin from its peroxidation as well as apoptotic cell death.

ACKNOWLEDGMENTS

We would like to thank all the colleagues who have helped in many invaluable ways in the production of this chapter, in particular, Dr. Hayashi, who helped us obtain N.M.R. data and gave advice throughout our studies.

REFERENCES

Abramovitch, D.A., Marsh, D., Powell, G.L., 1990, Activation of beef-heart cytochrome c oxidase by cardiolipin and analogs of cardiolipin. *Biochim. Biophys. Acta,* **1020:** 34-42

Arai, M., Imai, H., Sumi, D., Imanaka, T., Takano, T., Chiba, N., and Nakagawa, Y., 1996, Import into mitochondria of phospholipid hydroperoxide glutathione peroxidase requires a leader sequence. *Biochem. Biophys. Res. Commun.*, **277:** 433-439

Awasthi, Y.C., Chuang, T.F., Keenan, T.W., and Crane, F.L., 1971, Tightly bound cardiolipin in cytochrome *c* oxidase. *Biochim. Biophys. Acta,* **226:** 42-52

Beyer, K., and Klingenberg, M., 1985, ADP/ATP carrier protein from beef heart mitochondria has high amounts of tightly bound cardiolipin, as revealed by ^{31}P nuclear magnetic resonance. *Biochemistry,* **24:** 3821-3826

Beyer, K., and Nuscher, B., 1996, Specific cardiolipin binding interferes with labeling of sulfhydryl residues in the adenosine diphosphate/adenosine triphosphate carrier protein from beef heart mitochondria. *Biochemistry,* **35:** 15784-15790

Bossy-Wetzel, E., Newmeyer, D.D., and Green, D.R., 1998, Mitochondrial cytochrome c release in apoptosis occurs upstream of DEVD-specific caspase activation and independently of mitochondrial transmembrane depolarization. *EMBO J.* **17:** 37-49

Choi, S., and Swanson, J.M., 1995, Interaction of cytochrome *c* with cardiolipin: an infrared spectroscopic study. *Biophys. Chem.*, **54:** 271-278

Costantini, P., Belzacq, A. S., Vieira, H. L., Larochette, N., de Pablo M. A., Zamzami, N., Sunsin, S. A., Brenner, C., and Kroemer, G., 2000, Oxidation of a critical thiol residue of the adenine nucleotide translocator enforces Bcl-2-independent permeability transition pore opening and apoptosis. *Oncogene*, **19:** 307-314

de Jongh, H.H., Ritsema,T., and Killian, J.A., 1995, Lipid specificity for membrane mediated partial unfolding of cytochrome *c*. *FEBS Lett.,* **360:** 255-260

Duan, Y.-J., Komura, S., Fiszer-Szafarz B., Szafarz, D., and Yagi, K, 1988, Purification and characterization of a novel monomeric glutathione peroxidase from rat liver. *J. Biol. Chem.,* **263:** 19003-19008

Flohe, L., Eisele, B., and Wendel, A., 1971, Glutathione peroxidase. I. Isolation and determinations of molecular weight. *Hoppe Seylers Z. Physiol. Chem.*, **352:** 151-158

Gallet, P.F., Petit, J.M., Maftah, A., Zachowski, A., and Julien, R., 1997, Asymmetrical distribution of cardiolipin in yeast inner mitochondrial membrane triggered by carbon catabolite repression. *Biochem. J.,* **324:** 627-634

Gibson, D. D., Hornbrook, K. R., and McCay, P. B., 1980, Glutathione-dependent inhibition of lipid peroxidation by a soluble, heat-labile factor in animal tissues. *Biochim. Biophys. Acta*, **620:** 572-582

Godeas, C., Sandri, G., and Panfili, E., 1994, Distribution of phospholipid hydroperoxide glutathione peroxidase (PHGPx) in rat testis mitochondria. *Biochim. Biophys. Acta,* **1191:** 147-150

Green, D.R., and Reed, J.C., 1998, Mitochondria and apoptosis. *Science*, **281:** 1309-1312

Hoffmann, B., Stockl, A., Schlame, M., Beyer, K., and Klingenberg, M., 1994, The reconstituted ADP/ATP carrier activity has an absolute requirement for cardiolipin as shown in cysteine mutants. *J Biol Chem.* **269:** 1940-1944

Kisselev, P., Tuckey, R.C., Woods, S.T., Triantopoulos, T., Schwarz, D., 1999, Enzymatic properties of vesicle-reconstituted human cytochrome P450SCC (CYP11A1) differences in functioning of the mitochondrial electron-transfer chain using human and bovine adrenodoxin and activation by cardiolipin. *Eur. J. Biochem.*, **260:** 768-773

Kroemer, G., Zamzami, N., and Susin, S., 1997, Mitochondrial control of apoptosis. *Immunol. Today*, **18:** 44-51

Nomura, K., Imai, H., Koumura, T., Kobayashi, T., and Nakagawa, Y., 2000, Mitochondrial phospholipid hydroperoxide glutathione peroxidase inhibits the release of cytochrome *c* from mitochondria by suppressing the peroxidation of cardiolipin in hypoglycaemia-induced apoptosis. *Biochem. J.*, **351:** 183-193

Nuñez, G., Benedict, M., Hu, Y., and Inohara, N., 1998, Caspases: the proteases of the apoptotic pathway. *Oncogene,* **17:** 157-168

Pepe, S., 2000, Mitochondrial function in ischemeia and reperfusion of the ageing heart. *Clin. Exp. Pharmacol. Physiol.*, **27:** 745-750

Polyak, K., Xia, Y., Zweier, J.L., Kinzler, K.W., and Vogelstein, B., 1997, A model for p53-induced apoptosis. *Nature,* **389:** 300-305

Pushpa-Rekha, T. R., Burdsall, A. L., Oleksa, L. M., Chisoln, G. M., and Driscoll, D. M., 1995, Rat phospholipid-hydroperoxide glutathione peroxidase. cDNA cloning and identification of multiple transcription and translation start sites. *J. Biol. Chem.*, **270:** 26993-26999

Roveri, A., Maiorino, M., and Ursini, F., 1994, Enzymatic and immunological measurements of soluble and membrane-bound phospholipid-hydroperoxide glutathione peroxidase. *Methods Enzymol.* **233:** 202-212

Rytömaa, M., and Kinnunen, P.K.J., 1994, Evidence for two distinct acidic phospholipid-binding sites in cytochrome *c*. *J. Biol. Chem.*, **269:** 1770-1774.

Rytömaa, M., and Kinnunen, P.K.J., 1995, Reversibility of the binding of cytochrome c to liposomes. Implications for lipid-protein interactions. *J. Biol. Chem.*, **270:** 3197-3202

Salamon, Z., and Tollin, G., 1996, Surface plasmon resonance studies of complex formation between cytochrome c and bovine cytochrome *c* oxidase incorporated into a supported planar lipid bilayer. II. Binding of cytochrome *c* to oxidase-containing cardiolipin /phosphatidylcholine membranes. *Biophys. J.*, **71:** 858-867

Salamon, Z., and Tollin, G., 1997, Interaction of horse heart cytochrome *c* with lipid bilayer membranes: effects on redox potentials. *J. Bioenerg. Biomembr.* **29:** 211-221

Scarlett, J.L., and Murphy, M.P., 1997, Release of apoptogenic proteins from the mitochondrial intermembrane space during the mitochondrial permeability transition. *FEBS Lett.* **418:** 282-286

Schlame, M., Brody, S., and Hostetler, K., 1993, Mitochondrial cardiolipin in diverse eukaryotes. Comparison of biosynthetic reactions and molecular acyl species. *Eur. J. Biochem.* **212:** 727-735

Schluter, T., Struy, H., and Schonfeld P., 2000, Protection of mitochondrial integrity from oxidative stress by the triaminopyridine derivative flupirtine. *FEBS Lett.*, **481:** 42-46

Shidoji, Y., Nakamura, N., Moriwaki, H., and Muto, Y., 1997, Rapid loss in the mitochondrial membrane potential during geranygeranoic acid-induced apoptosis. *Biochem. Biophys. Res. Commun.* **230:** 58-63

Shidoji, Y., Hayashi, K., Komura, S., Ohishi, N., and Yagi, K., 1999, Loss of molecular interaction between cytochrome c and cardiolipin due to lipid peroxidation. *Biochem. Biophys. Res. Commun.* **264:** 343-347

Soussi, B, Idstrom, J.P, Schersten, T., and Bylund-Fellenius, A.C., 1990, Cytochrome *c* oxidase and cardiolipin alterations in response to skeletal muscle ischemeia and reperfusion. *Acta Physiol Scand* **138:** 107-14

Spooner, P.J., Duralski, A.A., Rankin, S.E., Pinheiro, T.J., Watts, A., 1993, Dynamics in a protein-lipid complex: Nuclear magnetic resonance measurements on the headgroup of cardiolipin when bound to cytochrome *c*. *Biophys. J.* **65:** 106-112

Suter, M., Remé, C., Grimm, C., Wenzel, A., Jäättela, M., Esser, P., Kociok, N., Leist, M., Richter, C., 2000, Age-related macular degeneratuin. The lipofuscin component N-retinyl-N-retinylidene ethanolamine detaches proapoptotic proteins from mitochondria and induces apoptosis in mammalian retinal pigment epithelial cells. *J. Biol. Chem.* **275:** 39625-39630

Ursini, F., Maiorino, M., and Gregolin, C., 1985, The selenoenzyme phospholipid hydroperoxide glutathione peroxidase. *Biochim. Biophys. Acta* **839:** 62-70

Ushmorov, A., Ratter, F., Lehmann, V., Droge, W., Schirrmacher, V., and Umansky, V., 1999, Nitric-oxide-induced apoptosis in human leukemic lines requires mitochondrial lipid degradation and cytochrome *c* release. *Blood*, **93:** 2342-2352

Wendel, A., 1981, Glutathione peroxidase. *Methods Enzymol.* **77:** 325-333

Yagi, K., Komura, S., Kojima, H., Sun, Q., Nagata, N., Ohishi, N., and Nishikimi, M., 1996, Expression of human phospholipid hydroperoxide glutathione peroxidase gene for protection of host cells from lipid hydroperoxide-mediated injury. *Biochem. Biophys. Res. Commun.*, **219:** 486-491

Yagi, K., Shidoji, Y., Komura, S., Kojima, H., and Ohishi, N., 1998, Dissipation of mitochondrial membrane potential by exogenous phospholipid monohydroperoxide and protection against this effect by transfection of cells with phospholipid hydroperoxide glutathione peroxidase gene. *Biochem. Biophys. Res. Commun.* **245:** 528-533

Yan, L. J., and Sohal, R. S., 1998, Mitochondrial adenine nucleotide translocase is modified oxidatively during aging. *Proc. Natl. Acad. Sci. USA* **95**: 12896-12901

Zou, H., Henzel, W., Liu, X., Lutschg, A., and Wang, X., 1997, Apaf-1, a human protein homologous to C. elegans CED-4, participates in cytochrome *c*-dependent activation of caspase-3. *Cell,* **90:** 405-413

Chapter 3

Plasma Membrane Phospholipid Asymmetry

Peter J. Quinn
Division of Life Sciences, King's College London, 150 Stamford Street, London SE1 9NN, UK

1. INTRODUCTION

The matrix in which the proteins of biological membranes are oriented and distributed is comprised of a bilayer of amphipathic lipids. In animal cell membranes the predominant lipid components are diacylglycerophospholipids and sphingolipids with lesser amounts of glyceryl ether phospholipids, plasmalogens and cholesterol. Biophysical studies have established that the bilayer is fluid and the interior hydrocarbon domain has a viscosity approximating that of olive oil. Because of this liquid-like character the polar lipids are able to rotate freely about their axis perpendicular to the plane of the membrane and to diffuse readily in the lateral plane. Movement from one leaflet of the bilayer to the other, however, is severely constrained and measured in half times of hours or days. The origin of this constraint is the free energy required to move a hydrated polar moiety from the aqueous interface into the hydrocarbon interior of the structure.

As a consequence of this restricted motion an asymmetric distribution of lipids can be created and maintained across biological membranes. The first evidence that the polar lipids are not randomly distributed between the two leaflets of the membrane lipid bilayer was obtained using reactivity of aminophospholipids to chemical probes such as formylmethionyl-(sulphonyl)methylphosphate and trinitrobenzenesulphonic acid (TNBS) both of which are impermeant to the membrane (Bretscher, 1972a; Gordesky and Marinetti, 1973). Later studies employed florescent amino reagents which offered greater utility (Koynova and Tenchov, 1983). Similar conclusions were drawn from studies of the susceptibility of membranes to phospholipase A_2 and sphingomyelinase (Verkleij *et al.*, 1973). These early studies concentrated on the erythrocyte membrane in which the methodology was not compromised by the presence of subcellular membranes. The distribution of the major membrane lipids in the inner and outer leaflets of the erythrocyte membrane are presented in Figure 1.

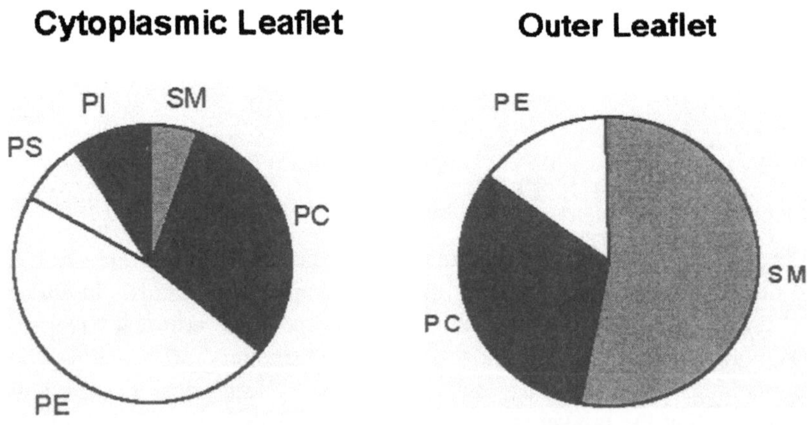

Figure 1. Percent distribution of major phospholipids between the outer and cytoplasmic leaflets of the human erythrocyte membrane. SM, sphingomyelin; PC, phosphatidylcholine; PE, phosphatidylethanolamine; PS, phosphatidylserine; PI, phosphatidylethanolamine

It can be seen from Figure 1 that the choline-containing phospholipids, phosphatidylcholine and sphingomyelin are localized predominantly in the outer monolayer of the plasma membrane. The aminophospholipids, comprising phosphatidylethanolamine and phosphatidylserine, by contrast, are enriched in the cytoplasmic leaflet of the membrane (Bretcher, 1972b; Rothman and Lenard, 1977; Op den Kamp, 1979). The transmembrane distribution of the minor membrane lipid components has been determined by reaction with lipid-specific antibodies (Gascard *et al.*, 1991) and lipid hydrolases (Bütikofer *et al.*, 1990). Such studies have shown that phosphatidic acid, phosphatidylinositol and phosphatidylinositol-4,5-*bis*-phosphate all resemble phosphatidylethanolamine in that about 80% of the phospholipids are localized in the cytoplasmic leaflet of the membrane.

Reviews of membrane phospholipid asymmetry in health (Zachowski, 1993) and disease (Zwaal and Schroit, 1997) have been published. This chapter will be concerned with the process whereby the asymmetric distribution of phospholipids is generated and how dissipation of the arrangement of phospholipids in the membrane triggers physiological responses by cells, principally apoptosis.

2. MEASUREMENT OF MEMBRANE PHOSPHOLIPID ASYMMETRY

Methods used to demonstrate the existence of membrane phospholipid asymmetry, such as chemical labelling and susceptibility to hydrolysis or modification by phospholipases and other enzymes, are unsuitable for dynamic studies because the rates of chemical and biochemical reactions are of a different order compared to the transmembrane translocation of the phospholipids. Indirect methods have therefore been developed to measure the translocation rate which are consequent on the loss of membrane phospholipid asymmetry. Thus time scales appropriate to rates of lipid scrambling under resting conditions or when the forces preserving the asymmetric phospholipid distribution are disturbed can be monitored. Generally the methods rely on detecting the appearance of phosphatidylserine on the surface of cells. Methods of demonstrating lipid translocation in mammalian cells has been the subject of a recent review (Bevers *et al.*, 1999).

2.1 Thrombin formation

The appearance of anionic phospholipids, particularly phosphatidylserine, on the cell surface activates prothrombinase complex culminating in the formation of thrombin (Bevers *et al.*, 1982; Connor *et al.*, 1989). The assay can be performed with pure coagulation proteins and specific chromogenic substrates to produce a very sensitive test to detect the appearance of phosphatidylserine on cell surfaces. Nevertheless, it has been shown that changes in the disposition of phosphatidylethanolamine and sphingomyelin may interfere with the ability of phosphatidylserine-containing membranes to activate prothrombinase (Smeets *et al.*, 1996).

2.2 Annexin V binding

Annexin V is a human placental anticoagulant protein of molecular weight 35kDa that binds to membranes and lipid bilayers containing phosphatidylserine in the presence of free calcium. Annexin V binding to cell surfaces said to result from transmembrane movement of phosphatidyserine is often used as a diagnostic indicator of apoptosis. Studies of a florescent derivative of annexin V showed binding to phospholipid model membranes with a stoichiometry of 84 phosphatidylserine molecules per molecule of protein in the presence of 1.2 mM Ca^{2+} with a Kd of 0.036nM (Tait and Gibson, 1992).

The binding of annexin to membranes is influenced by other membrane components. This was exemplified in studies of annexin V binding to phospholipid mixtures coated onto glass beads and separated by flow cytometry. It was found that incorporation of phosphatidylethanolamine into the phospholipid mixture reduced the threshold level for annexin V binding, sphingomyelin had no significant effect and cholesterol reduces binding to the phospholipid surface (Stuart *et al.*, 1998). Furthermore, calcium concentrations greater than 3mM were required to maximize binding of annexin V to the phospholipid.

The calcium-dependent binding of annexin V to phosphatidylserine-containing membranes has been monitored using derivatives that are florescent (Dachary-Prigent *et al.*, 1993), radioactive (Thiagarajan and Tait, 1990), biotinylated (van Engeland *et al.*, 1996) or covalently linked to magnetic nanoparticles (Sestier *et al.*, 1995; Geldwerth *et al.*, 1999). With biotinylated or magnetically labelled derivatives it is possible to separate annexin V binding cells in a population and to compare their behaviour and properties with cells that do not bind the protein. Time-dependent processes can also be investigated using such labels.

The reliability of annexin V binding as diagnostic evidence of apoptosis has been challenged recently in studies of annexin binding to human periferal lymphocytes, Jurkat T cells, U937 cells or HeLa cells subjected to apoptotic stimuli campared to oncotic stimuli (Lecoeur *et al.*, 2001). It was reported that oncotic cells, like apoptotic cells, bind annexin V without loss of membrane integrity as judged by propidium iodide uptake. IgE-dependent stimulation of rat, murine and human mast cells in which high-affinity IgE receptors, FcepsilonRI, are aggregasted on the surface of the cells and allergic mediator released, is found to be associated, in a dose-dependent manner, with appearance of phosphatidylserine on the cell surface as judged by calcium-dependent annexin V binding (Martin *et al.*, 2000). Blockage of stimulus-secretion coupling with pharmacological agents prevented appearance of phosphatidyserine on the cell surface and blocked calcium ionophore-induced phosphatidylserine appearance leading to the conclusion that the loss of phosphatidylserine asymmetry is a direct consequence of exocytosis rather than an early signaling event associated with IgE receptor redistribution. The loss of phosphatidylserine asymmetry across the plasma membrane was reversed irrespective of whether the stimulus was prolonged and there were no signs of apoptosis from examination of the nuclei indicating that the appearance of phosphatidylserine does not invariably initiate apoptosis.

Similar reversibility of annexin V binding to lymphocytes has been reported. Hammill *et al.* (1999) investigated the mechanism controlling apoptosis in a B-cell lymphoma cell line stimulated by cross-linking membrane immunoglogulin (mIg) receptors. It was demonstrated that receptor cross-linking resulted in a high proportion of the cells binding annexin V but once the signal was removed the cells were shown to be viable and could resume growth and release annexin V from the cell surface indicating that phospholipid asymmetry across the membrane had been restored. It was concluded that membrane phospholipid asymmetry as reflected in annexin V binding precedes commitment to apoptotic death in these cells.

Annexin V binding has been used to investigate the induction of apoptosis by a variety of factors. Induction of apoptosis in the human leukaemia cell line, HL60, by photosensitizer dyes (Purpurin-18) administered in low doses after exposure to visible light, for example, has been monitored by annexin V binding to the cell surface (Di Stefano *et al.*, 2001). It was found that when higher doses of the drug was employed necrosis of the cells resulted.

2.3 Phospholipid probes

The dynamics of lipid movement between the two leaflets of membrane lipid bilayers has been monitored using a variety of phospholipid analogs. Most of the analogs incorporate a reporter group attached to the C-2 of the glycerol moiety. One common substituent is a florescent-labelled fatty acid to form 1,2-(palmitoyl-N-4-nitrobenzo-2-oxa-1,3-diazole-amino-caproyl) phosphatidylserine (NBD-PS) (Martin and Pagano, 1987). Spin-labelled analogs have also been used (Seigneuret and Deveaux, 1984). Florescent and spin-labelled analogs can be readily incorporated into the outer monolayer of the membrane and then rapidly quenched by addition of membrane-impermeant reagents to provide information about the proportion of the incorporated analog resident in the outer leaflet and that which has migrated to inaccessible locations within the cell (McIntyre and Sleight, 1991; Williamson *et al.*, 1995). The major disadvantage of these analogs is that the reporter group introduces significant changes in lipid conformation and properties and the effect this may have on behaviour is problematic. This is not a factor in radioactive analogs which can be readily incorporated into the membrane by phospholipid transfer proteins but bilayer dynamics must be assessed by susceptibility to phospholipase enzymes (Tilley *et al.*,

1986). Methods of assessing transmembrane movement of sphingolipids have been reviewed by Sillence *et al.* (2000).

Measurement of transmembrane movement of phospholipids using florescent or other probe methods requires demonstration that incorporation of the phospholipid analogs occurs by a monomeric transfer from an exogenous resevoir into the outer leaflet of the plasma membrane rather than a fusion of bilayer vectors containing the probes with the plasma membrane. This problem has been addressed by Gadella *et al.* (1999) who established a protocol for incorporation of NBD-phospholipid analogs into plasma membrane of boar spermatozoa under physiological conditions. The total amount of fluorescence phospholipid incorporated and the proportion of translocated phospholipid fluorescence were determined by flow cytometric analysis before, and after fluorescence from the outer leaflet was abolished by addition of dithionite. Phospholipid uptake and internalization was monitored at 38°C and after 1 hr of labeling, about 96% of the incorporated phosphatidylserine, 80% of phosphatidylethanolamine, 18% of phosphatidylcholine, and 4% of sphingomyelin were found to have moved across the plasma membrane bilayer to the interior of the spermatozoa. Catabolism of incorporated florescent phospholipids was inhibited by addition of phenylmethylsulfonyl fluoride and cells with damaged membranes were identified with impermeant DNA stains and could therefore be excluded from the flow cytometric data. The inward movements of florescent phospholipids were found to be ATP-dependent and could be blocked with sulfhydryl reagents. It was also noted that movements of florescent phospholipids from the inner to the outer leaflet of the spermatozoa plasma membrane were not significant for intact, but were rapid and ATP-independent for florescent lipid metabolites.

3. GENERATION OF PHOSPHOLIPID ASYMMETRY

An account of how membrane lipids are distributed throughout the cell and topologically across membranes has been given by van Meer (2000). The process is initiated during membrane biogenesis and is sustained and augmented by phospholipid translocases as membranes are sorted and differentiated throughout the cell.

3.1 Vectorial biosynthesis

An asymmetric distribution of phospholipids is generated during membrane biogenesis. Most of the enzymes responsible for synthesis of phospholipids and their subsequent turnover are located at sites accessible to the cytoplasmic surface of the membrane. Evidence for this has come from various approaches including the action of regulatory factors like c-Fos. c-Fos is an inducible transcription factor that constitutes DNA-binding AP-1 complexes which regulate gene expression responsible for long-lasting cellular changes and act to modulate phospholipid biosynthesis. It has been shown that c-Fos interacts directly with lysophosphatidic acid acyl transferase and phosphatidic acid phosphatase, both key enzymes in phospholipid biosynthesis, on the cytoplasmic surface of the endoplasmic reticulum (Bussolino *et al.*, 2001). The biosynthesis of sphingolipids appears to be an exception to the rule. Sphingomyelin, for example, is synthesized on the luminal surface of the membrane in the Golgi phase and glycosidic residues of the glycosphingolipids are also attached to ceramide at this site.

Phospholipid turnover also takes place in an asymmetric manner. The enzymes responsible for phospholipid turnover in response to receptor-mediated phospholipase c activation are active from the cytoplasmic surface of the membrane. Likewise, diacylglycerol kinases converting the product of phospholipase c back into the key intermediate of phospholipid biosynthesis, phosphatidic acid, are also located on the cytoplasmic surface of the membrane (Sanjuan *et al.*, 2001).

3.2 Protein-lipid interactions

Once synthesized several factors influence the particular leaflet of the membrane lipid bilayer where the lipids reside. One is static interactions with intrinsic and extrinsic membrane proteins which, by virtue of their mechanism of biosynthesis, are also asymmetric with respect to the membrane. The interaction of the cytoplasmic protein, spectrin with the erythrocye membrane has been the subject of a number of studies. Coupling of spectrin to the transmembrane proteins, band 3 and glycophorin 3 via ankyrin and protein 4.1, respectively, has been well documented (van Dort *et al.*, 1998). Interaction of spectrin with membrane lipids is still somewhat conjectural but recent studies have characterized such interactions more precisely. O'Toole *et al.* (2000) have used a fluorescine derivative of phosphatidylethanolamine to investigate the binding affinity of spectrin to lipid bilayers comprised of phosphatidylcholine or a binary mixture of phosphatidylcholine and phosphatidylserine. They concluded on the basis of fluorescence intensity measurements that despite possessing a net negative charge at pH>5.6 the surface of the protein interfaced with the lipid presented a positively charged domain which presumably represented sites of interaction with negatively-charged lipids. Nevertheless, the dissociation constants of spectrin for binding to neutral and negatively-charged phospholipid interfaces could not be distinguished suggesting a multipoint attachment of the whole protein to the interface with specific domains on the protein preferentially interacting with phosphatidylserine.

An example of preferential interaction of membrane associated proteins with acidic phospholipids is dynamin. Dynamin acts at the cytoplasmic surface of the plasma membrane to effect receptor mediated endocytosis. The protein locates at the neck of deeply invaginated coated pits where it induces membrane constriction and the pinching off of endocytotic vesicles resulting from a conformational change driven by GTP hydrolysis (Schmid *et al.*, 1998). Studies of the binding of dynamin to different phospholipid monolayers has revealed that the protein binds very strongly to acidic phospholipids, particularly phosphatidylinositol and phosphatidic acid. The binding to these phospholipids is said to induce penetration of the protein into the hydrophobic domain of the lipid, causing destabilization of the membrane and fission (Burger *et al.*, 2000).

While examples such as these provide evidence that strong interactions of negatively-charged membrane lipids with membrane proteins the role in maintaining asymmetric distributions of lipids across biological membranes is unclear. In any event such effects are likely to be of minor importance relative to actively mediated phospholipid translocation processes.

3.3 Transmembrane phospholipid translocation

Active translocation of phospholipids across the plasma membrane has been demonstrated both from the inner to the outer leaflet and from the outer to the inner leaflet. The translocation processes specifically transport phosphatidyserine and phosphatidylethanolamine from the cytoplasmic to the outer surface of the membrane while choline phosphatides are transported from the outer to the cytoplasmic surface. The rate of translocation, in general, is greater for the amino phospholipids compared with the choline phospholipids.

3.3.1 Translocation of amino phospholipids

The aminophospholipid translocase is an ATPase II-type enzyme that requires Mg^{2+} and is activated by phosphatidylserine and to a lesser extent by phosphatidylethanolamine and is sensitive to the sulphydryl group reagent, N-ethylmaleimide. Because the enzyme is inhibited by vanadate ions it is categorised as a P-type ATPase. The properties with respect to substrate specificity, biochemical and physical properties of the enzyme and strategies for identification and isolation from various sources have been reviewed by Daleke and Lyles (2000). The protein has a molecular weight of about 116kDa and has been isolated from the plasma membrane of erythrocytes (Morrot *et al.*, 1990), clathrin-coated vesicles (Xie *et al.*, 1989), synaptic vesicles (Hicks and Parsons, 1992) and chromaffin granules (Zachowski *et al.*, 1989).

Sequence data of ATPase II from human, mouse and bovine tissues indicates that the protein has several P-type consensus sequences and a membrane topology that includes 10 putative transmembrane domains. An ATP binding site and a phosphoenzyme formation site are located within the largest cytosolic loop, whereas a sequence implicated in the coupling to transport activity was identified in another hydrophilic loop. The sequence of mammalian ATPase II is homologous to a yeast ATPase encoded by the *drs2* gene, of which the null mutant of a yeast strain lacks a specific phosphatidylserine internalization activity that is otherwise present in wild type yeast strains (Tang *et al.*, 1996). Four isoforms of the ATPase II enzyme have recently been identified in bovine brain (Ding *et al.*, 2000). The substrate specificities and selectivities were characterized and the results are summarised in Table 1. This shows relative specificities of phosphatidylcholine and phosphatidylethanolamine compared with phosphatidylserine for activation of the ATPase by different isoforms of recombinant ATPase II. It was found that there was no significant difference between the presence or absence of phosphatidylcholine suggesting that this phospholipid is not an activator of ATPase activity. The ratio of activation phosphatidlyserine/phosphatidylcholine shown in the Table is therefore a reliable indication of the relative specific activity of the different ATPase II isoforms. The activity is in the order $\alpha 2 > \alpha 1 > \beta 2 > \beta 1$. A different order is observed in relative selectivity of phosphatidylserine compared with phosphatidylethanolamine in which the order is $\alpha 2 > \beta 2 > \alpha 1 > \beta 1$.

Table 1. Relative phospholipid specificities for activation of different isoforms of recombinant ATPase II.

Relative Specificity	α1	α2	β3	β4
PS/PC	53.0	68.0	25.0	32.5
PS/PE	2.7	10.9	1.3	3.5

PS, phosphatidylserine; PC, phosphatidylcholine; PE, phosphatidylethanolamine. Data from Ding *et al.* (2000).

The involvement of Drs2 protein in the transbilayer movement and distribution of phospholipids in the plasma membrane of the *S. cerevisiae end4Δ* mutant has been investigated by (Marx *et al.*, 1999). In this mutant, both growth and the internalization step of endocytosis are blocked at a restrictive temperature of >34°C (Raths *et al.*, 1993). It is known that several proteins such as ATP-binding cassette multidrug transporters Pdr5 (Egner *et al.*, 1995) and Ste6 (Kölling and Hollenberg, 1994) accumulate in the plasma membrane of the yeast after incubation under non-permissive conditions and it was thought that this may have consequences for the organization and dynamics of the plasma-membrane lipid phase. It was pointed out that the Ste6 protein has been shown to transport synthetic alkyl phospholipids (Reutz, *et al.*, 1997) but it is not known whether the protein-mediated transbilayer movement and the transbilayer asymmetry of lipids in the plasma membrane of the *S. cerevisiae end4Δ* mutant are altered. The transbilayer movement, of NBD analogs of choline and serine phosphatides was measured in an *S. cerevisiae end4Δdrs2Δ* strain, to determine whether internalization of NBD-labelled analogs by transbilayer movement depended on a functional *drs2* gene in endocytosis-deficient yeast cells. It was reported that the exposure of endogenous aminophospholipids to the exoplasmic leaflet of the mutant cells is altered with respect to wild-type cells.

3.3.2 Translocation of choline phospholipids

There is now convincing evidence that, in the erythrocyte membrane, a family of membrane P-glycoproteins are involved in transmembrane phospholipid movement. The translocase responsible for non-specific movement of phospholipid from the cytoplasmic leaflet to the surface of the erythrocye membrane has been shown to be identical with the multi drug resistance protein 1 (Kamp and Haest, 1998; Dekkers *et al.*, 1998, 2000). The proteins appear to translocate particular phospholipids on different cell types. Sphingolipids, for example, are preferentially transported across the plasma membrane of LLC-PK1 kidney epithelial cells transfected with MRP1 cDNA (Raggers *et al.*, 1999). By contrast, phosphatidylcholine was the preferred phospholipid translocated by MPR 3 in bile canicular cells where the protein predominates (Smit *et al.*, 1993) and the plasma membrane of fibroblasts of transgenic mice (Smith *et al.*, 1994). Other MRP P-glycoproteins have different specificities for translocation of phospholipids across membranes (van Helvoort *et al.*, 1996, 1997).

4. DISSIPATION OF TRANSMEMBRANE PHOSPHOLIPID ASYMMETRY

The loss of membrane phospholipid asymmetry is known to be an important signaling mechanism. As will be discussed below, the consequences of appearance of specific phospholipids on opposite sides of the membrane triggers a variety of cellular responses.

4.1 Phospholipid scramblase

A family of membrane proteins referred to as phospholipid scramblases (PLSCR) have been implicated in dissipating phospholipid asymmetry in a process that depends on elevated cytoplasmic calcium concentration (Zhao *et al.*, 1998a). Four genes have been identified that code for phospholipid scramblases in human and mouse and all have been conserved through evolution (Wiedmer *et al.*, 2000). The amino acid sequence deduced for one of the human phospholipid scramblases, HuPLSCR1, indicates that the scramblase is a Type-2 membrane protein of molecular weight 35kDa (318 amino acid residues) with a transmembrane helical domain located towards the C-terminus (Zhou *et al.*, 1997). The N-terminal cytoplasmic domain was found to possess a putative phosphorylation site and a calcium binding segment. There is evidence that regulation of scramblase activity may be mediated by phosphorylation in the presence of PKCδ the action of which results in the surface exposure of phosphatidylserine in apoptosing cells (Frasch *et al.*, 2000). Regulation of scramblase activity may also occur via palmitolylation at cysteine thiol residues of the protein (Zhao *et al.*, 1998b).

Interferon has been shown to markedly upregulate transcription of PLSCR1 but the resulting increased expression of the scramblease was not associated with any perturbation of plasma membrane phospholipid asymmetry or enhancement of phospholipid scrambling in response to increased cytoplasmic calcium as judged by exposure of phosphatidylserine on the cell surface (Zhou *et al.*, 2000). Studies of a mouse mutant strain in which the N-terminal region of the scramblase is truncated have provided evidence that the c-Abl proto-oncogene tyrosine kinase binds to this region of the scramblase and phosphorylates two tyrosyl residues (Sun *et al.*, 2001). This suggests that phosphorylation by this reaction may represent an additional regulatory mechanism to control phospholipid scramblase activity.

4.2 The role of cytoplasmic calcium

Cytoplasmic Ca^{2+} concentration modulates membrane phospholipid asymmetry by activating the scramblase and inhibiting the Mg^{2+}-ATPase aminophospholipid translocase. The sensitivity of the phospholipid scramblase to cytoplasmic Ca^{2+} in resealed human erythrocyte ghost membranes has been investigated using florescent probes sensitive to $[Ca^{2+}]$ and measuring the translocation of florescent phospholipid analogs across the membrane (Woon *et al.*, 1999). It was found that phospholipid scrambling was half maximally activated at about 75mM cytoplasmic free Ca^{2+}.

4.3 Membrane phospholipid homeostasis

The asymmetric distribution of phospholipids across the plasma membrane has implications for membrane phospholipid turnover and phospholipid homeostasis as reflected in their susceptibility to soluble phospholipases. This is evidenced by studies reported by Murakami *et al.* (1999) who examined the functional properties of distinct $sPLA_2$ isozymes, namely types IIA, V, and X, by overexpressing them in human embryonic kidney 293 cells. They obtained evidence that $sPLA_2$-X acts on cells in a manner different from the heparin-binding group II subfamily of $sPLA_2$s. The heparin-binding $sPLA_2$s-IIA and -V, but not the heparin-nonbinding $sPLA_2$-X, have the ability to up-regulate the cytokine-induced COX-2 expression, which leads to efficient delayed prostaglandin biosynthesis from the released arachidonic acid substrate.

Experiments involving coexpression of phospholipid scramblase with $sPLA_2$ revealed that altered membrane asymmetry leads to increased cellular susceptibility to the group II subfamily of $sPLA_2$s. Nevertheless, Murakami *et al.* (1999) identified several lines of evidence to suggest that phospholipid scramblase-mediated alteration in plasma membrane asymmetry culminating in increased cellular susceptibility to the group II subfamily of $sPLA_2$s does not entirely mimic their actions on cells after IL-1 signaling. It was noteworthy that coexpression of type IIA or V, but not X, $sPLA_2$ and phospholipid scramblase resulted in a marked reduction in cell growth, suggesting an antiproliferative action of particular classes of $sPLA_2$.

5. CELL FUNCTIONS MEDIATED BY MEMBRANE PHOSPHOLIPID DISTRIBUTION

The creation of a solute gradient across biological membranes in unstimulated cells and receptor-mediated dissipation of the gradient is a mechanism frequently employed in transmembrane signal transduction. The creation of an asymmetric distribution of membrane phospholipids in an ATP-driven mechanism and randomisation in response to specific signals shares many common features with this process. Some of these signal transducing systems, including contraction coupling in cardiac muscle and membrane phospholipid turnover, have been reviewed recently (Verkleij and Post, 2000). Several physiological events have been implicated in the dissipation of membrane phospholipid asymmetry including adaptation of plants to low temperature, the transduction of light signals, endocytosis and apoptosis.

5.1 Acclimation of plants to low temperature

The maintenance of membrane lipid asymmetry has been linked to adaptation of plants to cold. A protein believed to be associated with cold acclimation in Arabidopsis, ALA1 (aminophospholipid ATPase 1), a novel P-type ATPase has been found to have close sequence homology with those of yeast Drs2 and bovine ATPase II both of which are putative aminophospholipid translocases (Gomes *et al.*, 2000). ALA1 was shown to complement the deficiency in phosphatidylserine internalization into intact yeast cells of the *drs2Δ* mutant, and expression of ALA1 results in increased translocation of aminophospholipids in vesicles reconstituted from yeast membranes. ALA1 also complements the cold sensitivity of the *drs2Δ* yeast mutant. Present evidence suggests that ALA1 is involved in generating membrane lipid asymmetry and probably encodes an aminophospholipid translocase in Arabidopsis plasma membrane. Down regulation of ALA1 in Arabidopsis results in cold susceptible plants that are much smaller than those of the wild type.

5.2 Light transduction cycle of rhodopsin

Asymmetric distribution of phospholipids across the retinal rod outer segment disk membrane has been shown to be associated with light reception by rhodopsin. It is known that the major phospholipids of this membrane, phosphatidylcholine and phosphatidylethanolamine, are symmetrically distributed across the membrane in the dark but not phosphatidylserine which is located predominantly on the cytoplasmic leaflet (Wu and Hubbell, 1993; Hessel *et al.*, 2000). It is thought that the asymmetry of phosphatidylserine is dictated by the charge asymmetry of the rhodopsin molecule (Tsiu *et al.*, 1990; Hubbell, 1990). It has been reported recently using spin-lable phospholipid analogs that phosphatidylserine is released from its binding to the rhodopsin molecule immediately upon illumination whereupon it redistributes between the two leaflets of the disc membrane (Hessel *et al.*, 2001). Transmembrane movement of phosphatidylcholine and phosphatidylethanolamine was also observed but the kinetics were more complex and unrelated to that of phosphatidylserine. The transient changes in phospholipid transmembrane distribution were believed to be associated with the disappearance of active rhodopsin species and their subsequent regeneration in the signal transduction cycle.

5.3 Endocytosis

Membrane phospholipid asymmetry has been implicated as a mechanism of inducing membrane curvature required for endocytotic vesicles to be pinched off from the plasma membrane. In this process the action of the aminophospholipid translocase in moving aminophospholipids from the outer to the inner leaflet of the membrane is said to increase the proportion of total phospholipids in the inner leaflet. As a consequence a local curvature is generated sufficient to cause budding of small endocytotic vesicles (Farge, 1995). The involvement of the aminophospholipid translocase in this process has been demonstrated by observing a 2 to 4-fold increase in the number of endocytotic vesicles in living K562 cells following addition of exogenous aminophospholipids which are translocated to the cytoplasmic leaflet (Farge *et al.*, 1999; Rauch and Farge, 2000).

A function of membrane phospholipid asymmetry at the subcellular membrane level has also been implied from the location of P-type ATPase coded by *drs2* gene identical to aminophospholipid translocase in golgi vesicles in yeast (Chen *et al.*, 1999). A mutant *drs2*Δ strain of *Saccharomyces cerevisiae* was found to have a defect in late Golgi function and exhibit several phenotypes at a nonpermissive temperature that suggested a loss of clathrin function in the yeast Golgi *in vivo*. These conclusions were consistant with immunofluorescence and subcellular fractionation studies which indicated that Drs2p localizes to late Golgi membranes. This suggested that the primary site of Drs2p function is in the Golgi complex rather than the plasma membrane, and infered a link between Drs2p and the ARF-dependent recruitment of clathrin to Golgi membranes.

5.4 Apoptosis

Reviews of the role of aminophospholipid translocase and scramblase (Schlegel *et al.*, 2000) and the consequences of the appearance of phosphatidylserine on the cell surface (Williamson *et al.*, 2001) in apoptosis of thymocytes have been published. The precise relationship between membrane phospholipid asymmetry and apoptosis is currently a topic of considerable interest.

The temporal sequences of apoptotic events has received much attention, particularly with respect to the initiation process. A number of studies have been designed to investigate the relationship between changes in plasma membrane phospholipid asymmetry and other signs of apoptosis. Chan *et al.* (1998), for example, examined the time-course of annexin V binding to the plasma membrane of three cell types (i) rat thymocytes treated with methylprednisolone (ii) the human histiocytic lymphoma cell line, U937 induced to undergo apoptosis by treatment with topoisomerase inhibitor (iii) PC12 rat pheochromocytoma cells apoptosing in serum depleted growth medium and correlated this with DNA strand breaks as judged by *in situ* tailing (IST or TUNEL) reaction. They concluded that, at least in the three systems examined, exposure of phosphatidylserine on the cell surface preceeded strand breaks in DNA of cells induced to undergo apoptosis.

The loss of phospholipid asymmetry in HL-60 cells and exposure of phosphatidylserine on the cell surface does not appear to be a mandatory requirement to trigger apoptosis in these cells. When HL-60 cells were irradiated with UV light after polyamines such as putrescine and spermidine levels had been reduced by treatment of the cells with an inhibitor of ornithine decarboxylase, difluoromethylornithine (Bratton *et al.*, 1999) they did not bind annexin V indicating that phosphatidylserine was not localized on the surface of the cells. Nevertheless, the cells displayed typical signs of apoptosis including DNA fragmentation, caspase 3 activation and decreased transmembrane flip-flop activity as judged by uptake of a florescent analog of phosphatidycholine. Apoptotic human HT-1080 fibrosarcoma cells produced by UV irradiation of cells depleted of polyamines were not phagocytosed by macrophages compared with apoptotic cells that bound annexin V which were found to be phagocytosed normally (Fadok *et al.*, 2001). It was suggested that external phosphatidylserine acted as a specific recognition factor for macrophage engulfment. The authors tested this hypothesis by enriching the plasma membrane of cells induced to undergo apoptosis by irradiation with UV after depletion of polyamines with either D- or L-sterioisomers of 1,2-diacyl-*sn*-glycerophosphoserine. Phagocytosis was restored only if the L-sterioisomer was present in the plasma membrane and not the D-sterioisomer suggesting a specific receptor for the natural sterioisomer of phosphatidylserine is present on macrophages that triggers phagocytosis.

The role of external phosphatidylserine in opsonization has also been investigated by examination of phagocytosis of apoptotic cells using the florescent dye, merocyanine 540 (Callahan *et al.*, 2000). It was found that the dye binds not only to thymocytes which express phosphatidylserine on their surface but also to apparently normal macrophages suggesting that these cells also display phosphatidylserine on their surface. Studies of annexin V binding to macrophages confirmed that phosphatidylserine was normally expressed on the macrophage surface and that annexin V binding blocked phagocytosis of apoptotic thymocytes. Control experiments showed engulfment of latex particles or Fc receptor-mediated phagocytosis of opsonized erythrocytes was unaffected under these conditions. It was concluded that phosphatidylserine is constitutively expressed on the surface of macrophages and is required for phagocytosis of apoptotic cells displaying phosphatidylserine on their surface.

One of the consequences of lipid scrambling across the plasma membrane is that, apart from exposure of phosphatidylserine on the cell surface, sphingomyelin and phosphatidylcholine are presented on the cytoplasmic surface. Tepper *et al.* (2000) examined the relationship between the two lipid events in the apoptotic execution phase. They used scrambling-competent Jurkat T and SKW6.4 B cells and scrambling-deficient Raji B cells to follow the fate florescent (NBD-analogs of sphingomyelin and phosphatidylcholine. They found that one of the consequences of the loss of phospholipid asymmetry, sphingomyelin moves from the outer to the inner leaflet of the plasma membrane, where it serves as a substrate for an intracellular sphingomyelinase. This process serves to modulate the availability of substrate during apoptosis and thereby controls ceramide formation. They went on to show that the breakdown of sphingomyelin causes a concomitant efflux of cholesterol which results in significant alterations in the biophysical properties of the plasma membrane representing a prerequisite for membrane blebbing and vesiculation at the surface of the apoptotic cell.

6. CONCLUSIONS

The maintenance of an asymmetric distribution of phospholipids across the plasma membrane with choline phospholipids predominating on the external surface and amino phospholipids confined to the cytoplasmic leaflet of the membrane has now been well established. The participation of aminophospholipid translocases in this process has been clearly demonstrated and the translocases themselves have been characterized.

One of the more significant events associated with dissipation of phospholipid asymmetry is the appearance of phosphatidylserine on the surface of the cell. This provides a clear signal for altering the hemeostatic balance since several procoagulant and anticoagulant reactions depend on interfaces containing phosphatidyserine. Furthermore, the appearance of phosphatidylserine on the cell surface appears to be the primary reaction mediating recognition and clearance of apoptotic cells by macrophages so as to prevent release of inflammatory cell contents associated with lysis.

The control of the processes of phospholipid translocation and scrambling by cytoplasmic calcium concentration is known to be one factor in regulating the appearance of phosphatidylserine on the cell surface. There is now evidence that other factors are involved in modulating membrane phospholipid distribution and for preserving phospholipid homeostasis. Current research is directed to clarify the role of these agents and to establish any connections with the initiation of apoptosis.

REFERENCES

Bevers, E.M., Comfurius, P., Dekkers, D.W.C. and Zwaal, R.F.A., 1999, Lipid translocation across the plasma membrane of mammalian cells. *Biochim. Biophys. Acta*, **1439**: 317-330.

Bevers, E.M., Comfurius, P., van Rijn, J.L.M.L., Hemker, H.C. and Zwaal, R.F.A., 1982, Generation of platelet prothrobmin converting activity and the exposure of phosphatidylserine at the platelt outer surface. *Eur. J. Biochem.*, **122**: 429-436.

Bratton, D.L., Fadok, V.A., Richter, D.A., Kailey, J.M., Frasch, S.C., Nakamura, T. and Henson, P.M., 1999, Polyamine regulation of plasma membrane phospholipid flip-flop during apoptosis. *J. Biol. Chem.*, **274**: 28113-28120.

Bretcher, M.S., 1972a, Asymmetrical lipid bilayer structure for biological membranes. *Nature New Biol.*, **236**: 11-12.

Bretcher, M.S., 1972b, Phosphatidyl-ethanolamine: differential labelling in intact cells and cell ghosts of human erythrocytes by a membrane-impermeable reagent. *J. Mol. Biol.*, **71**: 523-528.

Burger, K.N., Demel, R.A., Schmid, S.L. and de Kruijff, B., 2000, Dynamin is membrane-active: lipid insertion is induced by phosphoinositides and phosphatidic acid. *Biochemistry*, **39**: 12485-12493.

Bussolino, D.F., Guido, M.E., Gil, G.A., Borioli, G.A., Renner, M.L., Grabois, V.R., Conde, C.B. and Caputto, B.L., 2001, c-Fos associates with the endoplasmic reticulum and activates phospholipid metabolism. *FASEB J.*, **15**: 556-558.

Bütikofer, P., Lin, Z.W., Chiu, D.T-Y. and Kuypers, F.A., 1990, Transbilayer distribution and mobility of phosphatidylinositol in human red blood cells. *J. Biol. Chem.*, **265**: 16035-16038.

Callahan, M.K., Williamson, P. and Schlegel, R.A., 2000, Surface expression of phosphatidylserine on macrophages is required for phagocytosis of apoptotic thymocytes. *Cell Death Differ.*, 7: 645-53.

Chan, A., Reiter, R., Wiese, S., Fertig, G. and Gold, R., 1998, Plasma membrane phospholipid asymmetry precedes DNA fragmentation in different apoptotic cell models. *Histochem. Cell Biol.*, **110**: 553-558.

Chen, C.Y., Ingram, M.F., Rosal, P.H. and Graham,'T.R., 1999, Role for Drs2p, a P-type ATPase and potential aminophospholipid translocase, in yeast late Golgi function. *J. Cell. Biol.*, **147**: 1223-1236.

Connor, J., Bucana, C., Fidler, I.J. and Schroit, A.J., 1989, Differentiation-dependent expression of phosphatidylserine in mammalian plasma membranes: quantitative assessment of outer leaflet lipid by prothrombinase complex formation. *Proc. Natl. Acad. Sci., U.S.A.* **86**: 3184-3188.

Dachary-Prigent, J., Freyssinet, J.M., Pasquet, J.M., Carron, J.C. and Nurden, A.T., 1993, Annexin V as a probe of aminophospholipid exposure and platelet membrane vesiculation: a flow cytometry study showing a role for free sulfhydryl groups. Blood, 81: 2554-2565

Daleke, D.L. and Lyles, J.V., 2000, Identification and purification of aminophospholipid flippases. *Biochim. Biophys. Acta*, **1486**: 108-127.

Dekkers, D. W. C., Comfurius, P., Schroit, A. J., Bevers, E. M. and Zwaal, R. F. A., 1998, Transbilayer movement of NBD-labeled phospholipids in red blood cell membranes: outward-directed transport by the multidrug resistance protein 1 (MRP1). *Biochemistry*, **37**: 14833–14837.

Dekkers, D.W.C., Comfurius, P., van Gool, R.G.J., Bevers, E.M. and Zwaal, R.F.A., 2000, Multidrug resistance protein 1 regulates lipid asymmetry in erythrocyte membranes. *Biochem. J.*, **350**: 531-535.

Di Stefano, A., Ettorre, A., Sbrana, S., Giovani, C. and Neri, P., 2001, Purpurin-18 in combination with light leads to apoptosis or necrosis in HL60 leukemia cells. Photochem Photobiol. **73**: 290-296.

Ding, J, Wu, Z, Crider, B.P., Ma, Y., Li, X., Slaughter, C., Gong, L. and Xie, X.S., 2000, Identification and functional expression of four isoforms of ATPase II, the putative aminophospholipid translocase. Effect of isoform variation on the ATPase activity and phospholipid specificity. *J. Biol. Chem.*, **275**: 23378-23386.

Egner, R., Mahé, Y., Pandjaitan, R. and Kuchler, K., 1995, Endocytosis and vacuolar degradation of the plasma membrane-localized Pdr5 ATP-binding cassette multidrug transporter in *Saccharomyces cerevisiae*. *Mol. Cell. Biol.*, **15**: 5879–5887.

Fadok, V.A., de Cathelineau, A., Daleke, D.L., Henson, P.M. and Bratton, D.L., 2001, Loss of phospholipid asymmetry and surface exposure of phosphatidylserine is required for phagocytosis of apoptotic cells by macrophages and fibroblasts. *J. Biol. Chem.*, **276**: 1071-1077.

Farge, E., 1995, Increased vesicle endocytosis due to an increase in the plasma membrane phosphatidylserine concentration. *Biophys. J.*, **69**: 2501-2506.

Farge, E., Ojcius, D.M., Subtil, A. and Dautry-Varsat, A., 1999, Enhancement of endocytosis due to aminophospholipid transport across the plasma membrane of living cells. *Am. J. Physiol. Cell Physiol.*, **276**: C725-C733.

Frasch, S.C., Henson, P.M., Kailey, J.M., Richter, D.A., Janes, M.S., Fadok, V.A. and Bratton, D.L., 2000, Regulation of phospholipid scramblase activity during apoptosis and cell activation by protein kinase Cdelta. *J. Biol. Chem.* **275**: 23065-23073.

Gadella, B.M., Miller, N.G., Colenbrander, B., van Golde, L.M. and Harrison, R, A. 1999, Flow cytometric detection of transbilayer movement of florescent phospholipid analogs across the boar sperm plasma membrane: elimination of labeling artifacts. *Mol. Reprod. Dev.***53**: 108-125.

Gascard, P., Tran, P., Sauvage, M., Sulpice, J-C., Fukami, K., Takenawa, T., Claret, M. and Giraud, F., 1991, Asymmetric distribution of phosphoinositides and phosphatidic acid in the human erythrocyte membrane. *Biochim. Biophys. Acta*, **1069**: 27-36.

Geldwerth, D., Helley, D., de Jong, K., Sabolovic, D., Sestier, C., Roger, J., Pons, J.N., Freyssinet, J.M., Devaux, P.F. and Kuypers, F.A., 1999, Detection of phosphatidylserine surface exposure on human erythrocytes using annexin V-ferrofluid. *Biochem. Biophys. Res. Commun.*, **258**: 199-203.

Gomes, E., Jakobsen, M.K., Axelsen, K.B., Geisler, M. and Palmgren, M.G., 2000, Chilling tolerance in Arabidopsis involves ALA1, a member of a new family of aminophospholipid translocases. *Plant Cell*, **12**: 2441-2454.

Gordesky, S.F. and Marinetti, G.V., 1973, The asymmetric arrangement of phospholipids in the human erythrocyte membrane. *Biochem. Biophys. Res. Commun.*, **50**: 1027-1031.

Hammill, A.K., Uhr, J.W. and Scheuermann, R.H., 1999, Annexin V staining due to loss of membrane asymmetry can be reversible and precede commitment to apoptotic death. *Exptl. Cell Res.*, **251**: 16-21.

Hessel, E., Herrmann, A., Muller, P., Schnetkamp, P.P. and Hofmann, K.P., 2000, The transbilayer distribution of phospholipids in disc membranes is a dynamic equilibrium evidence for rapid flip and flop movement. *Eur. J. Biochem.*, **267**: 1473-1483.

Hessel, E., Muller, P., Herrmann, A. and Hofmann, K.P., 2001, Light-induced reorganization of phospholipids in rod disc membranes. *J. Biol. Chem.*, **276**: 2538-2543.

Hicks, B.W. and Parsons, S.M., 1992, Characterization of the P-type and V-type ATPases of cholinergic synaptic vesicles and coupling of nucleotide hydrolysis to acetylcholine transport. *J. Neurochem.*, **58**: 1211-1220.

Hubbell, W.L., 1990, Transbilayer coupling mechanism for the formation of lipid asymmetry in biological membranes. Application to the photoreceptor disc membrane. *Biophys. J.* **57**: 99-108.

Kamp, D. and Haest, C.M.W., 1989, Evidence for the role of multidrug resistance protein (MRP) in the outward translocation of NBD-phospholipids in the erythrocyte membrane. *Biochim. Biophys. Acta*, **1372**: 91-101.

Kölling, R. and Hollenberg, C.P., 1994, The ABC-Transporter Ste6 accumulates in the plasma membrane in a ubiquitinated form in endocytosis mutants. *EMBO J.*, **13**: 3261–3271.

Koynova, R.D. and Tenchov, B.G., 1983, Effect of ion concentration on phosphatidylethanolamine distribution in mixed vesicles. *Biochim. Biophys. Acta*, **727**: 351-356.

Lecoeur, H., Prevost, M.C. and Gougeon, M.L., 2001, Oncosis is associated with exposure of phosphatidylserine residues on the outside layer of the plasma membrane: A reconsideration of the specificity of the annexin V/propidium iodide assay. *Cytometry*, **44**: 65-72.

Martin, O.C. and Pagano, R.E., 1987, Transbilayer movement of florescent analogs of phosphatidylserine and phosphatidylethanolamine at the plasma membrane of cultured cells. *J. Biol. Chem.*, **262**: 5890-5898.

Martin, S., Pombo, I., Poncet, P., David, B., Arock, M. and Blank, U., 2000, Immunologic stimulation of mast cells leads to the reversible exposure of phosphatidylserine in the absence of apoptosis. *Int. Arch. Allergy Immunol.*, **123**: 249-258.

Marx, U., Polakowski, T., Pomorski, T., Lang, C., Nelson, H., Nelson, N. and Herrmann, A., 1999, Rapid transbilayer movement of florescent phospholipid analogs in the plasma membrane of endocytosis-deficient yeast cells does not require the Drs2 protein. *Eur. J. Biochem.* **263**: 254-263.

McIntyre, J.C. and Sleight, R.G., 1991, Fluorescence assay for phospholipid membrane asymmetry. *Biochemistry*, **30**: 11819-11827.

Moriyama, Y., and Nelson, N., 1988, Purification and properties of a vanadate- and N-ethylmaleimide- sensitive ATPase from chromaffin granule membranes. *J. Biol. Chem.*, **263**: 8521-8527.

Morrot, G., Zachowski, A., and Devaux, P.F., 1990, Partial purification and characterization of the human erythrocyte Mg2(+)-ATPase. A candidate aminophospholipid translocase. *FEBS Lett.*, **266**: 29-32.

Murakami, M., Kambe, T., Shimbara, S., Higashino, K., Hanasaki, K., Arita, H., Horiguchi, M., Arita, M., Arai, H., Inoue, K. and Kudo, I., 1999, Different functional aspects of the group II subfamily (Types IIA and V) and type X secretory phospholipase A(2)s in regulating arachidonic acid release and prostaglandin generation. Implications of cyclooxygenase-2 induction and phospholipid scramblase-mediated cellular membrane perturbation. *J. Biol. Chem.*, **274**: 31435-31444.

Op den Kamp, J.A.F., 1979, Lipid asymmetry in membranes. *Annu. Rev. Biochem.*, **48**: 47-71.

O'Toole, P.J., Morrison, I.E. and Cherry, R.J., 2000, Investigations of spectrin-lipid interactions using fluoresceinphosphatidylethanolamine as a membrane probe. *Biochim. Biophys. Acta*, **1466**: 39-46.

Raggers, R. J., van Helvoort, A., Evers, R. and van Meer, G., 1999, The human multidrug resistance protein MRP1 translocates sphingolipid analogs across the plasma membrane. *J. Cell Sci.*, **112**: 415–422.

Raths, S., Rohrer, J., Crausaz, F. and Riezman, H., 1993, Two mutants defective in receptor-mediated and fluid-phase endocytosis in *Saccharomyces cerevisiae*. *J.Cell Biol.*, **120**: 55–65.

Rauch, C. and Farge, E., 2000, Endocytosis switch controlled by transmembrane osmotic pressure and phospholipid number asymmetry. *Biophys. J.*, **78**: 3036-3047.

Rothman, J.E. and Lenard, J., 1977, Membrane asymmetry. The nature of membrane asymmetry provides clues to the puzzle of how membranes are assembled. *Science*, **195**: 743-753.

Ruetz, S., Brault, M., Dalton, W. and Gros, P., 1997, Functional interactions between synthetic alkyl phospholipids and the ABC transporters P-glycoprotein, Ste-6, MRP, and Pgh-1. *Biochemistry*, **36**: 8180–8188.

Sanjuan, M.A., Jones, D.R., Izquierdo, M. and Merida, I., 2001, Role of diacylglycerol kinase alpha in the attenuation of receptor signaling. *J. Cell Biol.*, **153**: 207-220.

Schlegel, R.A., Callahan, M.K. and Williamson, P., 2000, The central role of phosphatidylserine in the phagocytosis of apoptotic thymocytes. *Ann. N.Y. Acad. Sci.*, **926**: 217-225.

Schmid, S.L., McNiven, M.A. and De Camilli, P., 1998, Dynamin and its partners: a progress report. *Curr. Opin. Cell Biol.*, **10**: 504-512.

Seigneuret, M. and Deveaux, P.F., 1984, ATP-dependent asymmetric distribution of spin-labelled phospholipids in the erythrocyte membrane: relation to shape changes. *Proc. Natl. Acad. Sci., U.S.A.*, **81**: 3751-3755.

Sestier, C., Sabolovic, D,. Geldwerth, D., Roger, J., Pons, J.N. and Halbreich, A., 1995, Use of annexin V-ferrofluid to enumerate erythrocytes damaged in various pathologies or during storage *in vitro*. *C. R. Acad. Sci., Paris*, **318**: 1141-1146.

Sillence, D.J., Raggers, R.J. and van Meer, G., 2000, Assays for transmembrane movement of sphingolipids. *Methods Enzymol.* **312**: 562-579.

Smeets, E.F., Comfurius, P., Bevers, E.M. and Zwaal, R.F.A., 1996, Contribution of different phospholipid classes to the prothrombin converting capacity of sonicated lipid vesicles. *Thromb. Res.*, **81**: 419-426.

Smit, J. J., Schinkel, A. H., Oude Elferink, R. P., Groen, A. K., Wagenaar, E., van Deemter, L., Mol, C. A., Ottenhoff, R., van der Lugt, N. M. and van Roon, M. A., 1993, Homozygous disruption of the murine mdr2 P-glycoprotein gene leads to a complete absence of phospholipid from bile and to liver disease. *Cell*, **75**: 451–462.

Smith, A. J., Timmermans-Hereijgers, J. L. P. M., Roelofsen, B., Wirtz, K. W. A., van Blitterswijk, W. J., Smit, J. J. M., Schinkel, A. H. and Borst, P., 1994, The human MDR3 P-glycoprotein promotes translocation of phosphatidylcholine through the plasma membrane of fibroblasts from transgenic mice. *FEBS Lett.*, **354**: 263–266.

Stuart, M.C.A., Reutelingsperger, C.P.M. and Frederik, P.M., 1998, Binding of annexin V to bilayers with various phospholipid compositions using glass beads in a flow cytometer, *Cytometry*, **33**: 414-419.

Sun, J., Zhao, J., Schwartz, M.A., Wang, J.Y.J., Wiedmer, T. and Sims, P.J., 2001, c-Abl tyrosine kinase binds and phosphorylates phospholipid scramblase 1. *J. Biol. Chem.*, **276**: 28984-28990.

Tait, J.F. and Gibson, D., 1992, Phospholipid binding of annexin V: effects of calcium and membrane phosphatidylserine content. *Arch. Biochem. Biophys.*, **298**: 187-191.

Tang, X., Halleck, M.S., Schlegel, R.A., and Williamson, P., 1996, A subfamily of P-type ATPases with aminophospholipid transporting activity. *Science*, **272**: 1495-1497

Tepper, A.D., Ruurs, P., Wiedmer, T., Sims, P.J., Borst, J. and van Blitterswijk, W.J., 2000, Sphingomyelin hydrolysis to ceramide during the execution phase of apoptosis results from phospholipid scrambling and alters cell-surface morphology. *J. Cell Biol.*, **150**: F5-7.

Thiagarajan, P. and Tait J.F., 1990, Binding of annexin V/placental anticoagulant protein I to platelets. Evidence for phosphatidylserine exposure in the procoagulant response of activated platelets. *J. Biol. Chem.*, **265**: 17420-17423.

Tilley, L., Cribier, S., Roelofsen, B., Op den Kamp, J.A.F. and van Deenen, L.L.M., 1986, ATP-dependent translocation of amino phospholipids across the human erythrocyte membrane. *FEBS Lett.*, **194**: 21-27.

Tsui, F.C., Sundberg, S.A. and Hubbell, W.L., 1990, Distribution of charge on photoreceptor disc membranes and implications for charged lipid asymmetry. *Biophys. J.* **57**: 85-97.

Van Dort, H.M., Moriyama, R. and Low, P.S., 1998, Effect of band 3 subunit equilibrium on the kinetics and affinity of ankyrin binding to erythrocyte membrane vesicles. *J. Biol. Chem.*, **273**: 14819-14826.

van Engeland, M., Ramaekers, F.C., Schutte, B. and Reutelingsperger, C.P., 1996, A novel assay to measure loss of plasma membrane asymmetry during apoptosis of adherent cells in culture. *Cytometry*, **24**: 131-139.

van Helvoort, A., Giudici, M. L., Thielemans, M. and van Meer, G., 1997, Transport of sphingomyelin to the cell surface is inhibited by brefeldin A and in mitosis, where C6-NBD-sphingomyelin is translocated across the plasma membrane by a multidrug transporter activity. *J. Cell Sci.* **110**: 75–83.

van Helvoort, A., Smith, A.J., Sprong, H., Fritzsche, I., Schinkel, A.H., Borst, P. and van Meer, G., 1996, MDR1 P-glycoprotein is a lipid translocase of broad specificity, while MDR3 P-glycoprotein specifically translocates phosphatidylcholine. *Cell*, **87**: 507-517.

van Helvoort, A., Smith, A. J., Sprong, H., Fritzsche, I., Schinkel, A. H., Borst, P. and van Meer, G., 2000, Cellular organelles: how lipids get there, and back. *Trends Cell Biol.*, **10**: 550-552.

van Meer, G., 1996, MDR1 P-glycoprotein is a lipid translocase of broad specificity, while MDR3 P-glycoprotein specifically translocates phosphatidylcholine. *Cell*; **87**, 507–517.

van Meer, G., 2000, Cellular organelles: how lipids get there, and back. *Trends Cell Biol.*, **10**: 550-552.

Verkleij, A.J. and Post, J.A., 2000, Membrane phospholipid asymmetry and signal transduction. *J. Memb. Biol.* **178**: 1-10.

Verkleij, A.J., Zwaal, R.F., Roelofsen, B., Comfurius, P., Kastelijn, D. and van Deenen, L.L.M., 1973, The asymmetric distribution of phospholipids in the human red cell membrane. A combined study using phospholipases and freeze-etch electron microscopy. *Biochim. Biophys. Acta*, **323**: 178-193.

Wiedmer, T., Zhou, Q., Kwoh, D.Y. and Sims, P.J., 2000, Identification of three new members of the phospholipid scramblase gene family. *Biochim. Biophys. Acta*, **1467**: 244-253.

Williamson, P., Bevers, E.M., Smeets, E.F., Comfurius, P., Schlegel, R.A. and Zwaal, R.F., 1995, Continuous analysis of the mechanism of activated transbilayer lipid movement in platelets. *Biochemistry*, **34**: 10448-10455.

Williamson, P., van den Eijnde, S. and Schlegel, R.A., 2001, Phosphatidylserine exposure and phagocytosis of apoptotic cells. *Methods Cell Biol.*, **66**: 339-364.

Woon, L.A., Holland, J.W., Kable, E.P. and Roufogalis, B.D., 1999, Ca^{2+} sensitivity of phospholipid scrambling in human red cell ghosts. *Cell Calcium*, **25**: 313-320.

Wu, G. and Hubbel, W.L., 1993, Phospholipid asymmetry and transmembrane diffusion in photoreceptor disc membranes. *Biochemistry*, **32**: 879-888.

Xie, X.-S., Stone, D.K., and Racker, E., 1989, Purification of a vanadate-sensitive ATPase from clathrin-coated vesicles of bovine brain. *J. Biol. Chem.*, **264**: 1710-1714.

Zachowski, A., 1993. Phospholipids in animal eukaryotic membranes: transverse asymmetry and movement. *Biochem. J.*, **294**: 1-14

Zachowski, A., Henry, J.P. and Devaux, P.F., 1989, Control of transmembrane lipid asymmetry in chromaffin granules by an ATP-dependent protein. *Nature*, **340**: 75-76

Zhao, J., Zhou, Q., Wiedmer, T. and Sims, P.J., 1998a, Level of expression of phospholipid scramblase regulates induced movement of phosphatidylserine to the cell surface. *J. Biol. Chem.* **273**: 6603-6606.

Zhao, J., Zhou, Q., Wiedmer, T. and Sims, P.J., 1998b, Palmitoylation of phospholipid scramblase is required for normal function in promoting Ca^{2+}-activated transbilayer movement of membrane phospholipids. *Biochemistry*, **37**: 6361-6366.

Zhou, Q., Zhao, J., Stout, J.G., Luhm, R.A., Wiedmer, T. and Sims, P.J., 1997, Molecular cloning of human plasma membrane phospholipid scramblase. A protein mediating transbilayer movement of plasma membrane phospholipids. *J. Biol. Chem.* **272**: 18240-18244.

Zhou, Q., Zhao, J., Al-Zoghaibi, F., Zhou, A., Wiedmer, T., Silverman, R.H. and Sims, P.J., 2000, Transcriptional control of the human plasma membrane phospholipid scramblase 1 gene is mediated by interferon-alpha. *Blood*, **95**: 2593-2599.

Zwaal, R.F.A., and Schroit, A.J., 1997. Pathophysiologic implications of membrane phospholipid asymmetry in blood cells. *Blood*, **89**: 1121-1132.

Chapter 4

Apoptosis by Phosphatidylserine in Mammalian Cells

Kazuo Emoto and Masato Umeda
Department of Molecular Biodynamics, The Tokyo Metropolitan Institute of Medical Science, 3-18-22 Honkomagome, Bunkyo-ku, Tokyo 113-8613, Japan; email: emoto@rinshoken.or.jp

1. INTRODUCTION

Phosphatidylserine (PS) is an essential phospholipid for the growth of mammalian cells, comprising ≈10% of the total membrane phospholipids of various mammalian tissues and culture cells. Although PS biosynthesis in mammalian cells is tightly regulated in the steady state, metabolism and transport of PS appears to be significantly changed during apoptosis. Furthermore, an aberrant increase of cellular PS content induces apoptosis. In this review, we will first describe the current understanding of PS metabolism and transport in mammalian cells, and then discuss how alterations in PS metabolism may affect apoptosis.

2. REGULATION OF PHOSPHATIDYLSERINE BIOSYNTHESIS AND TRANSPORT IN MAMMALIAN CELLS

2.1 Biosynthesis of phosphatidylserine in mammalian cells

Phosphatidylserine (PS) in mammalian cells is synthesized through an exchange of free L-serine with the base moiety of pre-existing phospholipids, phosphatidylcholine (PC) and phosphatidylethanolamine (PE) (Borkenhangen *et al.*, 1961; Marggraf *et al.*, 1974; Kuge and Nishijima, 1997; Nishijima *et al.*, 1997) (Figure 1). Studies on PS biosynthesis in Chinese hamster ovary (CHO) cells have suggested that the serine base-exchange is catalyzed by at least two distinct enzymes; one is PS synthase I (PSS I) which uses PC and PE as phosphatidyl donors for the serine base-exchange. The other is PS synthase II (PSS II) which uses PE but not PC as a phosphatidyl donor (Kuge *et al.*, 1997; Saito *et al.*, 1998). PSS I and II exhibit a ≈ 32% amino acid sequence identity. Hydrophobicity analysis

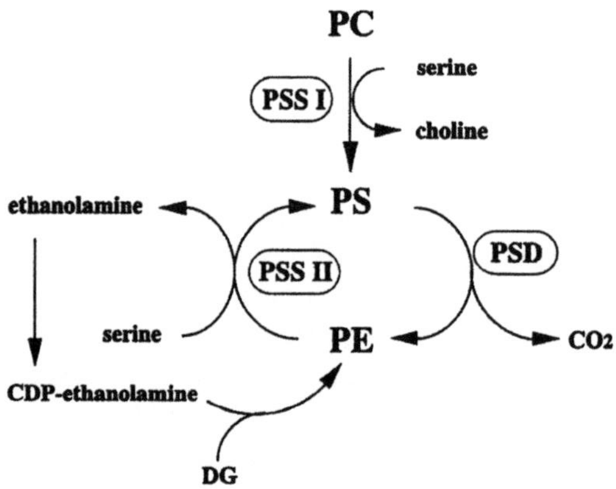

Figure 1. Synthetic pathway for PS and PE in mammalian cells. The major steps occuring in the synthesis and interconversion of PS and PE are shown. The PS synthases condense serine with a phosphatidyl moiety derived from PC and PE. The nascent PS can be converted to PE by decarboxylation. PE can also be formed by transfer of a phosphoethanolamine moiety from CDP-ethanolamine to diacylglycerol via the Kennedy pathway. The abbreviations used are: PC, phosphatidylcholine; PS, phosphatidylserine; PE, phosphatidylethanolamine; DG, diacylglycerol; PSD, phosphatidylserine decarboxylase; PSS, PS synthase.

revealed that PSS I and II are highly hydrophobic proteins containing nine potential membrane-spanning domains (Kuge and Nishijima; 1997). PSS II has a Met-Arg-Arg-Ala-Glu sequence at N-terminus, which corresponds to the endoplasmic reticulum (ER)-targeting sequence (Kuge *et al.*, 1997). Although PSS I does not have such a typical ER targeting motif as PSS I, it has a unique Gly-Val-Gly-Lys-Lys sequence at its C-terminus, which is analogous to another ER targeting motif (Jackson *et al.*, 1990). It is, therefore, likely that PSSI and II are localized in ER membrane. These distributions are consistent with the previous report that the serine base-exchange activity was recovered largely in the ER fraction (Jesema *et al.*, 1978).

Vance has recently isolated a unique mitochondria-associated membrane (MAM) from the crude rat mitochondrial fraction and found that MAM fraction has a high specific activity of serine base-exchange (Vance, 1990). Immunoblotting using specific antibodies for either PSS I and II showed that both enzymes are predominantly distributed to MAM (Stone and Vance, 2000). MAM have many, but not all, properties of ER. In particular, MAM are enriched in several lipid biosynthetic enzymes such as acyl-CoA: cholesterol acyltransferase (Rusinol *et al.*, 1994), diacylglycerol acyltransferase (Rusinol *et al.*, 1994), and some enzymes involved in biosynthesis of glycosylphosphatidylinositol anchors of proteins (Vidugiriene *et al*, 1999). Interestingly, MAM has been shown to be involved in transport of PS into mitochondria, in which PS is decarboxylated for PE formation (Shiao *et al*, 1995).

2.2 Control of phosphatidylserine biosynthesis

It has been shown that cellular PS content is tightly regulated in CHO cells. When CHO cells were cultured with exogenous PS, PS was efficiently incorporated into cells and utilized for membrane biogenesis, however, the total PS content was not significantly changed in the presence of exogenous PS (Nishijima *et al.*, 1986). Biochemical studies indicated that de novo PS biosynthesis was remarkably inhibited upon addition of PS to the culture medium. This inhibition of de novo PS biosynthesis by the added PS appears not to be caused by a reduction in the cellular level of PSS I and II, because the serine base-exchange activities in homogenates of cells grown with or without exogenous PS were not changed. In addition, overexpression of PSS I in CHO cells did not lead to accelerate the PS biosynthetic rate of intact cells, although cell extracts of the PSS I-overproducing cells showed a 6-fold higher serine base-exchange activity than that of CHO cells (Kuge *et al.*, 1991). It is, therefore, likely that the product inhibition of PSS I activity is involved in the regulation of PS biosynthesis.

Hasegawa *et al.* (1989) isolated a CHO cell mutant in which PS biosynthesis was highly resistant to the depression of exogenous PS. Recently, this PS- resistant PS biosynthesis in the mutant was traced to a point mutation in PSS I gene, resulting in the replacement of Arg-95 to lysine (Kuge *et al.*, 1998). Expression of the mutant PSS I (R95K) in CHO cells induced PS-resistant PS biosynthesis. Furthermore, PSS I (R95K)-expressing cells showed a ≈ 2-fold increase in cellular PS level compared with that in CHO cells. These results indicated that the product inhibition of PSS I activity is essential for the maintenance of a normal PS level in CHO cells. They have further shown that PSS II activity in CHO cells is also regulated by the same product inhibition as shown in PSS I (Kuge *et al.*, 1999).

The molecular mechanism underlying the inhibition remains unclear. It could comprise the simple product inhibition due to binding of PS to either catalytic site or the regulatory site of PSS I. It is also possible that a putative regulator molecule may bind PSS I to repress the PSS I activity in the presence of excess PS and that Arg-95 of PSS I may be essential for this binding. Determination of the three-dimensional structure of PSS I may provide a new insight into the role of Arg-95 of PSS I in the feedback regulation mechanism.

2.3 Intracellular transport of phosphatidylserine

In mammalian cells, the final stage of PS biosynthesis occurs in ER and MAM (Trotter and Voelker, 1994; Daum and Vance, 1997; Voelker, 2000). The other membranes in the cell, such as mitochondria, nucleus, and plasma membrane, are therefore assembled from PS exported from ER and MAM (Figure 2). Phospholipid synthesis in mitochondria is restricted to the formation of phosphatidylglycerol, cardiolipin, and PE, and other lipids such as PC and PS must be imported from sites of cellular lipid synthesis, ER or MAM (Daum, 1985; Vance, 1991). PS imported to the outer mitochondrial membrane is then translocated to the inner mitochondrial membrane, where it is converted to PE by PS decarboxylase (PSD) (Dennis and Kennedy, 1972; Voelker, 1990). It has been shown that the translocation of PS to mitochondria followed by its decarboxylation is a major pathway for the synthesis of PE in some cultured mammalian cells (Voelker, 1984; Kuge *et al.*, 1986; Voelker and Frazier, 1986), suggesting that significant amounts of PE found in cell membranes are derived from mitochondria.

Since the enzymes involved in PS synthesis are located in ER or MAM, and PSD is exclusively located at the inner mitochondrial membrane, the conversion of PS to PE by PSD has been used as an indicator of PS translocation into the inner mitochondrial membrane (Dennis and Kennedy, 1972; Voelker, 1990). Recent studies have shown that the transport of newly synthesized PS to the outer mitochondrial membrane requires no cytosolic proteins and is probably mediated by direct contact region between MAM and mitochondria (Voelker, 1989; Voelker, 1993; Shiano *et al.*, 1995). It is also suggested that the translocation of PS from the outer to inner mitochondrial membrane occurs through the contact sites where the two mitochondrial membranes are closely apposed and linked in a stable manner, since agents that disrupt the contact sites such as 1,4-dinitrophenol and adriamycin inhibit the PS transport (Hovius *et al.*, 1992; Voelker, 1991).

Ardail *et al.* (1991) reported that when radiolabeled PS was introduced into isolated mitochondria in the presence of hydroxylamine, an inhibitor of PSD, the imported PS accumulated in the contact site. In addition, PE newly synthesized through the decarboxylation of PS in the inner mitochondrial membrane is exported very rapidly to the outer membrane, without mixing with inner membrane PE (Ardail *et al.*, 1991). We recently isolated a CHO cell mutant with a decreased cellular PE level by screening for variants resistant to the cytotoxicity of a tetracyclic peptide, which forms a tight equimolar complex with PE on biological membranes and subsequently induces cytolysis (Aoki *et al.*, 1994). We have further shown that this mutant has a specific defect in the intramitochondrial transport of PS but is not affected in the PC transport (Emoto *et al.*, 1999). These results strongly suggest the presence of a specific machinery for the intramitochondrial transport of PS, which is tightly coupled to their biosynthesis.

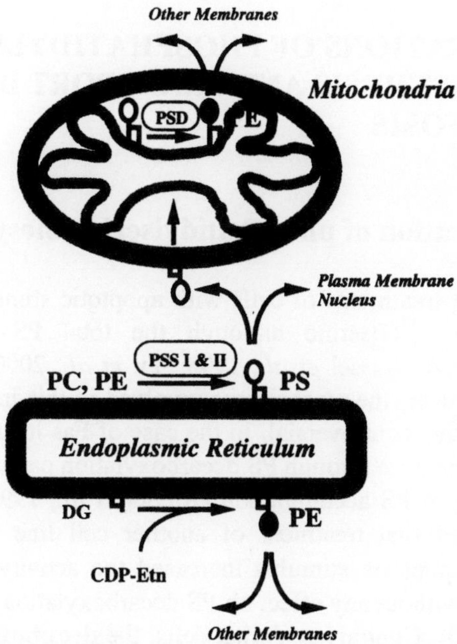

Figure 2 . General topological feature of PS translocation and decarboxylation in mammalian cells. PS is synthesized by PSS I and II in endoplasmic reticulum (ER) or mitochondria-associated membrane (MAM). The nascent PS is transported other membranes such as plasma membrane, nucleus, and mitochondria. The PS transported to the mitochondrial outer membrane is then translocated to the inner membrane, in which PS is converted to PE by PS decarboxylase (PSD). The PE formed in mitochondria is dynamic and can be exported to other organelles for membrane biogenesis.

PE synthesized in mitochondria is then exported from mitochondria to function in the biosynthesis of other organelles including ER (Voelker, 1989) and plasma membrane (Vance *et al.*, 1991). PE transport from mitochondria to plasma membrane is insensitive to the Golgi disrupting toxin brefeldin A (Vance *et al.*, 1991), indicating that the PE transport pathway is likely to bypass the Golgi apparatus. PE is produced by three distinct pathways: (i) the PS decarboxylation pathway, (ii) the CDP-ethanolamine pathway in which phosphoethanolamine is transferred to 1,2-sn-diacylglycerol in ER, and (iii) the base-exchange pathway in which free ethanolamie is exchanged with preexisting phospholipids in ER (Kent, 1995; Voelker, 1997). Vance *et al.* (1991) showed that PE derived from PS decarboxylation pathway appears on the cell surface much faster than that from other pathways, suggesting that PE synthesized mitochondria is preferentially transported to the cell surface.

3. ALTERATIONS OF PHOSPHATIDYLSERINE BIOSYNTHESIS AND TRANSPORT DURING APOPTOSIS

3.1 Acceleration of phosphatidylserine biosynthesis

It is shown that treatments of cells with apoptotic stimulus enhance the PS synthesis from [^{3}H]serine although the total PS content is not significantly changed (Aussel *et al.*, 1998; Yu *et al.*, 2000; Buratta *et al.*, 2000). So far, however, the mechanisms involved in this increase of the PS synthetic rate remains controversial. In the case of Fas-induced apoptosis in Jurkat cells, PE formation through PS decarboxylation pathway was strongly inhibited, resulting in PS accumulation (Aussel *et al.*, 1998). On the other hand, it is reported that treatment of another cell line U937 or mouse thymocytes with apoptotic stimulus increased the activity of serine base-exchange enzyme without any effect on PS decarboxylation (Yu *et al.*, 2000; Buratta *et al.*, 2000). Contrary to Jurkat cells, the decarboxylation of newly synthesized PS to PE was almost negligible in mouse thymoctes (Buratta *et al.*, 2000). Thus, although the increase of PS biosynthesis appears a common event of early apotosis, each apoptotic stimulus may have a specific effect on particular enzymes of PS metabolism.

3.2 Phosphatidylserine translocation on plasma membrane

In many eukaryotic plasma membranes, PS resides in the inner leaflet (Schroit and Zwaal, 1991; Zachowski, 1993). This transbilayer distribution of membrane lipids is not a static situation but a result of balance between the inward and outward translocation of phospholipids across the membranes. Recent studies showed that the transbilayer lipid asymmetry is regulated by several lipid transporter proteins, such as aminophospholipid translocase (Daleke and Lyles, 2000), ATP-binding cassette transporter family (van Helvoort *et al.*, 1996; Klein *et al.*, 1999), and phospholipid scramblase (Zhou *et al.*, 1997; Zhao *et al.*, 1998). An increment of intracellular Ca^{2+} due to cell activation, cell injury, and apoptosis affects the activities of these transporters, resulting in exposure of PS (Koopman *et al.*, 1994; Verhoven *et al.*, 1995) and PE (Emoto *et al.*, 1997) on the cell surface.

The exposed PS promotes the clearance of injured or apoptotic cells from the circulation by macrophages and phagocytotic cells (Nishikawa *et al.*, 1990; Ellis *et al.*, 1991; Shiratsuchi *et al.*, 1997, Nakano *et al.*, 1997). The phagocytosis of apoptotic cells by phagocytotic cells can be inhibited stereospecifically by PS, but not by other anionic phospholipids including phosphatidyl-D-serine (Fadok *et al.*, 1992; Pradhan *et al.*, 1997), suggesting that a PS-specific receptor on the cell surface must mediate the phagocytosis. Recently, Fadok *et al.* (2000) have identified a PS-specific receptor on the surface of activated human macrophages. The PS receptor is predicted to be a transmembane protein with one membrane-spanning region. Since PS receptor exhibits no significant homology with PKCs nor blood coagulation factors in the amino acid level, PS receptor is likely to recognize 1,2-sn-phosphatidyl-L-serine by a distinct manner from those of PKCs and blood coagulation factors.

In the early stage of apoptosis, newly synthesized PS is preferentially transferred to the plasma membrane and then shed as a small vesicle in cells undergoing apoptosis (Aussel *et al.*, 1998; Yu *et al.*, 2000; Buratta *et al.*, 2000). It is, therefore, possible that stimulation of PS synthesis may contribute to the enhancement of PS exposure on the cell surface of apoptotic cells.

In contrast to PS, the function of PE exposure during apoptosis is largely unclear. We have recently found that exposure of PE on cell surface at the cleavage furrow is essential for completion of cell division (Emoto *et al.*, 1996). We have further shown that redistribution of PE on the cleavage furrow contributes to regulation of actin filament reorganization in cell division (Umeda and Emoto, 1999; Emoto and Umeda, 2000). Interestingly, the transbilayer redistribution of PE in apoptotic cells is specifically observed on the bleb membrane in which acute actin filament reorganization is induced (Emoto *et al.*, 1997; Mills *et al.*, 1998). Thus, it is possible that PE redistribution may play a role in reorganization of cytoskeletons to form the apoptotic membrane blebbing.

4. INDUCTION OF APOPTOSIS BY PHOSPHATIDYLSERINE

4.1 Exogenously added phosphatidylserine induces apoptosis in CHO cells

It is previously shown that various lipid metabolites, such as ceramide (Hannun and Obeid, 1995) and dolichyl phosphate (Yasugi, *et al.*, 1995) induce apoptosis. We found that PS is also an inducer of apoptosis (Uchida *et al.*, 1998). Treatment of CHO-K1 cells with PS caused cell death in a dose-dependent manner. This cytotoxic effect was specific for PS and other phospholipids, such as PC, PE, phosphatidylinositol (PI), and phosphatidic acid (PA), had no effect on cell viability. It is unlikely that PS-induced cell death was due to lysophosphatidylserine (lyso-PS), as either a contaminant of the PS preparation or the result of conversion of PS by phospholipase A2, since even an higher concentration of lyso-PS was required to induce cell death. PS-treatment induced chromatin condensation and segregation and an extensive DNA fragmentation with a pattern characteristic of internucleosomal fragmentation on the agarose gel electrophoresis. Furthermore, PS-induced cell death was effectively inhibited by ionic zinc or a Ca^{2+}-chelater, such as ethyleneglycol-bis(B-aminoethyl)-N,N,N',N',-tetraacetic acid (EGTA), both of which are known to inhibit endonuclease. These observations indicate that PS induced apoptosis in CHO-K1 cells. The apoptosis-inducing activity of PS was also observed in other cell lines, such as Balb 3T3 (mouse fibroblast cell line) and GOTO (human neuroblastoma cell line).

4.2 Structure-activity relationship in apoptosis inducing activity

The PS-induced apoptosis was highly specific for PS. Synthetic PS analogs such as 1,2-diacyl-*sn*-glycero-3-phospho-D-serine (D-PS), 1,2-diacyl-*sn*-glycero-3-phospho-homoserine (P-H-S), 1,2-diacyl-*sn*-glycero-3-phospho-serine methyl esther (PS-O-Me), and 2,3-diacyl-*sn*-glycero-1-

phospho-L-serine (sn-2,3-PS) failed to induce the apoptosis (Figure 3). Endo *et al.* (1996) have shown that exogenous PS is internalized mainly by a protein called flippase and that the transbilayer movement of PS on the plasma membrane is an important step in the cytotoxic action of PS. This

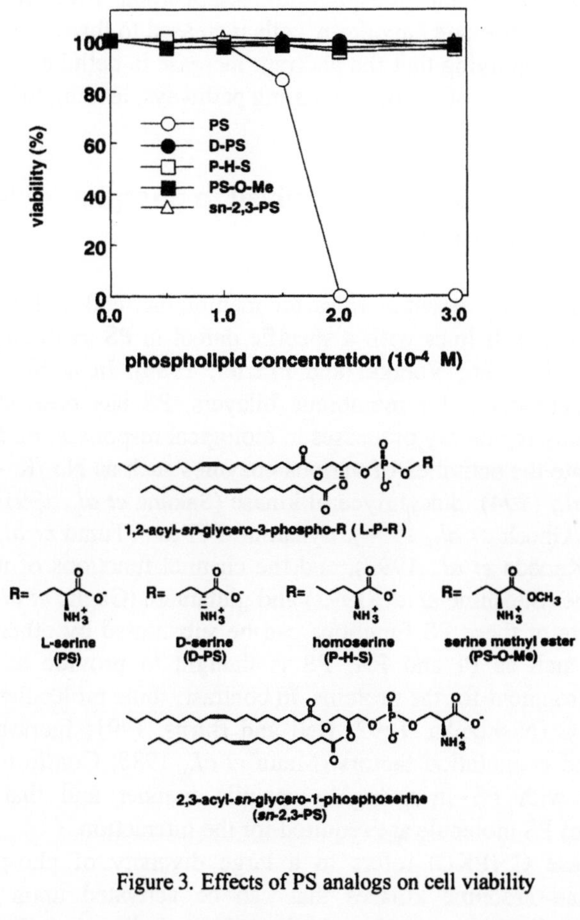

Figure 3. Effects of PS analogs on cell viability

observation raised the possibility that flippase on plasma membranes may selectively internalize PS, while other phospholipids including the synthetic PS analogs, may not be internalized as effectively as PS. To examine this possibility, we studied the uptake of fluorescence-labeled PS analogs of PS (apoptotic) and D-PS (non-apoptotic). Fluorescence-labeled PS and D-PS were taken up equally and then transferred to intracellular membranes, suggesting that the PS-specific induction of apoptosis was not the result of its specific internalization. These results suggested that certain molecules which may recognize the stereo-specific configuration of PS are involved in the apoptotic process triggered by PS.

PS content in CHO-K1 cells to be strictly regulated and that exogenously added PS was incorporated into cells and utilized for membrane biogenesis, resulting in the suppression of endogenous PS formation (Nishijima *et al.*, 1986). Analysis of cellular phospholipid composition revealed the PS content level in the PS-treated apoptotic cells increased to three times of that of in control cells, implying that the aberrant increase in cellular PS content may cause malregulation of cellular signaling pathways, leading to apoptosis of the cell.

4.3 Possible mechanisms involved in phosphatidylserine-induced apoptosis

PS is shown to play essential roles for mammalian cell viability, since mammalian mutant cell lines with a specific defect in PS synthesis cannot grow (Kuge *et al.*, 1986; Voelker and Frazier, 1986). In addition to the function as a component for membrane bilayers, PS has been shown to contribute to many regulatory processes in biological responses. PS has been shown to regulate the activities of various enzymes such as Na^{+}/K^{+}-ATPase (Stekhoven *et al.*, 1994), diacylglycerol kinase (Sakane *et al.*, 1991), B-Raf protein kinase (Ghosh *et al.*, 1994), dynamin GTPase (Tuma *et al.*, 1993), and GAPDH (Kaneda *et al.*, 1997), and the channel functions of receptors for acetylcholine (Sunshine *et al.*, 1992) and glutamate (Gagne *et al.*, 1996). Since large parts of these PS functions can be substituted by other anionic phospholipids such as PI and PA, PS is thought to provide an anionic membrane environment for the proteins. In contrast, some molecules such as protein kinase C (Nishizuka, 1992; Bell and Burns, 1991; Igarashi *et al.*, 1995) and blood coagulation factors (Mann *et al.*, 1988; Comfurius *et al.*, 1994) interact with PS in a highly specific manner and that precise configurations of PS molecule are required for the interaction.

Protein kinase C (PKC) refers to a large diversity of phospholipid-dependent serine-threonine kinases that can be activated upon external stimulation of cells by a number of ligands including growth factors, hormones, and neurotransmitters (Newton, 1995; Nishizuka, 1995; Mellor and Parker, 1998). PKC in mammalian cells consists of at least 11 related isoenzymes that contain four conserved domains named C1-C4 (Cousens *et al.*, 1986). According to their structure and cofactor regulation, PKCs can be classified into three groups. The classical isoforms (α, βI, βII, and γ) can be distinguished from other groups because functioning is regulated by diacylglycerol and also cooperatively by calcium and PS. In the classical PKC isoenzymes, the calcium-dependent binding to membranes presents a high specific to 1,2-*sn*-phosphatidyl-L serine (Boni and Rando, 1985; Lee and Bell, 1989; Johnson *et al.*, 1998). It is also shown that this binding is mediated by the C2 domain (Igarashi *et al.*, 1995; Edwards and Newton, 1997; Medkova and Cho, 1998), which is conserved in large variety of proteins involved in intracellular signaling and membrane trafficking (Nalefski and Falke, 1996; Rizo and Sudhof, 1998). Crystal structures of C2 domains from synaptotagmin I (Sutton *et al.*, 1995), phospholipase C-δ (Essen *et al.*, 1996), phospholipase A_2 (Perisic *et al.*, 1998; Bittova *et al.*, 1999; Dessen *et al.*, 1999), and PKCß (Sutton and Sprang, 1998) have revealed that a homologous ß-sandwich fold that serve as the scaffold for a generally bipartile Ca^{2+} binding site formed by a pair of loops that project from the opposing ß-sheets. Recently, the three-dimensional structures of PKCα-C2 domain in the presence of 1,2-dicaproyl-*sn*-phosphatidyl-L-serine was determined by X-ray crystallography (Verdaguer *et al.*, 1999). This structural information suggests that one calcium ion directly mediates the PS-specific recognition. Although PKC would be a most possible candidate affected by an increase of cellular PS content, PKC inhibitors such as H-7 and calphostin C slightly enhanced PS-induced apoptois while an activator of PKC, phorbol 12-myristate 13-acetate (PMA), inhibited apoptosis (Figure 4), suggesting that PKC inactivation, rather than activation, is involved in this apoptotic process. Ceramide is a crucial mediator of apoptosis induced by various ligands such as tumor necrosis factor α, γ-interferon, interleukin-1, and Fas activation.

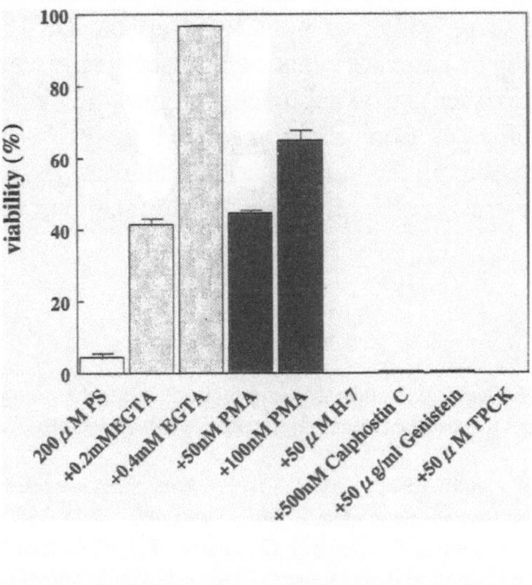

Figure 4. Effects of Various compounds on PS-induced apoptosis

This ceramide-mediated apoptosis was shown to be inhibited by the simultaneous addition of PKC activators (Ni *et al.*, 1994; Obeid *et al.*, 1993), implying that PS may activate the ceramide-mediated apoptotic pathway. However, the inhibitors of interleukin-1 converting enzyme (ICE)-like proteases (Caspase), such as tosyl-L-lysine chloromethyl ketone (TLCK), and tosyl-L-phenylalanine chloromethyl ketone (TPCK) which inhibit ceramide-mediated apoptosis, had no effect on PS-induced apoptosis (Figure 4). Thus, PS-induced apoptotic pathway appears to be distinct from that mediated by ceramide. Further studies are required to clarify the molecular mechanisms underlying the PS-induced apoptosis.

5. CONCLUSIONS

Much progress has been gained in understanding the molecular mechanisms responsible for PS biosynthesis in mammalian cells. It has been also shown that PS biosynthesis is tightly regulated, at least in part, by the feedback control system. These observations indicate that the PS concentration inside the cell is an important parameter that requires careful control and strongly suggest that PS plays specific roles *in vivo*. Apoptotic stimuli appear to rapidly modulate the regulation of PS biosynthesis and the intracellular transport of PS, however, the molecular mechanisms underlying these alterations remain poorly understood. The exact roles played by PS biosynthetic enzymes such as PSSI, II, and PSD during apoptosis need to be elucidated by both biochemical and genetic experiments. It also became known that PS is an effective inducer of apoptosis. This process likely involves signaling cascades that are specifically activated by PS. Furthermore, pharmacological approaches suggest that PS-induced apoptosis appears to be distinct from those mediated by PKC, a well-known PS effector. Elucidation of the molecular mechanisms by which PS induces apoptosis would provide a novel aspect of PS function *in vivo*.

REFERENCES

Aoki, Y., Uenaka, T., Aoki, J., Umeda, M., and Inoue, K., 1994, A novel peptide probe for studying the transbilayer movement of phosphatidylethanolamine. *J. Biochem.* **116**: 291-297.

Ardail, D., Lerme, F., and Louisot P., 1991, Involvement of contact sites in phosphatidylserine import into liver mitochondria. *J. Biol. Chem.*, **266**: 7978-7981.

Ardail, D., Gasnier, F., Lerme, F., Simonot, C., Louisot, P., and Gateau-Roesch, O., 1993, Involvement of mitochondrial contact sites in the subcellular compartmentalization of phospholipid biosynthetic enzymes. *J. Biol. Chem.* **268**: 25985-25992.

Aussel, C., Pelassy, C., Breittmayer, J. P., 1998, CD95 (Fas/APO-1) induces an increased phosphatidylserine synthesis that precedes its externalization during programmed cell death. *FEBS Lett.*, **431**: 195-199.

Bell, R.M., and Burns, D.J., 1991, Lipid activation of protein kinase C. *J. Biol. Chem.* **266**: 4661-4664.

Bittova, L., Sumandea, M., and Cho, W., 1999, A structure-function study of the C2 domain of cytosolic phospholipase A2. *J. Biol. Chem.* **274**: 9665-9672.

Boni, L.T., and Rando, R.R., 1985, The nature of protein kinase C activation by physically defined phospholipid vesicles and diacylglycerols. *J. Biol. Chem.* **260**: 10819-10825.

Borkenhangen, J.D., Kennedy, E.P., and Fielding, L., 1961, Enzymatic formation and decarboxylation of phosphatidylserine. *J. Biol. Chem.*, **236**: PC28-29.

Buratta, S., Migliorati, G., Marchetti, C., Mambrini, R., Riccardi, C., and Mozzi, R., 2000, Dexamethasone increases the incorporation of [3H]serine into phosphatidylserine and the activity of serine base exchange enzyme in mouse thymocytes: a possible relation between serine base exchange enzyme and apoptosis. *Mol. Cell. Biochem.* **211**: 61-67.

Comfurius, P., Smeets, E.F., Willems, G.M., Bevers, E.M., and Zwaal,R.F.A., 1994, Assembly of the prothrombinase complex on the lipid vesicles dependent on the stereochemical configuration of polar headgroup of phosphatidylserine. *Biochemistry* **33**: 10319-10324.

Coussens, L., Parker, P.J., Rhee, L. , Yang-Feng, T.L., Chen, E., Waterfield, M.D., Francke, U., and Ullrich, A., 1986, Multiple, distinct forms of bovine and human protein kinase C suggest diversity in cellular signaling pathways. *Science* **233**: 859-866.

Daleke, D.L., and Lyles, J.V., 2000, Identification and purification of aminophospholipid flippases. *Biochim. Biophys. Acta* **1486**: 108-127.

Daum ,G., 1985, Lipids of mitochondria. *Biochim. Biophys. Acta* **822**: 1-42

Daum, G., and Vance, J.E., 1997, Import of lipids into mitochondria. *Prog. Lipid Res.* **36**: 103-130.

Dennis, E.A., and Kennedy, E.P., 1972, Intracellular sites of lipid synthesis and the biogenesis of mitochondria. *J. Lipid Res.* **13**: 263-267.

Dessen, A. , Tang, J. , Schmidt, H. , Stahl, M. , Clark, J.D. , Seehra, J. and Somers, W.S., 1999, Crystal structure of human cytosolic phospholipase A2 reveals a novel topology and catalytic mechanism. *Cell* **97**: 349-360.

Edwards, A.S., and Newton, A.C., 1997, Regulation of protein kinase C II by its C2 domain. *Biochemistry* **36**: 15615-15623.

Ellis, R.E., Yuan, J.Y., and Horvitz, H.R., 1991, Mechanisms and functions of cell death. *Annu. Rev. Cell. Biol.* **7**: 663-698.

Emoto, K., Kobayashi, T., Yamaji, A., Aizawa, H., Yahara, I., Inoue, K., and Umeda, M., 1996, Redistribution of phosphatidylethanolamine at the cleavage furrow of dividing cells during cytokinesis. *Proc. Natl. Acad. Sci. USA*, **93**: 12867-12872.

Emoto, K., Toyama-Sorimachi, N., Karasuyama, H., Inoue, K., and Umeda, M., 1997, Exposure of phosphatidylethanolamine on the surface of apoptotic cells. *Exp. Cell Res.*, **232**: 430-434.

Emoto, K., Kuge, O., Nishijima, M., and Umeda, M., 1999, Isolation of a Chinese hamster ovary cell mutant defective in intramitochondrial transport of phosphatidylserine. *Proc. Natl. Acad. Sci. USA*, **96**: 12400-12405.

Emoto, K., and Umeda, M., 2000, An essential role for a membrane phospholipid in cytokinesis: Regulation of contractile ring disassembly by redistribution of phosphatidylethanolamine. *J. Cell Biol.*, **149**: 1215-1224.

Endo, T. A., Kobayashi, T., and Ohki, K., 1996, A Chinese hamster ovary mutant resistant to phosphatidylserine is defective in transbilayer movement of cell surface phosphatidylserine. *Exp. Cell Res.* **228**: 341-346.

Essen, L.-O., Perisic, O., Cheung, R., Katan, M., and Williams, R.L., 1996, Crystal structure of a mammalian phosphoinositide-specific phospholipase C. *Nature* **380**: 595-602.

Fadok, V.A., Voelker, D.R., Campbell, P.A., Cohen, J.J., Bratton, D.L., and Henson, P.M., 1992, Exposure of phosphatidylserine on the surface of apoptotic lymphocytes triggers specific recognition and removal by macrophages. *J. Immunol.* **148**: 2207-2216.

Fadok, V.A., Bratton, D.L., Rose, D.M., Pearson, A., Ezekewitz, R.A., and Henson, P.M., 2000, A receptor for phosphatidylserine-specific clearance of apoptotic cells. *Nature* **405**: 85-90.

Gagne, J., Giguere, C., Tocco, G., Ohayon, M., Thompson, R.F., Braudry, M., and Massicotte,G., 1996, Effect of phosphatidylserine on the binding properties of glutamate receptors in brain sections from adult and neonatal rats. *Brain Res.* **740**: 337-345.

Ghosh, S., Xie, W.Q., Quest, A.F.G., Strum, J.C. and Bell, R.M., 1994, The cystein-rich region of raf-1kinase contains zinc, translocates to liposomes, and is adjacent to a segment that binds GTP-ras. *J.Biol. Chem.***269**: 10000-10007.

Hannun, Y., A., and Obeid, L., M., 1995, Ceramide: an intracellular signal for apoptosis. *Trends Biochem. Sci.* **20**: 73-77.

Hasegawa, K., Kuge, O., Nishijima, M., and Akamatsu, Y., 1989, Isolation and characterization of a Chinese hamster ovary cell mutant with altered regulation of phosphatidylserine biosynthesis. *J. Biol. Chem.* **264**: 19887-19892.

Hovius, R., Faber, B., Brigot, B., Nicolay, K., and de Kruijff, B., 1992, On the mechanism of the mitochondrial decarboxylation of phosphatidylserine. *J. Biol. Chem.* **267**: 16790-16795.

Igarashi, K., Kaneda, M., Yamaji, A., Saido, T.C., Kikkawa, U., Ono, Y., Inoue, K., and Umeda, M., 1995, A novel phosphatidylserine-binding peptide motif defined by an anti-idiotypic monoclonal antibody. Localization of phosphatidylserine-specific binding sites on protein kinase C and phosphatidylserine decarboxylase. *J. Biol. Chem.* **270**: 29075-29078.

Jackson, M.R., Nilsson, T., and Peterson, P.A., 1990, Identification of a consensus motif for retention of transmembrane proteins in the endoplasmic reticulum. *EMBO J.* **9**: 3153-3162.

Jelsema, C. L., Morre, D.J., 1978, Distribution of phospholipid biosynthetic enzymes among cell components of rat liver. *J. Biol. Chem.* **253**: 7960-7971.

Johnson, J. E., Giorgione, J., and Newton, A.C., 2000, The C1 and C2 domains of protein kinase C are independent membrane targeting modules, with specificity for phosphatidylserine conferred by the C1 domain. *Biochemistry* **39**: 11360-11369.

Kaneda, M., Takeuchi, K., Inoue, K., and Umeda, M., 1997, Localization of the phosphatidylserine-binding site of glyceraldehyde-3-phosphate dehydrogenase responsible for membrane fusion. *J. Biochem.* **122**: 1233-1240.

Kent, C., 1995, Eukaryotic phospholipid biosynthesis., *Annu. Rev. Biochem.* **64**: 315-343.

Koopman, G., Reutelingsperger, C.P., Kuijten, G.A., Keehnen, R.M., Pals, S.T., and van Oers M.H., 1994, Annexin V for flow cytometric detection of phosphatidylserine expression on B cells undergoing apoptosis. *Blood* **84**: 1415-1420.

Kuge, O., Nishijima, M., and Akamatsu, Y., 1986, Phosphatidylserine biosynthesis in cultured Chinese hamster ovary cells. III. Genetic evidence for utilization of phosphatidylcholine and phosphatidylethanolamine as precursors. *J. Biol. Chem.* **261**: 5795-5798.

Kuge, O., and Nishijima M., 1997, Phosphatidylserine synthase I and II of mammalian cells. *Biochim. Biophys. Acta.* **1348**: 151-156.

Kuge, O., Saito, K., and Nishijima, M., 1997, Cloning of a Chinese hamster ovary (CHO) cDNA encoding phosphatidylserine synthase (PSS) II, overexpression of which suppresses the phosphatidylserine biosynthetic defect of a PSS I-lacking mutant of CHO-K1 cells. *J. Biol. Chem.* **272**: 19133-19139

Kuge, O., Hasegawa, K., Saito, K., and Nishijima, M., 1998, Control of phosphatidylserine biosynthesis through phosphatidylserine-mediated inhibition of phosphatidylserine synthase I in Chinese hamster ovary cells. *Proc. Natl. Acad. Sci. U S A.* **95**: 4199-4203.

Kuge, O., Saito, K., and Nishijima, M., 1999, Control of phosphatidylserine synthase II activity in Chinese hamster ovary cells. *J. Biol. Chem.* **274**: 23844-23849.

Lee, M.-H., and Bell, R.M., 1992, Supplementation of the phosphatidyl-L-serine requirement of protein kinase C with nonactivating phospholipids. *Biochemistry* **31**: 5176-5182.

Mann, K.G., Jenny, R.J., and Krishnaswamy, S., 1988, Cofactor proteins in the assembly and expression of blood clotting enzyme complexes. *Annu. Rev. Biochem.* **57**: 915-956.

Marggraf, W.D., and Anderer, F.A., 1974, Alternative pathways in the biosynthesis of phosphatidylserine in mouse cells. *Hoppe-Seyler's Z. Physiol. Chem.* **335** 1299-1304.

Mellor, H., and Parker, P.J., 1998, The extended protein kinase C superfamily. *Biochem. J.* **332:** 281-292.

Medkova, M., and Cho, W., 1998, Mutagenesis of the C2 domain of protein kinase C. *J. Biol. Chem.* **273:** 17544-17552.

Mills, J.C., Stone, N.L., Erhardt, J., and Pittman, R.N., 1998, Apoptotic membrane blebbing is regulated by myosin light chain phosphorylation. *J. Cell Biol.* **140**: 627-636.

Nakano, T., Ishimoto, Y., Kishino, J., Umeda, M., Inoue. K., Nagata, K., Ohashi, K., Mizuno, K., and Arita H., 1997, Cell adhesion to phosphatidylserine mediated by a product of growth arrest-specific gene 6. *J. Biol. Chem.* **272**: 29411-29414.

Nalefski, E.A., and Falke, J.J., 1996, The C2 domain calcium-binding motif: structural and functional diversity. *Protein Sci.* **5**: 2375-2390.

Newton, A.C., 1995, Protein kinase C: structure, function, and regulation. *J. Biol. Chem.* **270**: 28495-28498.

Ni, R., Tomita, Y., Matsuda, K., Ichihara, A., Ishimura, K., Ogasawara, J., and Nagata,S. 1994, Fas-mediated apoptosis in primary cultured mouse hepatocytes. *Exp.Cell Res.* **215**: 332-337.

Nishijima, M., Kuge, O., and Akamatsu, Y., 1986, Phosphatidylserine biosynthesis in cultured Chinese hamster ovary cells. *J. Biol. Chem.,* **261**: 5784-5789.

Nishijima, M., Kuge, O., and Hanada, K., 1997, Mammalian cell mutants of membrane phospholipid biogenesis. *Trends Cell Biol.* **7**: 324-329.

Nishikawa, K., Arai, H., and Inoue, K., 1990, Scavenger receptor-mediated uptake and metabolism of lipid vesicles containing acidic phospholipids by mouse peritoneal macrophages. *J. Biol. Chem.* **265**: 5226-5231.

Nishizuka,Y., 1992, Intracellular signaling by hydrolysis of phospholipids and activation of protein kinase C. *Science* **258**: 607-611.

Nishizuka, Y., 1995, Protein kinase C and lipid signaling for sustained cellular responses. *FASEB J.* **9**: 484-496.

Obeid, L.M., Linardic, C.M., Karolak, L.A., and Hannun,Y.A., 1993, Programmed cell death induced by ceramide. *Science* **259**: 1769-1771.

Perisic, O., Fong, S., Lynch, D.E., Bycroft, M., and Williams, R.L., 1998, Crystal structure of a calcium-phospholipid binding domain from cytosolic phospholipase A2. *J. Biol. Chem.* **273**: 1596-1604.

Pradhan, D., Krahling, S., Williamson, .P, and Schlegel, R.A., 1997, Multiple systems for recognition of apoptotic lymphocytes by macrophages. *Mol. Biol. Cell.* **8**: 767-778.

Rizo, J. and Sudhof, T.C., 1998, C2-domains, structure and function of a universal Ca2+-binding domain. *J. Biol. Chem.* **273**: 15879-15882.

Rusinol, A.E., Cui, Z., Chen, M.H., and Vance, J.E., 1994, A unique mitochondria-associated membrane fraction from rat liver has a high capacity for lipid synthesis and contains pre-Golgi secretory proteins including nascent lipoproteins. *J. Biol. Chem.*, **269**: 27494-27502.

Saito, K., Kuge, O, and Nishijima, M., 1998, Genetic evidence that phosphatidylserine synthase II catalyzes the conversion of phosphatidylethanolamine to phosphatidylserine in Chinese hamster ovary cells. *J. Biol. Chem.* **273**: 17199-17205.

Sakane, F., Yamada, K., Imai, S., and Kanoh, H., 1991, Porcine 80-kDa diacylglycerol kinase is a calcium-binding and calcium/phospholipid-dependent enzyme and undergoes calcium-dependent translocation. *J.Biol.Chem.* **266**: 7096-7100.

Shiao, Y.J., Lupo, G., and Vance, J.E., 1995, Evidence that phosphatidylserine is imported into mitochondria via a mitochondria-associated membrane and that the majority of mitochondrial phosphatidylethanolamine is derived from decarboxylation of phosphatidylserine. *J. Biol. Chem.* **270**: 11190-11198.

Shiratsuchi, A., Umeda, M., Ohba, Y., and Nakanishi, Y., 1997, Recognition of phosphatidylserine on the surface of apoptotic spermatogenic cells and subsequent phagocytosis by Sertoli cells of the rat. *J. Biol. Chem.* **272**: 2354-2358.

Schroit, A.J., and Zwaal, R.F., 1991, Transbilayer movement of phospholipids in red cell and platelet membranes. *Biochim. Biophys. Acta* **1071**: 313-329.

Stekhoven, F.M., Tijmes, J., Umeda, M., Inoue, K., and De Pont J.J., 1994, Monoclonal antibody to phosphatidylserine inhibits Na^+/K^+-ATPase activity. *Biochim. Biophys. Acta.* **1194**: 155-165.

Stone, S.J., and Vance, J.E., 2000, Phosphatidylserine synthase-1 and -2 are localized to mitochondria-associated membranes. *J. Biol. Chem.* **275**: 34534-34540.

Sunshine, C., and McNamee, M.G., 1992, Lipid modulation of nicotinic acetylcholine receptor function: the role of neutral and negatively charged lipids. *Biochim. Biophys. Acta* **1108**: 240-246.

Sutton, R.B., Davletov, B.A., Berghuis, A.M., Sudhof, T.C., and Sprang, S.R., 1995, Structure of the first C2 domain of synaptotagmin I: a novel calcium/phospholipid-binding fold. *Cell* **80**: 929-938.

Sutton, R.B., and Sprang, S.R., 1998, Structure of the protein kinase C phospholipid-binding C2 domain complexed with Ca2+. *Structure* **6**: 1395-1405.

Trotter P., J., and Voelker, D., R., 1994, Lipid transport processes in eukaryotic cells. *Biochim. Biophys. Acta.* **1213**: 241-262.

Tuma, P.L., Stachniak, M.C. and Collins, C.A.J., 1993, Activation of dynamin GTPase by acidic phospholipids and endogenous rat brain vesicles. *J.Biol.Chem.* **268**: 17240-17246.

Uchida, K., Emoto, K., Daleke, D. L., Inoue, K., and Umeda, M., 1998, Induction of apoptosis by phosphatidylserine. *J. Biochem.*, **123**: 1073-1078.

Umeda, M., Igarashi, K., Nam, K.S., and Inoue, K., 1989, Effective production of monoclonal antibodies against phosphatidylserine: stereo-specific recognition of phosphatidylserine by monoclonal antibody. *J. Immunol.* **143**: 2273-2279.

Umeda, M., and Emoto, K., 1999, Membrane phospholipid dynamics during cytokinesis: regulation of actin filament assembly by redistribution of membrane surface phospholipid. *Chem. Phys. Lipids*, **101**: 81-91.

Vance, J.E., 1990, Phospholipid synthesis in a membrane fraction associated with mitochondria. *J. Biol. Chem.* **265**: 7248-7256.

Vance, J.E., 1991, Newly made phosphatidylserine and phosphatidylethanolamine are preferentially translocated between rat liver mitochondria and endoplasmic reticulum. *J. Biol. Chem.* **266**: 89-97.

Vance, J.E., Aasman, E.J., and Szarka, R., 1991, Brefeldin A does not inhibit the movement of phosphatidylethanolamine from its sites for synthesis to the cell surface. *J. Biol. Chem.* **266**: 8241-8247.

Verdaguer, N., Corbalan-Garcia, S., Ochoa, W.F., Fita, I., and Gomez-Fernandez, J.C., 1999, Ca^{2+} bridges the C2 membrane-binding domain of protein kinase Calpha directly to phosphatidylserine. *EMBO J.* **18**: 6329-6338.

Verhoven, B., Schlegel, R.A., and Williamson, P., 1995, Mechanisms of phosphatidylserine exposure, a phagocyte recognition signal, on apoptotic T lymphocytes. *J. Exp. Med.* **182**: 1597-1601.

Vidugiriene, J., Sharma, D.K., Smith, T.K., Baumann, N.A., and Menon, A.K., 1999, Segregation of glycosylphosphatidylinositol biosynthetic reactions in a subcompartment of the endoplasmic reticulum. *J. Biol. Chem.* **274**: 15203-15212.

Voelker, D.R., 1984, Phosphatidylserine functions as the major precursor of phosphatidylethanolamine in cultured BHK-21 cells. *Proc. Natl. Acad. Sci. U S A* **81**: 2669-2673.

Voelker, D.R., and Frazier, J.L., 1986, Isolation and characterization of a Chinese hamster ovary cell line requiring ethanolamine or phosphatidylserine for growth and exhibiting defective phosphatidylserine synthase activity. *J. Biol. Chem.* **261**: 1002-1008.

Voelker, D.R., 1989, Phosphatidylserine translocation to the mitochondrion is an ATP-dependent process in permeabilized animal cells. *Proc. Natl. Acad. Sci. U S A.* **86**: .9921-9925.

Voelker, D.R., 1990, Lipid transport pathways in mammalian cells. *Experientia* **46**: 569-579.

Voelker, D.R., 1991, Adriamycin disrupts phosphatidylserine import into the mitochondria of permeabilized CHO-K1 cells. *J. Biol. Chem.* **266**: 12185-12188.

Voelker, D.R., 1993, The ATP-dependent translocation of phosphatidylserine to the mitochondria is a process that is restricted to the autologous organelle. *J. Biol. Chem.* **268**: 7069-7074.

Voelker, D.R., 1997, Phosphatidylserine decarboxylase. *Biochim. Biophys. Acta.* **1348**: 236-244.

Voelker, D.R., 2000, Interorganelle transport of aminoglycerophospholipids. *Biochim. Biophys. Acta.* **1486**: 97-107.

Yasugi, E., Yokoyama, Y., Seyama, Y., Kano, K., Hayashi, Y., and Oshima, M., 1995, Dolichyl phosphate, a potent inducer of apoptosis in rat glioma C6 cells. *Biochem. Biophys. Res. Commun.* **216**: 848-853.

Yu, A., Byers, D. M., Ridgway, N. D., McMaster, C. R., and Cook, H. W., 2000, Preferential externalization of newly synthesized phosphatidylserine in apoptotic U937 cells is dependent on caspase-mediated pathways. *Biochim. Biophys. Acta.* **1487**: 296-308.

Zachowski, A., 1993, Phospholipids in animal eukaryotic membranes: transverse asymmetry and movement. *Biochem. J.*, **294**: 1-14.

Zhao, J., Zhou. Q., Wiedmer. T., and Sims, P.J., 1998, Level of expression of phospholipid scramblase regulates induced movement of phosphatidylserine to the cell surface. *J. Biol. Chem.* **273**: 6603-6606.

Zhou, Q., Zhao, J., Stout, J.G., Luhm, R.A., Wiedmer, T, and Sims, P.J., 1997, Molecular cloning of human plasma membrane phospholipid scramblase. A protein mediating transbilayer movement of plasma membrane phospholipids. *J. Biol. Chem.* **272**: 18240-18244.

Chapter 5

Phosphatidylserine Peroxidation During Apoptosis

A signaling pathway for phagocyte clearance

Y. Y. Tyurina[†], V. A. Tyurin[†], S. X. Liu, C. A. Smith, A. A. Shvedova[§], N. F. Schor[#] and V. E. Kagan[*]

Departments of Environmental and Occupational Health, [*]Pharmacology, and [#]Pediatrics and Neurology, University of Pittsburgh, Pennsylvania 15238, [†]Institute of Evolutionary Physiology & Biochemistry, St. Petersburg, Russia, and [§]Health Effects Laboratory Division, Pathology and Physiology Research Branch, NIOSH, Morgantown, West Virginia 26505, U.S.A.

1. INTRODUCTION

Two distinct pathways, apoptosis and necrosis, may be triggered contemporaneously or independently of each other to cause cell death. Necrotic cell death is generally a consequence of severe stress conditions resulting in increase in the cell volume, disruption of plasma membrane, release of intracellular contents and an inflammatory response. Apoptosis, or programmed cell death, is essential for normal tissue development and organogenesis, immunologic selection, response to tissue injury. It is characterized by chromatin condensation, DNA fragmentation, cytochrome c release, caspase activation and cell shrinkage. Our understanding of apoptosis initiation and execution pathways and mechanisms is far from being complete. In the nematode *Caenorhabditis elegans*, genetic studies resulted in the discovery of 15 genes that are essential for the apoptotic program. Some of these genes have been shown to be conserved across a wide range of species. These 15 genes have been divided into four groups based on temporal order of their action during apoptosis. The products of these genes are involved in: a) decision making (*ces-1* and *ces-2*); b) execution (*ced-3*, *ced-4*, *ced-9* and *egl-1*); c) engulfment of apoptotic cells (*ced-1*, *ced-2*, *ced-5*, *ced-6*, *ced-7*, *ced-10*, *ced-12*) and d) degradation of apoptotic cells within phagocytes (Wu and Horvitz, 1988 ab; Liu *et al.*, 1999).

Removal of apoptotic, damaged and other unwanted cells from the bloodstream or tissues, is very important for cell and tissue homeostasis. *In vivo*, apoptotic cells are efficiently removed by professional phagocytes, macrophages or neighboring cells acting as semi-professional phagocytes (Terpstra and Berkel, 2000; Chang *et al.*, 2000; Witting *et al.*, 2000). The mechanisms by which phagocytes recognize apoptotic cells are poorly understood. It is clear that the outer leaflet of plasma membrane in apoptotic cells must be different from that of healthy cells. Promotion of "safe" phagocytic clearance of cells undergoing apoptosis requires the appearance of cell surface “eat-me” signals on the apoptotic cells (Savill, 1998, Ren and Savill, 1998, Savill and Fadok, 2000). Some of the "eat-me" signals, such as changes in surface carbohydrate moieties of membrane proteins and exposure of phosphatidylserine (PS) on the surface of apoptotic cells, are well characterized (Dini *et al.*, 1992; Benner *et al.*, 1995; Martin *et al.*, 1995; Adayev *et al.*, 1998, Savill, 1997, 1998, Savill and Fadok, 2000; Witting *et al.*, 2000). Other signals involved in the removal of apoptotic cells are not well characterized. These particular signals execute removal through adhesive bridging with extracellular matrix proteins that appear on apoptotic target cells.

Execution of the apoptotic program is associated with departure of cytochrome c from mitochondria and disruption of electron transport resulting in generation of reactive oxygen species (ROS), particularly superoxide and hydrogen peroxide. Whether this ROS production during apoptosis is an inevitable but unimportant consequence of dysregulated electron transport, or fulfills important signaling functions during apoptosis, is not fully understood. Recently, formation of various ROS has been implicated as components of the final common pathway leading to the execution of apoptosis following exposure to tumor necrosis factor, growth factor withdrawal, various oxidants, and numerous other insults. It has been established that apoptosis and PS externalization are associated with ROS generation (Wood and Youle, 1994; Jacobson, 1996). Thus, appearance of PS in outer monolayer of the plasma membrane could be related to its oxidation in the plasma membrane of apoptotic cells. Thus, oxidized PS on the cell surface may be recognized as an "eat-me" signal by macrophages and serve to regulate and stimulate phagocytosis of apoptotic cells (Kagan *et al.*, 2000).

2. MEMBRANE PHOSPHOLIPID ASYMMETRY DURING APOPTOSIS

Active maintenance of membrane phospholipid asymmetry is universal in normal cell membranes. Phosphatidylcholine (PC) and sphingomyelin (SPH) are found predominantly in the outer monolayer of plasma membrane, whereas phosphatidylinositol (PI) and phosphatidylethanolamine (PE), are confined predominantly to its inner leaflet. Phosphatidylserine (PS) is, however, the only phospholipid that localizes almost entirely to the inner leaflet of plasma membrane (Zwaal and Schroit, 1999; Bevers *et al.*, 1999). Surface exposure of PS has important physiological and phathological implications for blood coagulation, apoptosis and cell-cell recognition (Bevers *et al.*, 1998). The loss of transmembrane asymmetry of plasma membrane phospholipids and consequent externalization of PS on the surface of apoptotic cells is considered to be the critical triggering event required for their recognition by some macrophages (Martin *et al.*, 1995; Hampton *et al.*, 1996; Fadok *et al.*, 1998, 2001). Several proteins have been identified for the function of maintaining asymmetric transbilayer distribution of phospholipids. Maintenance of PS in the inner leaflet of the plasma membrane is accomplished by ATP-dependent aminophospholipid translocase (APT), P-type ATPase (Tang *et al.*, 1996; Daleke and Lyles, 2000; Ding *et al.*, 2000). APT transports aminophospholipids, PS and PE, from the outer to the inner leaflet of the plasma membrane, with a preference for PS over PE (Daleke and Huestis, 1985; Bitbol and Devaux, 1988). Originally discovered in erythrocytes, the APT activity has been identified in plasma membranes of a number of cells (Daleke and Lyles, 2000). Mechanisms of PS exposure on the surface of apoptotic cells appear to involve down-regulation of APT (Verhoven *et al.*, 1999; Bratton *et al.*, 1997). APT activity has been shown to be sensitive to oxidation and SH-modification and can be regulated in a redox-dependent manner (Fabisiak *et al.*, 2000, Herrmann and Devaux, 1990). In our experiments, direct inhibition of APT, as measured by NBD-PS internalization, occurs at the same time as PS oxidation (Fabisiak *et al.*, 2000). Oxidized PS may fail to be recognized by APT, and thus, escape its surveillance function. Alternatively, reactive oxidation products of PS may covalently modify APT within its active catalytic site and serve to "poison" the enzyme (Kagan *et al.*, 2000).

The steady-state level of PS in the inner leaflet of plasma membrane is the result of two opposite processes, an inward translocation mediated by APT, and an outward translocation mediated by Mg^{2+}, ATP-dependent floppase (Haest *et al.*, 1997). Similar to APT, floppase activity could be abolished by ATP depletion and sulfhydryl oxidation (Bevers *et al.*, 1999).

It has been demonstrated that floppase in red blood cells is identical to the multidrug resistance protein (MDRP) (Kamp and Haest, 1998). MDRP1 has very broad substrate specificity and translocates a number of lipid-soluble compounds, including anti-tumor drugs, from the inner to the outer leaflet. Recently, MDRP1 P-glycoprotein has been shown to transport NBD-labeled phospholipids (Van Helvoort *et al.*, 1996) including NBD-PS (Kamp and Haest, 1998). In mammalian cells, transbilayer arrangement of lipids is also mediated by ATP-binding cassette (ABC), proteins related to MDRP (Luciani and Chimini 1996; Wu and Horvitz, 1998b). ABC-1 transporter (Hamon *et al.*, 2000), a homologue of the *C.elegans* protein CED-7, is likely to function in both apoptotic cells and macrophages (Luciani and Chimini 1996; Wu and Horvitz, 1998b). It has been shown that, in addition to its externalization during apoptosis in target cells, PS is expressed on the surface of macrophages as well and is functionally significant and required for the phagocytosis of PS-expressing target cells (Marguet *et al.*, 1999; Callahan *et al.*, 2000). Furthermore, ABC-1 induced modification of lipid distribution in macrophage membranes, yielding exofacial PS flopping, generates the biophysical microenvironment required for the docking of apoA-1 at the cell surface (Cambenoit *et al.*, 2001).

Another protein, phospholipid scramblase, mediates non-specific bi-directional transbilayer movement of membrane phospholipids, including PS, under conditions of elevated calcium and acidification of cytosol (Zhao *et al.*, 1998; Verhoven *et al.*, 1999, Fadok *et al.*, 1998). Activation of phospholipid scramblase is also important for the appearance of PS on the surface of apoptotic cells. It has been reported that up-regulation of ATP-independent phospholipid scramblase during apoptosis is involved in PS externalization (Verhoven *et al.*, 1999; Bratton *et al.*, 1997). A recent study reported that scramblase activation precedes the appearance of PS on the cell surface. This supports the hypothesis that increased scamblase activity is required for PS externalization (Frasch *et al.*, 2000). PS externalization has been demonstrated to occur in a number of cell types undergoing apoptosis, including neutrophils, thymocytes, lymphocytes, actrocytes, as well as various tumor cell lines (Adayev *et al.*, 1998; Chang *et al.*, 2000; Fadok *et al.*, 1992, 2001; Schlegel *et al.*, 2001; Verhoven *et al.*, 1999; Witting *et al.*, 2000). Externalization of PS, however, is not an obligatory component of the apoptotic phenotype. Fadeel *et al.*, (1999) detected PS exposure in Jurkat and CEM cells subjected to different apoptotic treatments, whereas P39, HL-60 and Raji cells failed to externalize PS in response to the same pro-apoptotic stimuli (Fadeel *et al.*, 1999).

3. ROLE OF PS IN RECOGNITION OF APOPTOTIC CELLS BY MACROPHAGES

Recognition and removal of PS-displaying cells has been extensively studied. It has been demonstrated that not all macrophages recognize externalized PS (Fadok *et al.*, 1992; Schwartz *et al.*, 1999). Those macrophages that recognize PS on apoptotic cell use at least two different recognition systems (Schliegel *et al.*, 2000). Binding and uptake of PS-displaying cells appears to involve a multiple receptor-mediated system that recognizes PS either directly or indirectly through intermediate proteins that form a molecular bridge between apoptotic cell and macrophage (Savill and Fadok, 2000). A recent study showed several receptors and molecules to be important in the uptake of PS-displaying apoptotic cells in mammals. CD36 is a member of the family of class B scavenger receptor s like SR-B1 and is the first well-defined receptor shown to bind to apoptotic cells through a PS-related mechanism (Rigotti *et al.*, 1995). CD36 has been implicated as a putative receptor for oxidized LDL (Nicolson *et al.*, 1995; Krieger, 1997; Stainberg, 1997; Febbraio *et al.*, 1999), and appears to be the major PS-binding protein on the surface of retinal pigment epithelial cells (Ryeom *et al.*, 1996), platelets (Miao *et al.*, 2001), endothelial cells (Nicolson *et al.*, 1995; Febbraio *et al.*, 1999), human macrophages (Fadok *et al.*, 1998), and monocytes (Tait and Smith, 1999) and is specific for PS among anionic phospholipids. Several human macrophages use the vitronectin receptor ($\alpha_V\beta_3$-integrin) associated with CD36 to phagocytize apoptotic cells (Fadok *et al.*, 1992, 1998; Savill *et al.*, 1992). Furthermore, CD36 seems to be an important cofactor for recognition mechanisms mediated by PS-specific receptors (Fadok *et al.*, 1998). Existing data suggest that human macrophages might phagocytize PS-displaying apoptotic cells using the surface CD14 receptor, a glycophosphatidylinositol anchored protein (Devitt *et al.*, 1998: Schlegel *et al.*, 1999) that may function separately or jointly with the integrin and phospholipid receptor. Phosphatidylinositol (PI) and PS are more likely than the other phospholipids tested to be natural CD14 ligands (Wang *et al.*, 1998; Akashi *et al.*, 2000). Phagocyte scavenger receptor CD68 is involved in the binding of oxidized LDL (Sambrano *et al.*, 1994; Chang *et al.*, 1999), PS vesicles and apoptotic PS-displaying thymocytes to mouse peritoneal macrophages (Ramprasad *et al.*, 1995, 1996; Fadok *et al.*, 1992). A lectin-like receptor for oxidized LDL, LOX-1, was cloned from endothelial cells (Sawamura *et al.*, 1997) and it was shown to bind and phagocytize aged red blood cells and apoptotic cells through a PS-dependent mechanism (Oka *et al.*, 1998). Recently, a cDNA encoding a novel class of scavenger receptor s, SR-PSOX, that binds PS and oxidized LDL was isolated (Shimaoka *et al.*, 2000). Several studies proposed the existence of a so-called PS receptor (Fadok *et al.*, 1992, 1998; Tait and Smith, 1999). Lately, Fadok and co-workers cloned a gene that appears to recognize PS on apoptotic cells. Transfection of this gene to B and T lymphocytes results in recognition and engulfment of apoptotic cells in a PS-specific manner. It was found that newly-identified stereo-specific PS receptor (PSR) is expressed by professional phagocytes, fibroblasts and epithelial cells (Fadok *et al.*, 2000).

Rapid clearance of the PS-displaying particle from circulation and phagocytosis of apoptotic cells is associated with PS-bridging complexes between macrophage and the PS displaying particle or apoptotic cell (Chonn *et al.*, 1995; Sugihara *et al.*, 1993; Fadok *et al.*, 1998; Balasubramanian, *et al.*, 1997 Balasubramanian and Schroit, 1998, Mevorach, 2000). On the phagocyte, CD36 receptor and vitronectin receptor ($\alpha_V\beta_3$-inegrin) have the ability to bind a bridging molecule thrombospondin, that is present in plasma and found in the basement membrane of endothelial cells (Savill *et al.*, 1992; Jimenez *et al.*, 2000; Bornstein 2001). It was suggested that PS is the partner for thrombospondin bridging on the apoptotic cells (Sugihara *et al.*, 1993; Fadok *et al.*, 1998) and PS-displaying erythrocytes (Manodori *et al.*, 2000). PS induces conformational changes in plasma glycoprotein, β_2-glycoprotein I, that exposes a specific epitope that is recognized by a macrophage receptor distinct from CD36, CD68 and CD14 (Balasubramanian and Schroit, 1998). Recently it has been shown that compliment receptors CR3 and CR4 are also important for rapid elimination of the PS-containing liposomes from blood circulation (Huong *et al.*, 2001) and uptake of apoptotic cells by human macrophages (Mevorach *et al.*, 1998). Activation of the compliment is mediated, at least in part, by PS and results in coating the apoptotic cell surface with C3bi (Mevorach, 2000).

4. ROLE OF LIPID OXIDATION IN MACROPHAGE RECOGNITION

Macrophages are able to recognize and effectively take up from the circulation oxidized LDL particles as well as a wide variety of apoptotic and oxidatively damaged cells when they are no longer functionally required (Sambrano *et al.*, 1994; Eda *et al.*, 1997; Mold and Morris, 2001). Oxidized LDL is taken up and degraded by mouse macrophages significantly faster than native lipoproteins (Babiy and Gebidcki, 1999). It is tempting to assume that both oxidized LDL and apoptotic cells may have common ligands on their surface, consisting of oxidatively modified moieties that are recognized by common macrophage receptors, including CD36, CD68 and LOX-1. Chang and co-workers demonstrated that when cells undergo apoptosis, they express oxidation-specific epitopes on their cell surface, including oxidized phospholipids and/or oxidized phospholipid-protein adducts, and that these serve as ligands for macrophage recognition and phagocytosis (Chang *et al.*, 1999). Cellular plasma membrane and LDL particles are both composed mainly of phospholipids containing unsaturated fatty acids (Schnurr *et al.*, 1996). Polyunsaturated phospholipids are mostly prone to oxidative attack, thus the phospholipid oxidation products should contain hydroperoxy-fatty acid residues in the *sn*-2 position at the initial stage of free radical oxidation (Kagan, 1988). It has been shown that the lipid moiety of oxidized LDL, derived from oxidation of arachidonoyl phospholipids (Watson *et al.*, 1995), is responsible for its interaction with the scavenger receptor (Terstra *et al.*, 1998, Boullier *et al.*, 2000; Gillotte *et al.*, 2000). Furthermore, minimally oxidized LDL and the oxidation product of 1-palmitoyl-2-arachidonoyl-*sn*-glysero-3-phosphatidyl-choline have been shown to stimulate endothelial cells to increase monocyte-endothelial interaction (Berliner *et al.*, 1990; Lee *et al.*, 2000), which is inhibited by oxidized but not non-oxidized polyunsaturated fatty acids (Scthi *et al.*, 1996). Subbanagounder *et al.* identified 6 new bioactive, oxidized phospholipids that have specific effects on the monocyte-endothelium interaction and demonstrated that the specificity is determined by the *sn*-2 groups of oxidized phospholipids (Subbanagounder *et al.*, 2000).

5. PS OXIDATION IN APOPTOSIS

Oxidative stress in general and lipid peroxidation in particular have been implicated in the final common pathway of apoptosis (Chandra *et al.*, 2000). Cytochrome c released from mitochondria may be involved in specific interaction with PS located in the cytosolic leaflet of plasma membrane and be responsible for selective PS oxidation during apoptosis (Kagan *et al.*, 2000; Tyurina *et al.*, 2000). Recently, we demonstrated that cytochrome c can effectively induce oxidation of PS in both liposomes and intact living HL-60 cells (Kagan *et al.*, 2000). Moreover, PS oxidation during apoptosis may, in some way, participate in its externalization within the membrane. We have recently reported that apoptosis following exposure to paraquat was associated with enhanced sensitivity of PS to oxidation that was attenuated by Bcl-2 (Fabisiak *et al.*, 1997).

To characterize phospholipid oxidation during apoptosis, cellular phospholipids were metabolically labeled at *sn*-2 position with a natural unsaturated florescent fatty acid containing four conjugated double bonds, *cis*-parinaric acid (*cis*-PnA) (Kagan *et al.*, 1998; Tyurina *et al.*, 2001). Oxidative destruction of any part of the conjugated double bond system of *cis*-PnA results in irreversible loss of its characteristic fluorescence (Kuypers *et al.*, 1987). The amounts of PnA-labeled phospholipid in each class usually do not exceed 2%. That is low enough to have minimal effects on cell viability and function yet sufficient to permit quantitative detection of oxidative stress (Kagan *et al.*, 1998).

Using this protocol we found that apoptosis induced by a number of different oxidants is associated with selective oxidation of specific phospholipid classes, most notably PS (Kagan *et al.*, 2000, Tyurina *et al.*, 2000). Several cell lines were used to investigate the role of PS oxidation in apoptosis. Preferential oxidation of PS was observed in human leukemia HL-60 cells (Fabisiak *et al.*, 1998, 2000; Kawai *et al.*, 2000), and normal human keratinocytes (Shvedova *et al.*, 2001). Similarly, in pheochromocytoma PC12 cells exposed to a radical-generating antineoplastic drug, neocarzinostatin, externalization of PS was potentiated by its selective oxidation in whole cells (Schor *et al.*, 1999). In contrast, this selective PS oxidation did not occur in liposomes prepared from mixtures of PnA-labeled phospholipids extracted from the cells and exposed to oxidants under the same conditions (Fabisiak *et al.*, 1998; Kagan *et al.*, 2000; Shvedova *et al.*, 2001). Recently, we also documented that PS oxidation during oxidant-induced apoptosis occurred within the plasma membrane, the largest source of oxidized PS compared to other cellular organells (Kawai *et al.*, 2000).

Since in all these cases we used oxidants to induce apoptosis, a relatively massive background oxidation of all PnA-labeled phospholipid classes was observed and masked apoptosis-specific PS oxidation. Therefore, we next determined whether a non-oxidant-induced apoptosis model could be used to reveal more explicitly the selective mode of PS oxidation inherent to the apoptosis execution program. To this end, we used agonistic anti-Fas antibody to induce apoptosis in Jurkat B cells according to a commonly used protocol (Fadeel *et al.*, 1999). Importantly, we found that anti-Fas-triggered apoptosis in Jurkat cells was characterized by early and selective oxidaiton of *cis*-PnA-PS (Kagan *et al.*, 2001) without any significant involvement of other classes of phospholipids.

We further used a quantitative procedure to determine the amounts of externalized PS on the surface of apoptotic cells based on its availability to react with fluorescamine, a non-permeating reagent readily interacting with primary amino-groups. Subsequent separation (by high performance thin-layer chromatography) and quantitation of fluorescamine-modified PS permits determination of absolute amounts of externalized PS on the outer leaflet of plasma membrane. We were also able to determine the amounts of PnA-phospholipids oxidized during apoptosis based on their disappearance in the course of apoptosis. Thus both PS oxidation and externalization could be assessed.

Table 1. Externalization and oxidation of PS during apoptosis

Cell line	Stimuli	PS externalised*	PnA-PS oxidized*
HL-60	AMVN	51.2	12.2
HL-60	tert-BuOOH	41.6	2.2
HL-60	H_2O_2	27.8	2.9
NHEK	Cumene OOH	84.6	5.7
Jurkat	α-Fas antibody	52.8	0.3

* 10^6 molecules/cell

Our data on PS oxidation and externalization summarized in Table 1 show that relatively high amounts of PS were externalized as a result of apoptosis induced by both oxidants and non-oxidants (anti-Fas antibody). Interestingly, between 5-20% of externalized PS was represented by oxidized PnA-PS in the cells when apoptosis was induced by oxidants. In contrast, only <1% of esterified PS was accounted for as oxidized PnA-PS in Jurkat cells triggered with anti-Fas-antibody. This may be indicative of a specific role that oxidized PS may play in its externalization. Given that PnA-PS represents only ≈2 mol% of total PS in cells, realistic amounts of the fraction of oxidized PS in the total pool of externalized PS may be significantly higher. Thus, oxidation of PS may be a significant contributor to its externalization during apoptosis and may consequently play a very important role in recognition of apoptotic cells by macrophages.

6. OXIDATION OF PS AND RECOGNITION OF APOPTOTIC CELLS BY MACROPHAGES

The absolute amounts of externalized PS may be essential for clearance of apoptotic cells. Surprisingly, it has been recently found that viable developing and mature B cells can bind significant amounts of annexin V (Dillon *et al.*, 2000, 2001). These normal PS-displaying cells do not exert any of the biochemical or morphological characteristics associated with apoptosis. One may suggest that the relatively low amounts of externalized PS in normal non-apoptotric cells may be insufficient for recognition by macrophage receptors. PS exposed on the surface of viable cells and may be sequestered in membrane microdomains the characteristics of which are essential for recognition by macrophages. Liposome binding experiments indicate that increasing the amounts of liposomal PS from 10 to 30 mol% leads to a faster disappearance of the liposomes from circulation (Kamps *et al.*, 1999). Moreover, oxidation of PS may be an additional factor contributing to the enhanced recognition by macrophages. In line with this, no uncontrolled uptake of slightly oxidized LDL (fewer than 150 hydroperoxide groups per particle) by macrophages was observed (Babiy and Gebidcki, 1999). Thus, both the amount of PS exposed on the cell surface and its oxidative modification may be critical for recognition and engulfment of apoptotic cells.

To test whether oxidation of PS in plasma membrane may be functionally linked not only with PS externalization but with subsequent phagocytosis as well,we performed experiments to compare phagocytosis of HL-60 cells in which we integrated PS or oxidized PS into plasma membrane. We found that recognition of the cells with externalized PS was strongly facilitated when oxidized PS was also present on the surface of target cells. Integration of PS containing-liposomes into HL-60 cells pretreated with N-ethylmaleimide to inhibit APT (and prevent PS internalization) stimulated their phagocytosis by J774A.1 macrophages (Figure 1, A, B). This effect was markedly enhanced when oxidized PS (15 mol%) was integrated into HL-60 target cells along with PS (Figure 1, C). Neither PC, nor oxidized PC were able to stimulate phagocytosis of HL-60 cells by J774A.1 macrophages.

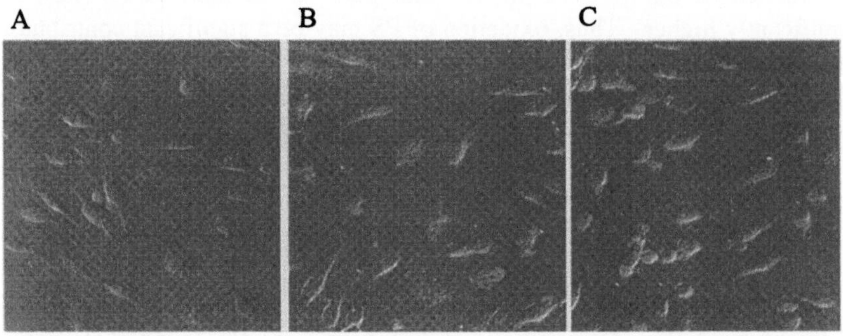

Figure 1. Phagocytosis of HL-60 cells by macrophages. Photographs depicting representative examples of phagocytosis of HL-60 cells by macrophage cell line. (A) naive (untreated); (B) HL-60 cells were pretreated with APT inhibitor, NEM, after which, PS was integrated into outer leaflet of plasma membrane; (C) HL-60 cells were pretreated with APT inhibitor, NEM, after which oxidized PS was integrated into outer leaflet of plasma membrane. HL-60 cells were labeled with CellTracker Orange (Molecular Probes, Eugene, OR) (2 mM in serum free medium at 37 °C for 15 min). Labeled cells were pre-treated with N-ethylmaleimide (10 μM, 5 min) and then incubated with liposomes containing either PS or oxidized PS at 37 °C for 30 min. HL-60 cells containing either externalized PS or oxidized PS were added to J774A.1 and phagocytosis was observed and photographed using Nikon fluorescence imaging system.

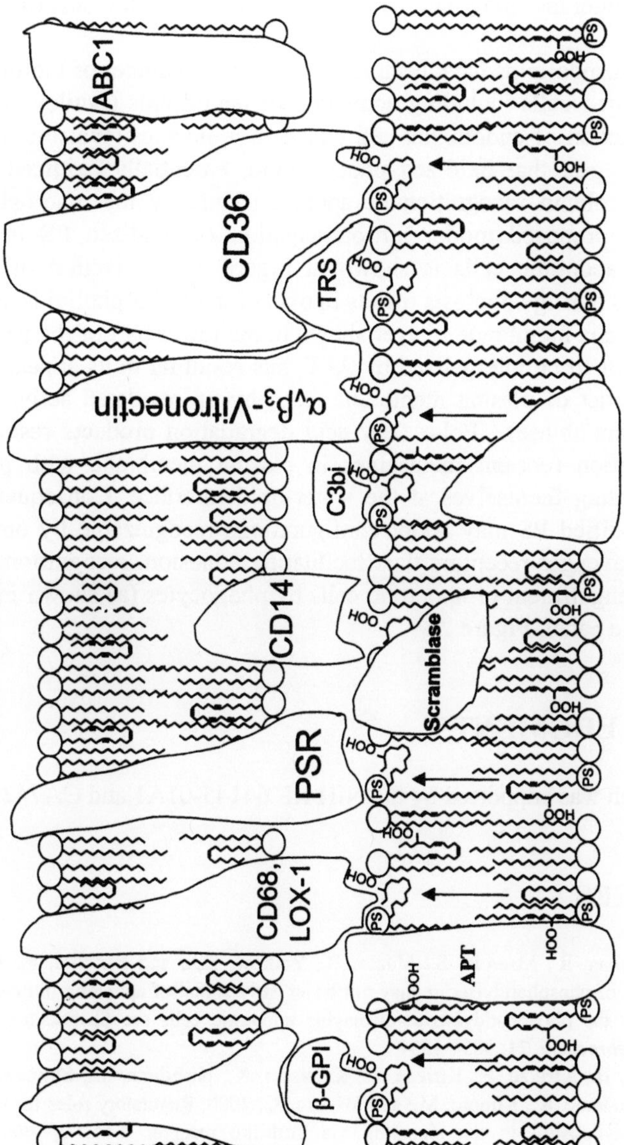

Figure 2. A model illustrating participation of oxidized PS in its externalization and recognition and engulfment of apoptotic cells.

Thus, selective oxidation and externalization of PS in plasma membrane are likely to create conditions where oxidized PS on the external surface of the plasma membrane may act as a preferred ligand for macrophage receptor.

Oxidized phospholipids have been shown to have a number of biological effects on cultured cells including activation of neutrophils (Smiley *et al.*, 1991) and induction of endothelial cell adhesion molecules (Watson *et al.*, 1995). We suggest that oxidized phospholipid, essentially oxidized PS, plays a major role in recognition of apoptotic cells by the macrophage receptor. Our proposed model for participation of oxidized PS in the engulfment of apoptotic cells is shown in Figure 2. Activation of free radical reactions during apoptosis results in oxidation of phospholipids, most notably PS (plasma membrane bilayer shown in the lower part of the panel). Oxidized PS fails to be internalized by APT thus resulting in its appearance in the outer leaflet of plasma membrane (e.g., by up-regulated activity of phospholipid scramblase). Polar fatty acid degradation products resulting from PS oxidation reorient themselves in plasma membrane with polar moieties presenting themselves at the water-lipid interface. Consequently, oxidatively modified PS may adopt configurations recognizable by one of macrophage scavenger receptors thus facilitating adhesion, recognition and PS-dependent engulfment of apoptotic cells by phagocytes (as shown in the upper part of the panel, Figure 2).

ACKNOWLEDGMENTS

This research was supported by the NIH HL 64145-01A1 and CA74289

REFERENCES

Adayev, T., Estephan, R., Mserole, S., Mazza, B., Yorkow, E.J., and Banerjel, P., 1998, Externalization of phosphatidylserine may not be an early signal of apoptotic in neuronal cells, but only the phoaphatidylserine-displaying apoptotic cells are phagocytosed by microglia, *J. Neurochem.* **71**: 1854-1864.

Akashi, S., Ogata, H., Kirikae, F., Kirikae, T., Kawasaki, K., Nishijima, M., Shimazu, R., Nagai, Y., Fukudome, K., Kimoto, M., and Miyake, K., 2000, Regulatory roles for CD14 and phosphatidylinositol in the signaling via Toll-like receptor 4-MD-2, *Biochem. Biophys. Res. Commun.* **268**: 172-177.

Babiy, A.V., and Gebicki, J.M., 1999, Decomposition of lipid hydroperoxides chances the uptake of low density lipoprotein by macrophages, *Acta Biochim. Pol.* **46**: 31-42.

Balasubramanian, K., and Schroit, A.J., 1998, Characterization of phosphatidylserine-sependent β2-glycoprotein I macrophage interactions, *J. Biol. Chem.* **273**: 29272-29277.

Berliner, J.A., Territo, M.C., Sevanian, A., Ramin, S., Kim, J.A., Bamshad, B., Esterson, M., and Fogelman, A.M., 1990, Minimally modified low density lipoprotein stimulates monocyte endothelial interactions, *J. Clin. Invest.* **85**: 1260-1266.

Bevers, E.M., Comfurius, P., Dekkers, D.W.C., and Zwaal, F.A., 1999, Lipid translocation across the plasma membrane of mammalian cells, *Biochim. Biophys. Acta* **1439**: 317-330.

Bitbol, M., and Devaux, P.F., 1988, Measurement ofoutward translocation of phospholipids across human erythrocyte membrane, *Proc. Natl. Acad. Sci. U.S.A.* **85**: 6783-6787.

Bornstein, P., 2001, Thrombospondins as matrix cellular modulators of cell function, *J. Clin. Invest.* **107**: 929-934.

Boullier, A., Gillotte, K.L., Horkko, S., Green, S.R., Friedman, P., Dennis, E.A., Witztum, J.L., Steinberg, D., and Quehenberger, O., 2000, The binding of oxidized low density lipoprotein to mouse CD36 in mediated in part by oxidized phospholipids that are associated with both the lipid and protein moieties of the lipoprotein, *J. Biol. Chem.* **275**: 9163-9169.

Bratton, G.W., Fadok, V.A., Richter, D.A., Kailey, J.M., Gurhrie, L.A., and Henson, P.M., 1997, Appearance of phosphatidylserine on apoptotic cells requires calcium-mediated nonspecific flip-flop and is enchanced by loss of the aminophospholipid translocase, *J. Biol. Chem.* **272**: 26159-26165.

Callahan, M.K., Williamson, P., and Schlegel, R.A., 2000, Surface expression of phosphatidylserine on macrophages is required for phagocytosis of apoptotic thymocytrs, *Cell Death Differ.* **7**: 645-653.

Chambenoit, O., Hamon, Y., Marguet, D., Rigneault, H., Rosseneu, M., and Chimini, G., 2001, Specific docking of apolipoprotein A-1 at the cell surface requires functional ABCA1 transporter, *J. Biol. Chem.* **276**: 9955-9960.

Chandra, J., Samali, A., and Orrenius, S., 2000, Triggering and modulation of apoptosis by oxidative stress, *Free Radical Biol. Med.* **29**: 323-333.

Chang, M.K., Bergmark, C., Laurila, A., Horkko, S., Han, K.H., Friedman, P., Dennis, E.A., and Witztum, J.L., 1999, Monoclonal antibodies against oxidized low-density lipoprotein bind to apoptotic cells and inhibit their phagocytosis by elicited macrophages: evidence that oxidation-specific epitopes mediate macrophage recognition, *Proc. Natl. Acad. Sci. U.S.A.* **96**: 6353-6358.

Chang, G.H., Barbaro, N.M., and Pieper, R.O., 2000, Phosphatidylserine-dependent phagocytosis of apoptotic cells by normal human microglia, astrocytes, and glioma cells, *Neuro-oncol.* **2**: 174-183.

Chonn, A., Semple, S.C., and Cullis, P.R., 1995, Beta 2 glycoprotein I is a major protein assocoated with very rapidly cleared liposomes *in vivo* suggesting a significant role in the immune clearance of "non-self" particles, *J. Biol. Chem.* **270**: 25845-25849.

Daleke, D.L., and Huestis, W.H., 1985, Incorporation and translocation of aminophospholipids in human erythrocytes, *Biochemistry* **24**: 5406-5416.

Daleke, D.L., and Lyles, J.V., 2000, Identification and purification of aminophospholipid flippases, *Biochim. Biophys. Acta* **1486**: 108-127.

Devitt, A., Moffatt, O.D., Raykundalia, C., Capra, J.D., Simmons, D.L., and Grecory, C.D., 1998, Human CD14 mediates recognition and phagocytosis of apoptotic cells, *Nature* **392**: 505-508.

Ding, J., Wu, Z., Crider, B.P., Ma, Y., Li, X., Slaughter, C., Gong, L., and Xie, X.S., 2000, Identification and functional expression of four isoforms of ATPase II, the putative aminophospholipid translocase, *J. Biol. Chem.* **275**: 23378-23386.

Dini, L., Autuori, F., Lentini, A., Oliverio, S., Piacentini, M., 1992, The clearance of apoptotic cells in the liver is mediated by the asialoglycoprotein recetpor, *FEBS Lett.* **296**: 174-178.

Dillon, S.R., Mancini, M., Rosen, A., and Schlissel, M.S., 2000, Annexin V binds to viable B cells and colocalized with a marker of lipid rafts upon C cell receptor activation, *J. Immunol.* **164**: 1322-1332.

Dillon, S.R., Constantinescu, A., and Schlissel, M.S., 2001, Annexin V binds to positively selected B cells, *J. Immunol.* **166**: 58-71.

Eda, S., Kikugawa, K., and Beppu, M., 1997, Oxidatively damaged erythrocytes are recognized by membrane proteins of macrophages, *Free Radic. Res.* **27**: 23-30.

Fabisiak, J.P., Kagan, V.E., Ritov, V.B., and Laso, J.S., 1997, Bcl-2 inhibits selective oxidation and externalization of phosphatidylserine during paraquat-induced apoptosis, *Am. J. Physiol. (Cell Physiol.)* **272**: C675-C684.

Fabisiak, J.P., Tyurina, Y.Y., Tyurin, V.A., Lazo, J.S., and Kagan, V.E., 1998, Random vresus selective membrane phospholipid oxidation in apoptotis: role of phosphatidylserine, *Biochemistry* **37**: 13781-13790.

Fabisiak, J.P., Tyurin, V.A., Tyurina, Y.Y., Sedlov, A., Lazo, J.S., and Kagan, V.E., 2000, Nitric oxide dissociates lipid oxidation from apoptosis and phosphatidylserine externalization during oxidative stress, *Biochemistry* **39**: 127-138.

Fadeel, B., Gleiss, B., Hogstrand, K., Chandra, J., Wiedmer, T., Sims, P., Henter, J.-I., Orrenius, S., and Samali, A., 1999, Phosphatidylserine exposure during apoptosis is a cell type-specific event and does not correlate with plasma membrane phospholipid scramblase expression, *Biochem. Biophys. Res. Commun.* **266**: 504-511.

Fadok, V.A., Voelker, D.R., Campbell, P.A., Cohen, J.J., Bratton, D.L., and Henson, P.M., 1992, Exposure of phosphatidylserine on the surface of apoptotic lymphocytes triggers specific recognition and removal by macrophages, *J. Immunol.* **148**: 2207-2216.

Fadok, V.A., Bratton, D.L., Frasch, S.C., Warner, M.L., and Henson, P.M., 1998, The role of phosphatidylserine in recognition of apoptotic cells by phagocytes, *Cell Death Differ.* **5**: 551-562.

Fadok, V.A., Warner, M.L., Bratton, D.L. and Henson, P.M., 1998, CD36 is required for phagocytosis of apoptotic cells by human macrophages that use either a phosphatidylserine receptor or the vitronectin receptor ($\alpha v\beta 3$)1, *J. Immunol.* **161**: 6250-6257.

Fadok, V.A., Bratton, D.L., David, M.R., Alan, P., Alan, B.E.R., and Henson, P.M., 2000, A receptor for phosphatidylserine-specific clearane of apoptotic cells, *Nature* **405**: 85-90.

Fadok, V.A., deCathelineau, A., Daleke, D.L., Henson, P.M., and Bratton, D.L., 2001, Loss of phospholipid asymmetry and surface exposure of phosphatidylserine is required for phagocytosis of apoptotic cells by macrophages and fibroblasts, *J. Biol. Chem.* **276**: 1071-1077.

Febbraio, A., Abumrad, N.A., Hajjar, D.P., Sharma, K., Cheng, W., Pearce, S.F., and Silverstein, R.L., 1999, A null mutation in murine CD36 reveals an important role in fatty acid and lipoprotein metabolism, *J. Biol. Chem.* **274**: 19055-19062.

Gillotte, K.L., Horkko, S., Witztum, J.L., and Steinberg, D., 2000, Oxidized phospholipids, linked to apolipoprotein B of oxidized LDL, are ligands for macrophage scavenger receptor, *J. Lipid Res.* **41**: 824-833.

Haest, C.W., Kamp, D., and Deuticke, B., 1997, Transbilayer reorientation of phospholipid probes in the human erythrocyte membrane, Lessions from studies on electroporated and resealed cells, *Biochim. Biophys. Acta* **1325**: 17-33.

Hamon, Y., Broccardo, C., Chambenoit, O., Luciani, M.F., Toti, F., Chaslin, S., Freyssinet, J.M., Devaux, P., Neish, J., Marguet, D., and Chimini, G., 2000, ABC1 promotes engulfment of apoptotic cells and transbilayer redistribution of phosphatidylserine, *Nat. Cell Biol.* **2**: 399-406.

Hampton, M.B., Vanags, D.M., Porn-Ares, I., and Orrenius, S., 1996, Involvement of extracellular calcium in phosphatidylserine exposure during apoptosis, *FEBS Lett.* **399**: 277-282.

Herrmann, A., and Devaux, P.F., 1990, Altereation of the aminophospholipid translocase activity during *in vivo* and artificial aging of human erythrocytes, *Biochim. Biophys. Acta* **1027**: 41-46.

Huong, T.M., Ishida, T., Harashima, H., and Kiwada, H., 2001, The compliment system enhances the clearance of phosphatidylserine (PS)-liposomes in rat and guinea pig, *Int. J. Pharm.* **215**: 197-205.

Jacobson, M.D., 1996, Reactive oxygen species and programmed cell death, *Trends Biochim. Sci.* **21**: 83-86.

Jimenez, B., Volpert, O.V., Ceawford, S.E., Febbraio, M., Silverstein, R.L., and Bouck, N., 2000, Signals leading to apoptosis-dependent inhibition of neovascularization by thrombospondine-1, *Natl. Med.* **6**: 41-48.

Krieger, M., 1997, The other side of scavenger receptors: pattern recognition for host defense, *Curr. Opin. Lipidol.* **8**: 275-280.

Kagan, V.E., 1988, Lipid peroxidation in biomembranes, pp1-146, CRC Press, Boca Raton, FL.

Kagan, V.E., Ritov, V.B., Tyurina, Y.Y., and Tyurin, V.A., 1998, Sensitive and specific florescent probing of oxidative stress in different classes of membrane phospholipids in live cells using metabolically integrated *cis*-parinaric acid, *Meth. Mol. Biol.* **108**: 71-87.

Kagan, V.E., Fabisiak, J.P., Shvedova, A.A., Tyurina, Y.Y., Tyurin, V.A., Schor, N.F., and Kawai, K., 2000, Oxidative signaling phathway for externalization of plasma membrane phosphatidylserine during apoptosis, *FEBS Lett.* **477**: 1-7.

Kagan, V.E., Gleiss,B., Tyurina, Y.Y., Tyurin, V.A., Liu, D., Serinkan, B., Arroyo, A., Magnusson, C.E., Chandra, J., Orrenius, S., Fadeel, B., 2001, Submitted to *Nature Cell Biology*.

Kawai, K., Tyurina, Y.Y, Tyurin, V.A., Kagan, V.E., and Fabisiak, J., 2000, Peroxidation and externalization of phosphatidylserine in plasma membrane of HL-60 cells during tert-butyl hydroperoxide-induced apoptosis: Role of cytochrome *c*, *The Toxicologist* **54** Suppl.: 776.

Kuypers, F.A., Van den Berg, J.J.M., Schalkwijk, C., Roelofsen, B., and Op den Kamp, J.A.F., 1987, Parinaric acid as a sensitive florescent probe for the determination of lipid peroxidation, *Biochim. Biophys. Acta* **921**: 266-274.

Kamp, D. and Haest, C.W.M., 1998, Evidence for a role of the multidrug resistance protein MRP in the outward translocation of NBD-phospholipids in the erythrocyte membrane, *Biochim. Biophys. Acta* **1372**: 91-101.

Kamps, J.A., Morselt, H.W., and Scherphof, G.L., 1999, Uptake of liposomes containing phosphatidylserine by liver cells *in vivo* and by sinusoidal liver cells in primary culture: *in vivo - in vitro* differences, *Biochem. Biophys. Res. Commun.* **256**: 57-62.

Lee, H., Shi, W., Tontonoz, P., Wang, S., Subbanagouder, G., Hedrick, C.C., Hama, S., Borromeo, C., Evans, R.M., Berliner, J.A., and Nagy, L., 2000, Role for peroxisome proliferator-activated receptor [alpha] in oxidized phospholipid-induced synthesis of monocyte chematactic protein-1 and interleukin-8 by endothelial cells, *Circ. Res.* **87**: 516-521.

Liu, Q.A., and Hengartner, M.O., 1999, The molecular mechanism of programmed cell death in *C. elegans*, *Ann NY Acad, Sci.* **887**: 92-104.

Luciani, M.F., and Chimini, G., 1996, The ATP binding cassette transporter ABC1, is required for the engulfment of corpses generated by apoptotic cell death, *EMBO J.* **15**: 226-235.

Marguet, D,, Luciani, M.F., Moynault, A., Williamson, P., and Chimini, G., 1999, Engulfment of apoptotic cells involves the redistribution of membrane phosphatidylserine on phagocyte and prey, *Nat. Cell Biol.* **1**: 454-456.

Martin, S.J., Reotelingsperger, C.P.M., McGahon, A.J., Rader, J.A., van Schie, R.C.A.A., LaFace, D.M., and Green, D,R., 1995, Early redistribution of plasma membrane phosphatidylserine is a general lymphocytes triggers specific regardless of the initiating stimulus: inhibition by overexpression of Bcl-2 and Abl, *J. Exp. Med.* **182**: 1545-1556.

Mevorach, D., 2000, Opsonization of apoptotic cells implications for uptake and autoimmunity, *Ann, NY Acad. Sci.* 926: 226-235.

Mevorach, D., Mascarenhas, J.O., Gershov, D., and Elkon, K.B., 1998, Compliment-dependent clearance of apoptotic cells by human macrophages, *J. Exp. Med.* **188**: 2113-2320.

Miao, W.M., Vasile, E., Lane, W.S., and Lawler, J., 2001, CD36 associated with CD9 and integrins on human blood platelets, *Blood* **97**: 1689-1696.

Manodori, A.B., Barabino, G.A., Lubin, B.H., and Kuypers, F.A., 2000, Adherence of phosphatidylserine-exposing erythrocytes to endothelial matrix thrombospondin, *Blood* **95**: 1293-1300.

Mold, C., and Morris, C.A., 2001, Complement activation by apoptotic endothelial cells following hypoxia/reoxygenation, *Immunology* **102**: 359-364.

Nicholson, A.C., Friede, S., Pearce, A., and Silverstein, R.L., 1995, Oxidized LDL binds to CD36 on human monocyte-derived macrophages and transfected cell lines: Evidence implicating the lipid moiety of the lipoprotein as the binding site, *Arterioscl. Tromb. Vasc. Biol.* **15**: 269-275.

Oka, K., Sawamura, T., Kikuta, K., Itokawa, S., Kume, N., Kita, T., and Masaki, T., 1999, Lectin-like low-density lipoprotein receptor 1 mediates phgocytosis of aged/apoptic cells in endothelial cells, *Proc. Natl. Acad. Sci. U.S.A.* **96**: 6353-6358.

Ramprasad, M.P., Fischer, W., Witztum, J.L., Sambrano, G.R., Quehenberger, O., and Steinberg, D., 1995, The 94- to 97-kDa mouse macrophage membrane protein that recognizes oxidized low density lipoprotein and phosphatidylserine-rich liposomes is identical to macrosialin, the mouse homologue of human CD68, *Proc. Natl. Acad. Sci. U.S.A.* **92**: 9580-9584.

Ramprasad, M.P., Terpstra, V., Kondratenko, N., Quehenberger, O., and Steinberg, D., 1996, Cell surface expression of mouse macrosialine and human CD68 and their role as macrophage receptors for oxidized low density lipoprotein, *Proc. Natl. Acad. Sci. U.S.A.* **93**: 14833-14838.

Rashba-Step, J., Tatoyan, A., Duncan, R,. Ann, D., Pushpa-Rehka, T.R., and Sevanian, A., 1997, Phospholipid peroxidation induces cytosolic phospholipase A_2 activity: membrane effects versus enzyme phosphorylation, *Arch. Biochem. Biophys.* **343**: 44-54.

Ren, Y, and Savill, J., 1998, Apoptosis: the importance of being eaten, *Cell Death Differ.* **5**: 563-568.

Rigotti, A., Acton, S.L., and Krieger, M., 1995, The class B scavenger receptors SR-B1 and CD36 are receptors for anionic phospholipids, *J. Biol. Chem.* **270**: 16221-16224.

Ryeom, S.W., Silverstain, R.L., Scotto, A., and Sparrow, J.R., 1996, Binding of anionic phospholipids to retinal pigment epithelium may be mediated by the scavenger receptor CD36, *J. Biol. Chem.* **271**: 20536-20539.

Sambrano, G.R., Parthasarathy, P.S., and Stainberg, D., 1994, Recognition of oxidatively damaged erythrocytes by a macrophage receptor with specificity for oxidized LDL, *Proc. Natl. Acad. Sci. U.S.A.* **91**: 3265–3269.

Sawamura, T., Kume, N., Aoyama, T., Moriwaki, H., Hoshikawa, H., Aiba, Y., Tanaka, T., Miwa, S., Katsura, Y., Kita, T., and Masaki, T., 1997, An endothelial receptor for oxidized low-density lipoprotein, *Nature* **386**: 73-77.

Savill, J., 1997, Recognition and phagocytosis of cell undergoing apoptosis, *Br. Med. Bull.* **53**: 491-508.

Savill, J., 1998, Apoptosis: Phagocytic docking without shocking, *Nature* 392: 442-443.

Savill, J.S., Hogg, N., Ren, Y., and Haselett, C., 1992, Thrombospondin cooperates with CD36 and the vitronectin receptor in macrophage recognition of neutrophils undergoing apoptosis, *J. Clin. Invest.* **90**: 1513-1522.

Savill, J., and Fadok, V., 2000, Corpse clearance defines the meaning of cell death, *Nature* **407**: 784-788.

Schlegel, R.A., Krahling, S., Callahan, M.K., and Williamson, P., 1999, CD14 is a component of multiple recognition system used by macrophages to phagocytose apoptotic lymphocytes, *Cell Death Differ.* **6**: 583-592.

Schlegel, R.A., Callahan, M.K., and Williamson, P., 2000, The central role of phosphatidylserine in the phagocytosis of apoptotic thymocytes, *Ann. NY Acad. Sci.* **926**: 217-225.

Schor, N.F., Tyurina, Y.Y., Fabisiak, J.P., Tyurin, V.A., Lazo, J.S., and Kagan, V.E., 1999, Selective oxidation and externalization of membrane phosphatidylserine: Bcl-2-induced potentiation of the final common pathway for apoptosis, *Brain Res.* 831: 125-130Schwartz, B.R., Karsan, A., Bombeli, T., and Harlan, J.M., 1999, A novel β1-integrin-dependent mechanism of leukocyte adherence to apoptotic cells. *J. Immunol.* **162**: 4842-4848.

Sethi, S., Eastman, A.Y., and Eaton, J.W., 1996, Inhibition of phagocyte-endothelium interaction by oxidized fatty acids: A natural anti-inflammatory mechanism? *J. Lab. Clin. Invest.* **128**: 27-38.

Shimaoka, T., Kume, N., Miniami, M,, Hayahida, K., Kataoka, H., Kita, T., and Yonehara, S., 2000, Molecular cloning of a novel scavenger receptor for oxidized low density lipoprotein, SDR-PSOX, on macrophages, *J. Biol. Chem.* **275**: 40663-40666.

Shvedova, A.A., Tyurina, Y.Y., Kawai, K., Tyurin, V.A., Kommineni, C., Fabisiak, J.P., and Kagan, V.E., 2001, Selective peroxidation and externalization of phosphatidylserine in normal human epidermal keratinocytes during oxidative stress induced by cumene hydroperoxide, *J. Inv. Derm.* (submitted for publication).

Smiley, P.L., Stremler, K.E., Prescott, S.M., Zimmerman, G.A., and McIntyre, T.M., 1991, Oxidatively fragmented phosphatidylcholines activate human neutrophils trough the receptor for platelet-activating factor, *J. Biol. Chem.* **266**: 11104-11110.

Steinberg, D., 1997, Low density lipoprotein oxidation and its pathobiological significance, *J. Biol. Chem.* **272**: 20963-20966.

Subbanagounder, G., Leitinger, N., Schwenke, D.C., Wong, J.W., Lee, H.R., Watson, A.D., Faull, K.F., Fogelman, A.M., and Berliner, J.A., 2000, Determination of bioactivity of oxidized phospholipids: Specific oxidized fatty acyl groups at the *sn*-2 position, *Arterioscler. Thromb. Vasc. Biol.* **20**: 2248-2254.

Sugihara, K., Sugihara, T., Mohandas, N., and Hebbel, R.P., 1992, Thrombospondin mediates adherence of CD36+ sickle reticulocytes to endothelial cells, *Blood* **80**: 2634-2642.

Tait, J.F., and Smith, C., 1999, Phosphatidylserine receptor: Role of CD36 in binding of anionic phospholipid vesicles to monocytic cells, *J. Biol. Chem.* **274**: 3048-3054.

Tang, X., Halleck, M.S., Schlegel, R.A., and Williamson, P., 1996, A subfamily of P-type ATPases with aminophospholipid transporting activity, *Science* **272**: 1495-1497.

Terpstra, V., and van Berkel, T.J., 2000, Scavenger receptors on liver Kupffer cells mediate the *in vivo* uptake of oxidatively damaged red blood cells in mice, *Blood* 95: 2157-2163.

Terpstra, V., Bird, D.A., and Stainberg, D., 1998, Evidence that the lipid moiety of oxidized low density lipoprotein plays a role in its interaction with macrophage receptor, *Proc. Natl. Acad. Sci. U.S.A.* **95**: 1806-1811.

Tyurina, Y.Y., Shvedova, A.A., Kawai, K., Tyurin, V.A., Kommineni, C., Quinn, P.J., Schor, N.F., Fabisiak, J.P., and Kagan, V.E., 2000, Phospholipid signaling in apoptosis: peroxidation and externalization of phosphatidylserine, *Toxicology* **148**: 93-101.

Tyurina, Y.Y., Tyurin, V.A., Shvedova, A.A., Fabisiak, J.P., and Kagan, V.E., 2001, Peroxidation of phosphatidylserine in mechanism of apoptotic signaling, *Meth. Enzymol.* (in press).

Van Helvoort, A., Smith, A.J., Sprong, H., Fritzsche, I., Schinke,l A.H., Borst, P., Van Meer, G., 1996, MDR1 P-glycoprotein is a lipid translocase of broad specificity, while MDRP3 P-glycoprotein specifically translocates phosphatidylcholine, *Cell* **87**: 507-517.

Verhoven, B., Krahling, S., Schlegel, R.A., and Williamson, P., 1999, Regulation of phosphatidylsserine exposure and phagocytosis of apoptotic T lympocytes, *Cell Death Differ.* **6**: 262-270.

Wang, P., Kitchens, R.L., Munford, R.S., 1998, Phosphatidylinositides bind to plasma membrane CD14 and can prevent monocyte activation by bacterial lipopolysacharide, *J. Biol. Chem.* **273**: 24309-24313.

Watson, A.D., Navab, M., Hama, S.Y., Sevanian, A., Rescott, S.M., Stafforini, D.M., McIntyre, T.M., Du, B.N., Fogelman, A.M., and Berliner, J.A., 1995, Effect of platelet activating factor-acetylhydrolase on the formation and action of minimally oxidized low density lipoprotein, *J. Clin. Invest.* **95**: 774-782.

Witting, A., Muller, P., Herrmann, A., Kettenmann, H.M., and Nolte, C., 2000, Phagocytic clearance of apoptotic neurons by microglia/brain macrophages *in vitro*: involvement of lectin-, integrin- and phisphatidylserine-mediated recognition, *J. Neurochem.* **75**: 1060-1070.

Wood, K.A., and Yuole, R.J., 1994, Apoptosis and free radicals, *Ann. NY Acad. Sci.* **738**: 400-407.

Wu, Y.C., and Horvitz, H.R., 1998, *C. elegans* phagocytosis and cell-migration protein CED-5 is similar to human COCK 180, *Nature* **392**: 501-504.

Wu, Y.C., and Horvitz, H.R., 1998, The *C. elegans* cell corpse engulfment gene ced-7 encodes a protein similar to ABC transporters, *Cell* **93**: 951-960.

Zhao, J., Zhou, Q., Wiedmer, T., Sims, P.J., 1998, Level of expression of phospholipid scramblase regulates induced movement of phosphatidylserine to the cell surface, *J. Biol. Chem.* **273**: 6603-6606.

Zwaal, F.A.A., and Schroit, A. 1997, Pathophysiologic implication of membrane phospholipid asymmetry in blood cells, *Blood* **89**: 1121-1132.

Chapter 6

Role of Nitric Oxide and Membrane Phospholipid Polyunsaturation in Oxidative Cell Death

C. Patrick Burns[*], Eric E. Kelley[*#], Brett A. Wagner[*], and Garry R. Buettner[#]
Departments of [*]Medicine and [#]Radiology (Free Radical and Radiation Biology Graduate Program), The University of Iowa College of Medicine, and The University of Iowa Holden Comprehensive Cancer Center, Iowa City, IA 52242

1. INTRODUCTION

There is great interest in the mechanisms of cell death since better understanding might lead to therapy that slows the rate of aging and prevents or treats human disease. Two major processes of cell death have been described, apoptosis and necrosis; other alternative pathways generally are variations of these (Formigli *et al.*, 2000; Sperandio *et al.*, 2000; Reed, 1999). Some of the intracellular events related to these types of death have been discovered (Reed, 2000). After exposure to noxious stimuli, the balance between antiapoptotic and proapoptotic influences can result in either survival or death. Many of these variable influences and the subsequent downstream concatenated events involve oxidation, which targets cellular components such as DNA, cellular proteins and membrane phospholipids. Our laboratory and others have studied the role of the redox-active cellular constituents nitric oxide ($^{\bullet}NO$)[1] and membrane phospholipid

[1] Abbreviations: BSO, D,L-buthionine-*S,R*-sulfoximine; $L^{\bullet}$, lipid alkyl radicals; LH, lipid; $LO^{\bullet}$, lipid alkoxyl radicals; $LOO^{\bullet}$, lipid peroxyl radicals; L-NAME, N^{ω}-nitro-L-arginine-methyl ester; MBI, methylene bridge index (mean number of *bis*-allylic methylene positions per fatty acid); $^{\bullet}NO$, nitric oxide; NOS, nitric oxide synthase; NO_2^-, nitrite; $^{\bullet}NO_2$, nitrogen dioxide; NO_2Cl, nitryl chloride; $O_2^{\bullet -}$, superoxide; $^{\bullet}OH$, hydroxyl radical; $OL^{\bullet}$, epoxyallylic radical; $OLOO^{\bullet}$, epoxyperoxyl radical; $O{=}NOO^-$, peroxynitrite; SNAP, *S*-nitroso-*N*-acetyl-D,L-penicillamine; SOD, superoxide dismutase; *contd. on p. 98,*

polyunsaturation in cellular demise. In this chapter we discuss the role of these two modulators of oxidation and oxidative cell death.

2. NITRIC OXIDE INHIBITS IRON-INDUCED LIPID PEROXIDATION AND OXIDATIVE CELL DEATH

2.1 Nitric oxide as a prooxidant and antiproliferative agent

•NO is a bioreactive free radical that is produced in cells by nitric oxide synthase (NOS). NOS catalyzes the five-electron oxidation of the guanidine nitrogen of L-arginine to generate •NO and L-citruline. There are three isoforms of NOS: (1) eNOS (endothelial NOS) is associated with endothelial cells, constitutively expressed, Ca^{2+}-dependent, and produces brief periods of •NO production (nM levels); (2) nNOS (neuronal NOS) is associated with neurons, constitutively expressed, Ca^{2+}-dependent, and also produces brief periods of •NO production (nM levels); and (3) iNOS (inducible NOS) is associated with phagocytic cells, induced by cytokines, Ca^{2+}-independent, and results in long lasting periods of •NO production (μM levels). It is now clear that this molecule plays a myriad of roles in cellular biology. It is a vasodilator (Moncada and Higgs, 1993; Waldman and Murad, 1988; Ignarro *et al.*, 1987), neurotransmitter (Moncada and Higgs, 1993; Bredt *et al.*, 1990), versatile signaling molecule, mediator of intracellular pathogen destruction by phagocytes (Moncada and Higgs, 1993) and redox reactant (Beckman *et al.*, 1990). It is also involved in gene expression (Bogdan, 2001), platelet aggregation (Broekman *et al.*, 1991), and male sexual function (Ignarro *et al.*, 1990).

Many of the features of the chemistry and biology of •NO are contradictory, suggesting a regulatory function. This contradiction is particularly obvious in its modulation of oxidative stress (Wink *et al.*, 1998; Beckman and Koppenol, 1996). •NO acts as a prooxidant in reacting with superoxide ($O_2^{\bullet -}$) to form peroxynitrite ($O{=}NOO^-$) (Ischiropoulos *et al.*, 1992; Beckman *et al.*, 1990), as shown in Figure 1, and it is generally assumed that $O{=}NOO^-$ is one of the key mediators of •NO toxicity. Another pathway for the formation of toxic or antiproliferative oxidants may be the conversion of nitrite (NO_2^-), an •NO metabolic product, to nitryl chloride (NO_2Cl) and nitrogen dioxide ($^{\bullet}NO_2$) (Eiserich *et al.*, 1998). These products

TBARS, thiobarbituric acid reactive substances; 22:6ω3, docosahexaenoic acid; 20:4ω6, arachidonic acid; 18:1ω9, oleic acid; 20:5ω3, eicosapentaenoic acid.

are formed in neutrophils by a myeloperoxidase-dependent pathway, and it has been shown in a cell-free system that $^{\bullet}$NO is a substrate for mammalian peroxidases including myeloperoxidase, eosinophil peroxidase and lactoperoxidase (Abu-Soud and Hazen, 2000). $^{\bullet}$NO has an antiproliferative effect such as on Drosophila development (Kuzin *et al.*, 1996), and cultured mesangial cells (Garg and Hassid, 1989). There is also evidence that $^{\bullet}$NO may increase the effectiveness of cancer chemotherapy agents (Wink *et al.*, 1997).

$$O_2^{\bullet -} + {}^{\bullet}NO \xrightarrow{k = 10^{10}\ M^{-1}s^{-1}} O{=}NOO^-$$

Figure 1. Reaction of $^{\bullet}$NO with superoxide ($O_2^{\bullet -}$) to form the toxic peroxynitrite ($O{=}NOO^-$).

2.2 Nitric oxide as an antioxidant and protective agent

$^{\bullet}$NO also has a protective function in oxidative stress and the balance between its pro- and antioxidant activity may depend upon the concentrations of reactants (Rubbo *et al.*, 1994). In chemical non-cellular systems, $^{\bullet}$NO inhibits free radical-mediated chain propagation reactions such as lipid peroxidation induced by H_2O_2 and ferrous complexes (Kanner *et al.*, 1991), by lipoxygenase (Rubbo *et al.*, 1995), low-density lipoprotein oxidation (Yamanaka *et al.*, 1996; Rubbo *et al.*, 1995; Hogg *et al.*, 1993), and linoleate lipid peroxyl radicals (LOO$^{\bullet}$) (O'Donnell *et al.*, 1997). This antioxidant activity is not trivial. For example, $^{\bullet}$NO reacts with LOO$^{\bullet}$ and LO$^{\bullet}$ (lipid alkoxyl radicals) with near diffusion-limited rates and the rate constant for $^{\bullet}$NO reactions with LOO$^{\bullet}$ is several orders of magnitude greater than with the important antioxidant α-tocopherol (Niki *et al.*, 1984). $^{\bullet}$NO and α-tocopherol can synergize to inhibit lipid peroxidation in membranes, and this occurs at biologically relevant concentrations (Rubbo *et al.*, 2000). This is due to $^{\bullet}$NO terminating propagation. Although $^{\bullet}$NO protects α-tocopherol from oxidation, it does not directly react with α-tocopherol radical (Rubbo *et al.*, 2000). In addition, the lipophilic solubility of $^{\bullet}$NO positions it preferentially in membranes, thereby magnifying the antioxidant effect since many lipid-derived free radicals are generated from polyunsaturated membrane lipids. $^{\bullet}$NO has also been shown to protect normal cells such as cultured heart cells, hepatocytes, intestinal epithelium, endothelial cells and fibroblasts (Gorbunov *et al.*, 1998; Chamulitrat, 1998; Gupta *et al.*, 1997; Sergent *et al.*, 1997; Struck *et al.*, 1995; Wink *et al.*, 1995; Wink *et al.*, 1993). Taken together, these observations suggest that the antioxidant effect results in a limiting of free radical propagation reactions and an extent of antioxidant activity that has a major effect in modulating the redox environment of the cell.

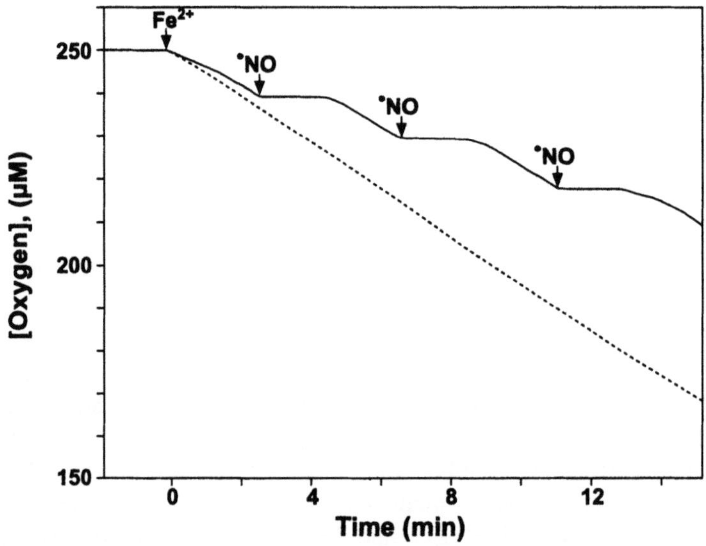

Figure 2. •NO inhibits iron-induced lipid peroxidation. The rate of O_2 consumption of HL-60 cells (5×10^6/ml) was determined using a YSI O_2 monitor. Fe^{2+} (20 μM) was added at the first arrow and subsequently •NO (0.45 μM) was added (other arrows). When •NO was added, the O_2 consumption was inhibited for a period of a few min, then it resumed at near its initial rate until the reintroduction of additional •NO. Also shown (lower dashed line) is a control of HL-60 cells subjected to Fe^{2+}-induced oxidative stress in the absence of •NO addition. The background rate of O_2 uptake of the HL-60 cell suspension before the addition of Fe^{2+} was 10 nM/sec. Upon the addition of 20 μM Fe^{2+}, this rate increased to 220 nM/sec. The addition of •NO resulted in a decrease in O_2 consumption to <10 nM/sec. (From Kelley, E.E., Wagner, B.A., Buettner, G.R., and Burns, C.P., 1999, *Arch. Biochem. Biophys.* **370**: 97-104).

The interest of our research group has been on the interactions of •NO and lipid peroxidation. The following reactions outline the three major components of lipid peroxidation:

INITIATION: $LH + oxidant^\bullet \rightarrow L^\bullet + oxidant\text{-}H$

PROPAGATION: $L^\bullet + O_2 \rightarrow LOO^\bullet$

$LOO^\bullet + LH \rightarrow L^\bullet + LOOH$

$LOOH + Fe^{2+} \rightarrow LO^\bullet + OH^- + Fe^{3+}$ (Chain Branching)

TERMINATION: $L^{\bullet} + L^{\bullet} \rightarrow$ non-radical products

$L^{\bullet} + LOO^{\bullet} \rightarrow$ non-radical products

$LOO^{\bullet}$ + antioxidant-H $\rightarrow$ LOOH + antioxidant$^{\bullet}$

($L^{\bullet}$, $LO^{\bullet}$, and $LOO^{\bullet}$ represent lipid alkyl, alkoxyl, and peroxyl radicals, respectively derived from lipid, LH, and lipid hydroperoxides, LOOH).

Ferrous iron (Fe^{2+}) can play a role by reacting with O_2 or H_2O_2 to generate an initiating oxidant$^{\bullet}$ (e.g. $^{\bullet}OH$) and by targeting LOOH during propagation. The $LO^{\bullet}$ produced in the reaction of LOOH with Fe^{2+} will rearrange to produce a carbon-centered epoxyallylic radical ($OL^{\bullet}$) that will react with an additional dioxygen to produce an epoxyperoxyl radical ($OLOO^{\bullet}$) (Qian *et al.*, 2000; Marnett and Wilcox, 1995; Wilcox and Marnett, 1993).

$$LO^{\bullet} \rightarrow OL^{\bullet}$$

$$OL^{\bullet} + O_2 \rightarrow OLOO^{\bullet}$$

Just like $L^{\bullet}$, $OL^{\bullet}$ will react immediately with O_2 to produce peroxyl radicals. Both $LOO^{\bullet}$ and $OLOO^{\bullet}$ will maintain the chain reaction by propagating the cycle as shown above.

In this review we will emphasize the manner in which $^{\bullet}NO$ limits O_2-related damage since that has been the focus of our work. Because of the known antioxidant effect of $^{\bullet}NO$ in chemical systems as well as normal cells, we investigated its effects on real time Fe^{2+}-induced oxidation in a human neoplastic cell line, leukemia HL-60. We have stimulated oxidation using Fe^{2+} that induces oxidative stress especially via reactions with cellular unsaturated fatty acids. In neoplastic cells we have found that Fe^{2+} induces lipid peroxidation as measured by increased O_2 consumption, ethane generation, thiobarbituric acid reactive substances (TBARS) and lipid-derived free radical generation (Wagner *et al.*, 1996; Wagner *et al.*, 1994; Burns and Wagner, 1991). In order to have a sensitive and reliable system that allowed the measurement of oxidation in real time, we utilized O_2 consumption measurement in a YSI Model 5300 Biological Oxygen Monitor (Yellow Springs, OH). When Fe^{2+} is added to a suspension of these cells in the presence of O_2, oxidative stress is initiated and the rate of O_2 consumption, which is proportional to the rate of lipid peroxidation in the cell, increases dramatically from its low basal metabolic level (Figure 2). After a few min, an aqueous $^{\bullet}NO$ solution was added.[2] We found that low

[2] $^{\bullet}NO$ was prepared by mixing 1 M $NaNO_2$ with 1 M HCl (1:1, v:v) in an O_2-free, bottle. The gas liberated from this was bubbled through oxygen-free distilled H_2O, into which it dissolves readily in the absence of oxygen, for 2 hr prior to diluting the solution. At

micromolar concentrations of •NO inhibited the Fe^{2+}-induced lipid peroxidation immediately upon addition as can be seen in Figure 2. This effect lasted a few min then the rate of O_2 consumption returned to its original rate. Subsequent additions of •NO inhibited the peroxidation in a similar manner, but each dissipated after a few min similar to the first addition. The effect was also dependent upon •NO concentration up to about 2 μM since the duration of inhibition could be increased to almost 9 min by adding higher concentrations of •NO. The relationship of •NO concentration and duration of inhibition appeared linear up to concentrations of ≈2 μM at which point the predominant reaction probably becomes •NO reacting directly with O_2 to form NO_2^-. In control experiments we showed that nitrogen-saturated media, L-arginine and L-NAME, an inhibitor of cellular production of •NO, had no effect indicating that the exogenous •NO is the direct inhibitor.

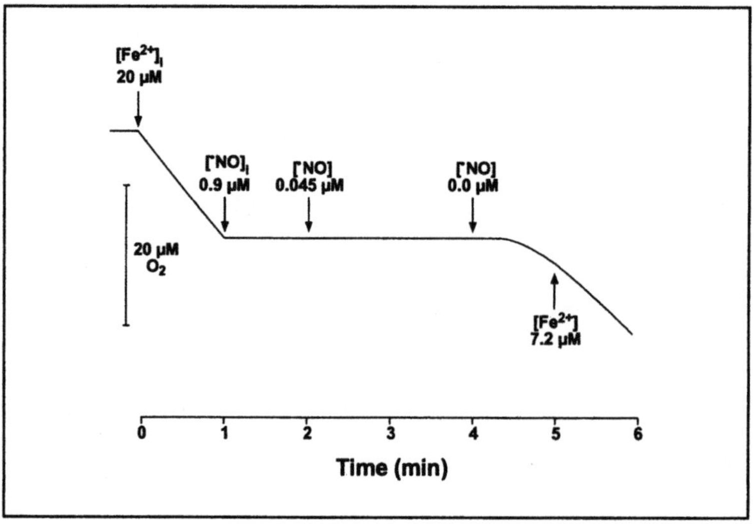

Figure 3. Changes in concentration of O_2, •NO and Fe^{2+} during cellular lipid peroxidation and its inhibition by •NO. Shown are the concentrations of •NO and Fe^{2+} at key time points. Peroxidation, as measured by O_2 uptake, was initiated with 20 μM Fe^{2+}. At 1 min after the addition of Fe^{2+}, 0.9 μM •NO was introduced. •NO was rapidly depleted and is below the limit of detection at about 4 min. At the time of •NO depletion, rapid O_2 uptake resumes. This reinitiation of O_2 consumption is due to Fe^{2+} that is still present at 7.2 μM or about 36% of its original value. (From Kelley, E.E., Wagner, B.A., Buettner, G.R., and Burns, C.P., 1999, *Arch. Biochem. Biophys.* **370**: 97-104).

standard temperature and pressure the saturated •NO concentration was about 1.8 mM as determined using a Sievers Model 280 Nitric Oxide Analyzer (Boulder, CO).

We thought that the resumption of oxidation after a few min was likely due to the depletion of •NO since subsequent additions of •NO inhibited oxidation with kinetics similar to the first addition. In order to prove this, we repeated the experiment, but this time determined the [•NO] at periodic time intervals. We initiated the experiment with 20 μM Fe^{2+} and then added 0.9 μM •NO 1 min later. At this 1 min time point, oxidation was inhibited (Figure 3). This inhibition continued until about 4.5 min, which is about when the [•NO] fell to below the limit of detection. At this time there was still sufficient Fe^{2+}(7.2 μM) to reinitiate oxidation. These results demonstrate that •NO is acting as a chain-breaking antioxidant during cellular lipid peroxidation.

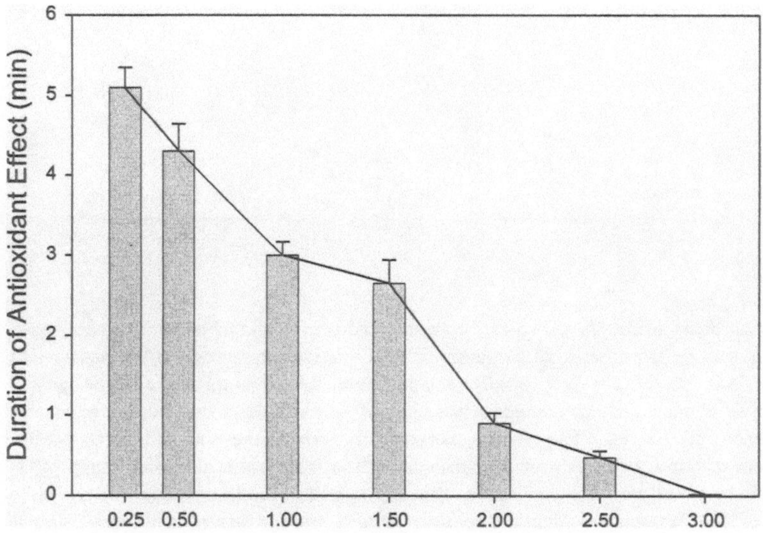

Figure 4. Effect of adding •NO prior to Fe^{2+}. •NO (0.9 μM) inhibits lipid peroxidation when added prior to Fe^{2+} addition but the effect diminishes as the time prior to introduction to Fe^{2+} increases. Each bar represents the mean of at least three independent determinations with the standard error. (Drawn using data abstracted from Kelley, E.E., Wagner, B.A., Buettner, G.R., and Burns, C.P., 1999, *Arch. Biochem. Biophys.* **370**: 97-104.

In our experiments described above, we added the •NO 1 min after the initiation of oxidation with Fe^{2+}. However, we realized that the sequence of addition of the initiator of oxidation and antioxidant •NO might be critical, so in additional studies we examined the importance of this sequence by adding the •NO prior to the Fe^{2+}. There was an antioxidant effect when the

$^{\bullet}NO$ is added immediately prior to the Fe^{2+} as shown in Figure 4. As the time interval prior to the Fe^{2+} increases, the antioxidant effect diminishes in a roughly linear manner until it is totally absent when the delay between $^{\bullet}NO$ and subsequent Fe^{2+} is 3 min or greater.

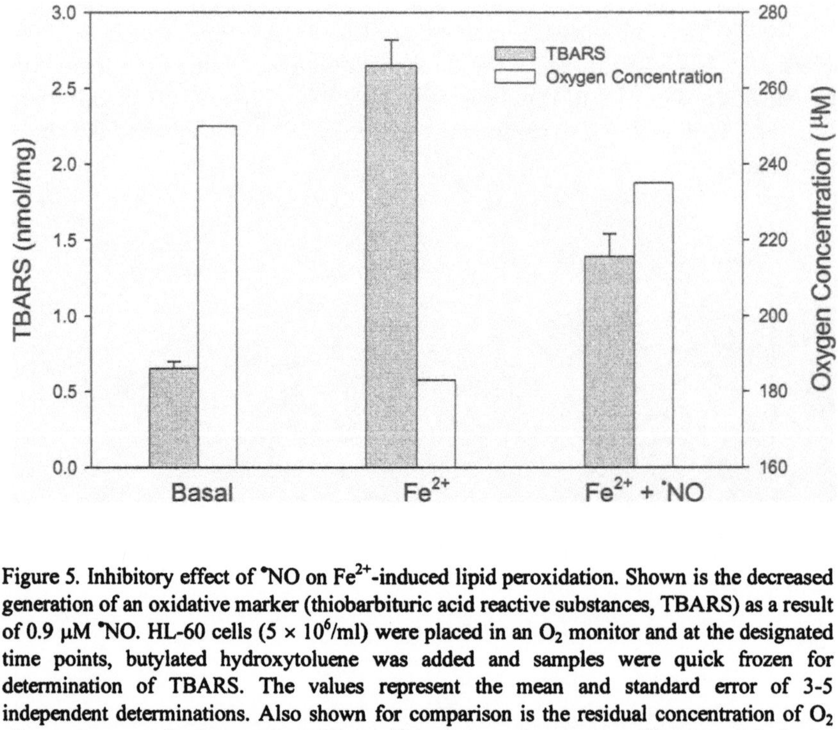

Figure 5. Inhibitory effect of $^{\bullet}NO$ on Fe^{2+}-induced lipid peroxidation. Shown is the decreased generation of an oxidative marker (thiobarbituric acid reactive substances, TBARS) as a result of 0.9 μM $^{\bullet}NO$. HL-60 cells (5×10^6/ml) were placed in an O_2 monitor and at the designated time points, butylated hydroxytoluene was added and samples were quick frozen for determination of TBARS. The values represent the mean and standard error of 3-5 independent determinations. Also shown for comparison is the residual concentration of O_2 after exposure to the the same conditions. This shows a decrease in utilization of O_2 in the presence of $^{\bullet}NO$. We conclude that $^{\bullet}NO$ reduces TBARS, and the percent inhibition is similar to the percent inhibition of O_2 consumption. (Modified from our data in Kelley, E.E., Wagner, B.A., Buettner, G.R., and Burns, C.P., 1999, *Arch. Biochem. Biophys.* **370**: 97-104).

To confirm that the effects of $^{\bullet}NO$ on O_2 consumption are due to inhibition of lipid peroxidation, we also examined the effect of $^{\bullet}NO$ on TBARS, a product of lipid peroxidation. Cells were oxidized by Fe^{2+}, and 0.9 μM $^{\bullet}NO$ was added 1 min later. Cells were collected after 5 min for assay. Figure 5 shows that Fe^{2+} increased lipid peroxidation, and $^{\bullet}NO$ inhibited it by 63% (after subtracting basal levels). It is noteworthy that the percentage inhibition of O_2 uptake by $^{\bullet}NO$ as measured by the change in O_2 concentration under similar conditions was similar (78%) lending verification to these complementary methods. These results confirm the relationship of O_2 consumption and lipid peroxidation in these experiments and demonstrate that ${}^{\bullet}NO$ stops lipid peroxidation in HL-60 cell membranes measured both by decreased generation of a lipid peroxidation product and by decreased utilization of a reactant (higher residual O_2 concentration).

Further evidence that the effect of ${}^{\bullet}NO$ involves oxidation comes from experiments with glutathione depletion. Cells were incubated with 100 μM D,L-buthionine-*S,R*-sulfoxime (BSO) for 24 hr prior to determination of the effect of ${}^{\bullet}NO$ on 20 μM Fe^{2+}-induced O_2 consumption. The duration of inhibition was significantly reduced compared to experiments done without BSO (Kelley *et al.*, 1999). This is consistent with our oxidative-based hypothesis since the lowering of cellular reducing potential diminishes the ability of cells to handle an oxidative stress. Therefore, ${}^{\bullet}NO$ was consumed faster perhaps due to pre-existing LOOH in cells. In further experiments, we also delivered ${}^{\bullet}NO$ using the donor *S*-nitroso-*N*-acetyl-D,L-penicillamine (SNAP); the inhibition was equal to or greater than bolus addition of ${}^{\bullet}NO$, depending on the concentration used (Kelley *et al.*, 1999).

The antioxidant activity afforded by ${}^{\bullet}NO$ is most important if it translates into a meaningful therapeutic event. To test the importance of this reaction pathway to cell survival, we compared viability of the cells exposed to Fe^{2+} in the presence or absence of ${}^{\bullet}NO$. When cells are exposed to 20 μM Fe^{2+}, there was appreciable loss of cell membrane integrity as measured by trypan blue dye exclusion (Kelley *et al.*, 1999); the addition of ${}^{\bullet}NO$ decreased cell membrane damage (Figure 6).

For the ${}^{\bullet}NO$ studies that show inhibition of lipid peroxidation, we used HL-60 cells made more oxidizable by adding 32 μM docosahexaenoic acid (22:6ω3[3]) to the growth media. When unenriched cells were examined, inhibition of oxidation was qualitatively identical, but the magnitude of the effect was less, and therefore more difficult to measure. The effect of lipid unsaturation will be developed below.

2.3 Mechanism of nitric oxide as an antioxidant

It is generally thought that ${}^{\bullet}NO$ interferes with lipid peroxidation by providing alternative ways to chain termination (Rubbo *et al.*, 1995; Rubbo *et al.*, 1994). For example, radical-radical reactions between ${}^{\bullet}NO$ and $LOO^{\bullet}$, $LO^{\bullet}$ and $L^{\bullet}$ radicals, as shown below, likely occur:

$$LOO^{\bullet} + {}^{\bullet}NO \rightarrow LOONO$$

$$LO^{\bullet} + {}^{\bullet}NO \rightarrow LONO$$

$$L^{\bullet} + {}^{\bullet}NO \rightarrow LNO$$

[3] Fatty acids are designated as number of carbon atoms:number of double bonds. ω refers to the position of the first double bond from the methyl terminus.

These reactions, if of sufficient magnitude, would result in the formation of non-radical, nitrogen-containing species, thereby stopping the propagation cycle of lipid peroxidation.

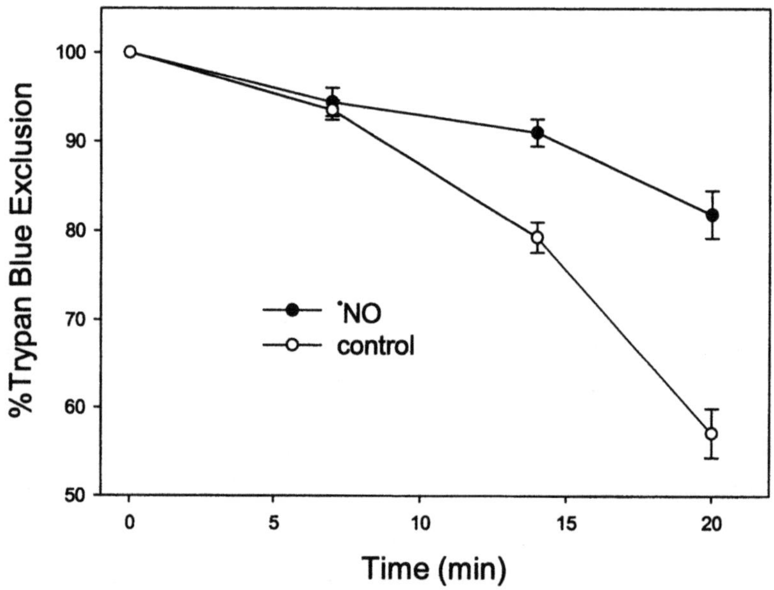

Figure 6. •NO protects cells from lipid peroxidation-induced cell damage. HL-60 cells at 5 × 10^6/ml were assayed for trypan blue uptake just before iron addition in the presence or absence of •NO (0.9 μM) addition, then at 7, 14, and 20 min after the initiation of peroxidation. Each data point represents the mean of at least three independent determinations and the bars the standard error. Membrane leakage was significantly different at 14 min ($p=0.01$) and 20 min ($p=0.005$). (Drawn using our data abstracted from Kelley, E.E., Wagner, B.A., Buettner, G.R., and Burns, C.P., 1999, *Arch. Biochem. Biophys.* **370**: 97-104.

3. ROLE OF MEMBRANE POLYUNSATURATION IN OXIDATIVE CELL DEATH

The extent of membrane lipid polyunsaturation is another modulator of cellular oxidative susceptibility; an example is mentioned above in the •NO section. This is not surprising, and it has a chemical rationale since the sites of attack of oxidizing free radical species in lipids are the double bonds in polyunsaturated phospholipid structures in membranes. Therefore, it follows, that cells with more polyunsaturated lipids in their membranes are likely to be more susceptible to oxidation.

3.1 Lipid biochemical features are changed by fatty acid modification

We have utilized a powerful experimental model in order to explore the relationship of polyunsaturation and oxidative susceptibility (Burns and Spector, 1987; Spector and Burns, 1987; Burns and North, 1986; Guffy *et al.*, 1984; Burns *et al.*, 1979). The model is based on the fact that the fatty acid composition of cell membrane phospholipids can change substantially when cells are exposed to different types of fat added to the media of cultured cells or, alternatively, to the diets of tumor-bearing animals. *In vitro* and dietary studies by our group have demonstrated that the membranes and cytoplasmic intracellular storage sites of mammalian neoplastic cells can be enriched with diverse fatty acids of various degrees of unsaturation from 1-6 double bonds (Burns and Wagner, 1993; Burns and Wagner, 1991; Burns and North, 1986; Burns *et al.*, 1979). For example, the model allows enrichment of cells with docosahexaenoic acid (22:6ω3), arachidonic acid (20:4ω6), or as "controls" oleic acid (18:1) and unmodified cells, to name a few. The fatty acid modification is specific since there is no change in: (1) membrane cholesterol, (2) total phospholipid content, (3) phospholipid head group composition, or (4) membrane acylglycerol content (Guffy *et al.*, 1982; Burns *et al.*, 1979; Burns *et al.*, 1977). The model also provides a way of examining the different influence of omega-3 and omega-6 fatty acids, families of fatty acids known to have differing effects on tumor growth (Burns and Wagner, 1993). Our work with cellular omega-3 fatty acid enrichment has been done both *in vitro* (Burns and North, 1986) and *in vivo* using diet modification (Burns and Wagner, 1993). There is considerable polyunsaturated fatty acid enrichment of the lipids of L1210 cells from animals fed omega-6-rich sunflower oil diets as compared to tumor cells from those fed saturated fatty acid-rich coconut oil diets (Burns *et al.*, 1979), and in the omega-6 fatty acid content of cells in tissue culture (Burns and Wagner, 1991). Both types of modification have been useful in oxidative studies.

3.2 Physical features are changed by fatty acid modification

This experimental model has promoted understanding of lipid transport, membrane remodeling and the relationship of membrane form and function

(Burns, 1988; Burns and Spector, 1987; Spector and Burns, 1987). When membranes contain more polyunsaturated fatty acyl chains, the kinetics of drug transport carriers, the conformation and functional properties of certain enzymes, receptors, and other compounds are modified (Burns and Spector, 1987; Spector and Burns, 1987). The modulation of the hydrophobic core without compensatory changes in other membrane lipids also results in changes in membrane order, which increases with polyunsaturation (Burns *et al.*, 1979). Increased polyunsaturation of membrane lipids has allowed us to examine the effect of a change in the structure of membranes on the interaction with ˙NO as described above, on the the effect of vitamin E in slowing the rate of lipid peroxidation without causing a delay in the time of onset as had been observed in cell-free systems (Wagner *et al.*, 1996), acceleration of retinoic acid induced HL-60 differentiation (Burns *et al.*, 1989), stimulation of lipid peroxidation by the novel anticancer drug edelfosine (Wagner *et al.*, 1993; Wagner *et al.*, 1992), heightened cytotoxicity of ilmofosine (Petersen *et al.*, 1992), increased passive diffusion of mitoxantrone into cells (Burns *et al.*, 1988; Burns *et al.*, 1987), increased cytotoxicity due to greater drug accumulation of doxorubicin (Burns *et al.*, 1988; Burns and North, 1986; Guffy *et al.*, 1984), lower transition temperture for carrier-mediated melphalan transport (Burns and Dudley, 1982) and lower K_M for methotrexate uptake (Burns *et al.*, 1979).

3.3 Fatty acid modification has no effect on tumor growth

Although some investigators have reported that fatty acid enrichment *per se* influences the growth of established tumors (reviewed in Burns and Wagner, 1993), we have found no effect on tumor growth rate at the concentrations of supplemental fatty acids that we use. This is true both for omega-6 fatty acids *in vitro* or *in vivo* and for omega-3 fatty acids in tissue culture. For example, when tumor cells were modified with omega-3 fatty acids *in vitro*, the culture doubling time and clonogenic efficiency were unchanged, as was the proportion of cells in S-phase (Guffy *et al.*, 1984). Likewise, after dietary modification with omega-6 fatty acids, neither the tumor growth rate nor the rate of DNA synthesis by the leukemia cells differed appreciably from controls (Burns *et al.*, 1978).

3.4 Many cell types can be modified

We have used human leukemias (HL-60 and K562), murine leukemia (L1210) and human breast cancer (MCF-7) for most of our studies.

However, similar changes in lipid composition are possible using a variety of other cell types (Burns and Wagner, 1993) such as an adenocarcinoma (Awad and Spector, 1976), and hepatoma (Wood *et al.*, 1975). The dietary fatty acid modification is not limited to tumor cells, since many tissues of the host are affected (Burns *et al.*, 1983). However, different tissues are modified to varying extents, and because of this it will likely be possible to develop protocols that provide therapeutic selectivity by producing greater or lesser enrichment of the neoplastic cells with a particular type of fatty acid as compared to normal tissues. We have also demonstrated that L1210 leukemia cells that were fatty acid modified *in vivo* and then placed into tissue culture maintain their experimentally-induced fatty acid composition for up to 4 days (Burns *et al.*, 1980). Therefore, this is a versatile model.

3.5 Fatty acid modification changes oxidative susceptibility

The changes in physical properties such as membrane order brought about by fatty acid modification are important. However, for the purposes of studying oxidative events, the change in number of double bonds is central to the strategy since the degree of unsaturation is the major determinant of susceptibility to peroxidation. Most importantly, the degree of unsaturation can be manipulated incrementally in our model.

It was known previously that in a noncellular homogeneous solution of lipids, the rate of peroxidation is linearly correlated to the number of *bis*-allylic positions contained in the lipids (Cosgrove *et al.*, 1987). However, it was not known whether this would also be the case in intact cells. This is particularly important since there is evidence that after an oxidative stress, resultant radicals originate within the cells and not from lipids released into the media (Wagner *et al.*, 1994). In order to investigate the relationship of the number of double bonds to cell oxidizability, we enriched the cultured murine leukemia L1210 with one of eight fatty acids of different degrees of unsaturation that were added to the growth media for 48 hr at concentrations of 32 μM (Wagner *et al.*, 1994). We then did a total cellular lipid analysis to determine the number of lipid carbon-carbon double bonds contained in each cell. This number was then compared to oxidizability that was determined by two complementary methods, O_2 uptake and the generation of lipid-derived radicals. The technique proved to be particularly powerful since the mole percent of polyunsaturated fatty acids ranged from 9% to 64%, thereby providing a wide spectrum of lipid unsaturation for study of related peroxidative events.

Fatty acid chain length had no effect on cellular oxidizability. We found a strong relationship between polyunsaturation and oxidizability. Unmodified and 18:1-modified cells had the lowest initial rates of oxidation; in contrast 20:5 and 22:6-enriched cells had the highest initial rates, so the greater the degree of unsaturation of the enriching lipid, the higher the oxidizability of the cell. There was a log-linear correlation of oxidizability (initial rates) with the mean number of double bonds per fatty acid in the cells. We were surprised to observe that rates of cellular oxidation increased *exponentially* with the number of *bis*-allylic positions contained in the lipids (Figure 7) since in homogeneous solution a linear relationship is observed.

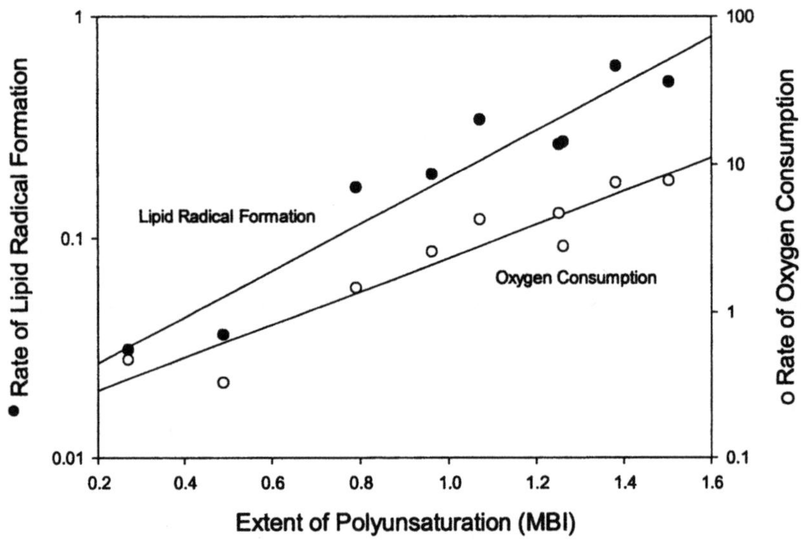

Figure 7. Relationship of oxidation and degree of polyunsaturation. Polyunsaturation is measured as the methylene bridge index (MBI), which is a more precise measure of extent of unsaturation and oxidizability than the double bond index. It is the mean number of *bis*-allylic methylene bridge positions per fatty acid (or fatty acyl chain) in a lipid ensemble. The rate of lipid radical formation measures formation of an oxidative product, while O_2 consumption (% O_2 lost per sec) is a measure of utilization of a reactant. (Drawn using our data abstracted from Wagner, B.A., Buettner, G.R., and Burns, C.P. 1994, *Biochemistry* **33**: 4449-4453).

The weakest carbon-hydrogen bonds are those at the *bis*-allylic methylene position, as shown in Figure 8. Therefore, this is the favored position for an attack by $LOO^{\bullet}$ (see the second propagation reaction shown earlier). Theoretically, the more *bis*-allylic methylene positions present in a lipid, as estimated by a higher methylene bridge index (MBI), the greater will be the oxidizability (Barclay, 1993; Cosgrove *et al.*, 1987; Porter, 1984).

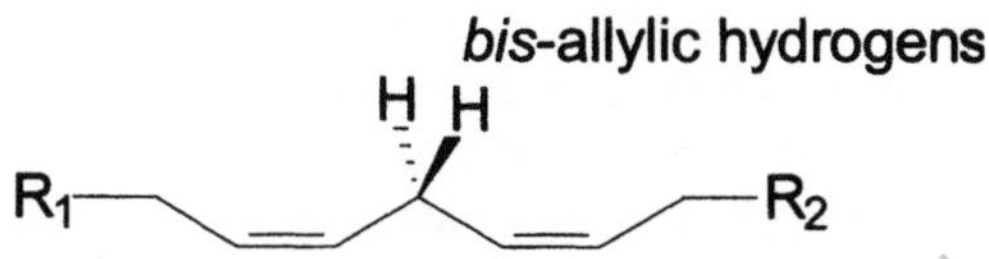

Figure 8. *Bis*-allylic carbon-hydrogen bond. The dissociation energy of this bond is lower than allylic and alkyl carbon-hydrogen bonds, and therefore the *bis*-allylic bond is favored for attack by $LOO^{\bullet}$ which explains the susceptibility of polyunsaturated lipids.

Our data on both disappearance of a reactant (O_2 consumption) and the formation of a product (lipid-derived radical) clearly support the conclusion that cellular oxidizability increases exponentially with MBI. This exponential relationship of cellular oxidizability and polyunsaturation could be due to the close packing of unsaturated lipids in the structured environment of the membrane.

Work in several laboratories including our own has shown that polyunsaturation correlates with change in susceptibility to oxidant-mediated cytotoxicity and decreased cell survival (Wagner *et al.*, 1993; Buettner, 1993; Wagner *et al.*, 1992; Spitz *et al.*, 1992; Hart *et al.*, 1991; Guffy *et al.*, 1984). The quantitative observation on the relationship of *bis*-allylic positions and oxidizability provides a rational basis for understanding the role of polyunsaturation in oxidation-related cell death.

We have utilized the generation of hydrocarbon gases released during lipid peroxidation as another measure to confirm the relationship of polyunsaturation and oxidizability. This technique measures the total amount of peroxidation, and it complements the studies of initial rate of cellular oxidation described above. Ethane from omega-3 fatty acids, and similarly pentane from omega-6 fatty acids, are volatile gases released during lipid peroxidation. They can be used as a sensitive measure of the process. These hydrocarbon gases are generated in a reaction that leads to β-scission of ethyl or pentyl radical from $LO^{\bullet}$ leaving an aldehyde. This will only take place as a result of oxygen free radical activity. To study this process we enriched L1210 leukemia cells with seven different fatty acids and then determined the generation of ethane by the different cells after exposure to an oxidative stimulus and compared that to the percentage of omega-3 fatty acids. Normal cells contained only low levels of omega-3 fatty acids and a low MBI. Figure 9 shows that there is a linear relationship between ethane generation and omega-3 fatty acid content of the cells enriched with various fatty acids.

The efficiency of peroxidation could be estimated for each modified cell type from the proportion of susceptible fatty acids that generate this hydrocarbon molecule. This assay proved to be sensitive, reproducible and ethane generation was eliminated by the antioxidants vitamin E and butylated hydroxyanisol. The generation of ethane was a function of the number of cells, time of incubation, and concentration of Fe^{2+} used as an oxidative stimulus. Cells that were unmodified, or those enriched only in the monounsaturated oleic acid (18:1ω9), generated little or no ethane even with high levels of oxidative stimulus and high concentrations of Fe^{2+}. Ethane generation proved to be more sensitive than the loss of polyunsaturated fatty acids, another frequently used measure of lipid peroxidation.

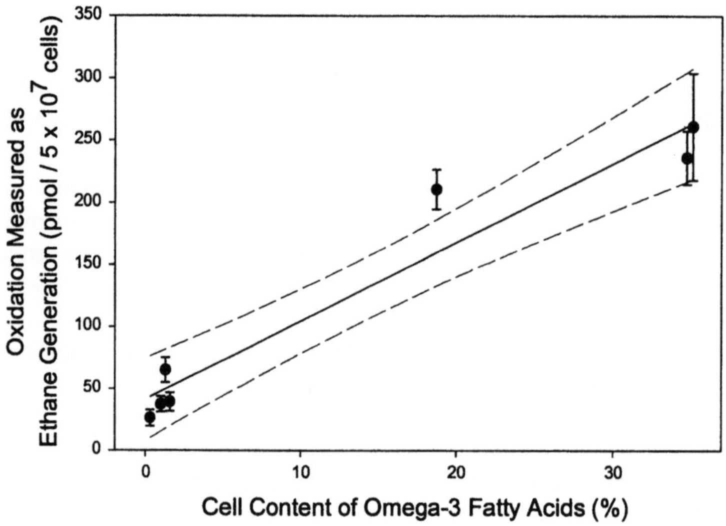

Figure 9. Relationship of oxidation measured by ethane generation from cellular omega-3 fatty acids and the cellular content of omega-3 fatty acids. Ethane generation into head spaces during cell incubation was collected in a closed plastic syringe previously flushed with low hydrocarbon compressed air. Headspace was collected and analyzed for ethane using gas-liquid chromatography. Ethane values are amounts generated over 2 hr by L1210 cells with varying degrees of polyunsaturation after exposure to the oxidative stimulus (20 μM $FeSO_4$). Polyunsaturation was obtained by growth for 72 hr in medium supplemented with fatty acids of various degrees of polyunsaturation (18:1, 18:3ω3, 18:3ω6, 20:4ω6, 20:5ω3, 22:4ω6, 22:6ω3). Shown are the regression line and 95% confidence limits. (Drawn using our data abstracted from Burns, C.P. and Wagner, B.A., 1991, *J. Lipid Research* **32**: 79-87).

From these studies we concluded that ethane is a highly useful measure of lipid peroxidation, and they confirmed that polyunsaturation, in this case with omega-3 fatty acids, is highly correlated with oxidizability. The correlation of brisk lipid peroxidation with the proportion of omega-3 fatty acids and a high MBI, means that an experimental or therapeutic gain might result from the ingestion of a specific type of fatty acid.

4. IMPLICATIONS OF NITRIC OXIDE AND MEMBRANE POLYUNSATURATION FOR HUMAN DISEASE

More needs to be learned about the role of $^{\bullet}$NO in oxidatively-mediated cell death before it can be used in clinical medicine. The factors that determine under which conditions it is an antioxidant vs prooxidant need to be elucidated since they are likely to include more than just the concentration of $^{\bullet}$NO (Rubbo *et al.*, 1994). For example, the augmenting or mitigating response may also depend upon the nature of the oxidative stress or diseased cell type.

In the case of $^{\bullet}$NO, clinical translation will likely be possible since it may be feasible to manipulate $^{\bullet}$NO in humans using $^{\bullet}$NO by inhalation, ingestion of oral nitroglycerin or $^{\bullet}$NO donors, or $^{\bullet}$NO-modulation with NOS inhibitors or tumor transfection in gene therapy. Of course, targeting of the tissue where the modulation of $^{\bullet}$NO is desired is the major challenge. Potential applications include neurologic, rheumatologic, pulmonary, cardiovascular, and neoplastic diseases. Regarding cancer, tumor $^{\bullet}$NO levels have been shown to correlate with tumor growth, metastasis and neovascularization (Garcia-Cardena and Folkman, 1998; Wink *et al.*, 1998; Jenkins *et al.*, 1995; Dong *et al.*, 1994) so it seems likely that therapeutic manipulation of $^{\bullet}$NO levels might influence tumor growth and metastasis. In fact, induction of NOS in animal models by gene transfection was found to be associated with suppression of tumor growth and regression of established metastasis (Dong *et al.*, 1994).

$^{\bullet}$NO might also be used in combination with standard cancer therapies since many of the current ones such as ionizing radiation (Saran and Bors, 1997; Riley, 1994; Borek *et al.*, 1986; Croute *et al.*, 1982), photodynamic therapy (Kelley *et al.*, 1997; Buettner *et al.*, 1993; Buettner and Need, 1985), ultraviolet radiation (Peus *et al.*, 1998; Bertling *et al.*, 1996) and selected chemotherapy drugs such as doxorubicin (Yang *et al.*, 1996; Keizer *et al.*, 1990), bleomycin (Giloni *et al.*, 1981), mitomycin (Komiyama *et al.*, 1986), edelfosine and other ether lipids (Wagner *et al.*, 1998; Wagner *et al.*, 1993;

Wagner *et al.*, 1992) involve oxidative mechanisms. Furthermore, resistance to vincristine has been associated with high cellular content of myeloperoxidase perhaps due to metabolism of the drug by the oxidant hypochlorous acid (Ozgen *et al.*, 2000). There is evidence that •NO might enhance these therapies. For example, it is known that it sensitizes hypoxic cancer cells to ionizing radiation (Mitchell *et al.*, 1993). There is also evidence that •NO may increase the efficacy of cancer chemotherapy agents. Wink *et al.* have shown that •NO can enhance the effect of cisplatin in V79 cells (1997). They postulated that reactive nitrogen oxide species inhibit DNA repair. Cook *et al.* reported a similar enhancement of the toxicity of melphalan (1997). It may also have a role in tumor angiogenesis (Gallo *et al.*, 1998).

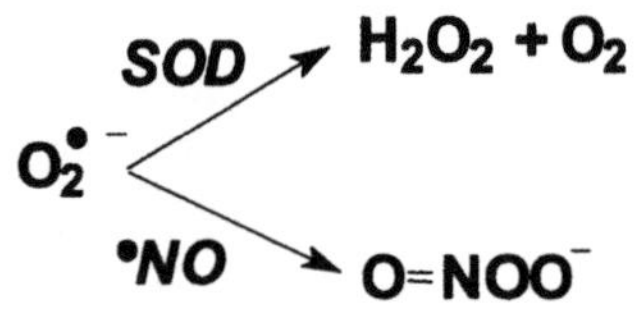

Figure 10. Competition of superoxide dismutase (SOD) and •NO for superoxide ($O_2^{\bullet -}$).

One possible way that •NO could affect anticancer drug action includes the formation of $O{=}NOO^-$, or possibly $^{\bullet}NO_2$, N_2O_3, or NO_2Cl. The mechanism of formation of $O{=}NOO^-$ can be seen in Figure 10. Selectivity of anticancer agents for tumor, and relatively less toxicity to normal tissues is a necessary characteristic of useful drugs. There are several reasons why the therapeutic use of •NO might confer a greater toxicity to tumor vs normal tissues. Superoxide dismutase (SOD) competes with •NO for superoxide ($O_2^{\bullet -}$) (Huie and Padmaja, 1993) as shown in Figure 10. It has been estimated that the metabolism of $O_2^{\bullet -}$ is distributed between SOD and •NO as 88% and 12% respectively, so a change in the concentration of SOD and •NO could change the relative rates of formation of H_2O_2 vs $O{=}NOO^-$ (Bovaris, 2000). Since SOD is low in most neoplastic tissue, these tumors may have a chemically competitive "disadvantage" in the formation of more oxidatively toxic $O{=}NOO^-$ when exposed to some anticancer drugs.

Membrane polyunsaturation also has potential in the treatment of human disease and trials of this approach are somewhat further along than with •NO. For example, highly polyunsaturated omega-3 fatty acids are being tested for their ability to ameliorate cancer cachexia in humans (Wigmore *et al.*, 2000; Burns *et al.*, 1999) and animals (Tisdale and Dhesi, 1990).

In addition, there is emerging evidence that the toxicity of selected polyunsaturated fatty acids such as those of the omega-3 family may have an antitumor effect alone (Beck *et al.*, 1991; Gonzalez *et al.*, 1991; Rose *et al.*, 1991; Gabor and Abraham, 1986; Karmali *et al.*, 1984) or when combined with cancer drugs (Hardman *et al.*, 1999; Petersen *et al.*, 1992; Burns and North, 1986). The degree of polyunsaturation of tumors can be increased by changing the diet fed to tumor-bearing animals (Burns and Wagner, 1993; Liepkalns and Spector, 1975) and similar changes can be produced *in vivo* in human tumors (Burns *et al.*, 1999). However, since meaningful cancer therapy requires fractional cell kill, we are working on the hypothesis that increased susceptibility to oxidative attack resulting from membrane lipid alteration alone will be insufficient for major clinical use. Therefore, we believe that the greatest promise is as an adjunct to chemotherapy. We have already shown that the anticancer effects of doxorubicin, and ether lipids can be augmented in experimental animal models or tissue culture by polyunsaturation of the tumor cells as mentioned earlier. In addition, a diet containing Fe^{2+}, an oxidatively active element via the Fenton reaction, might further influence the effect of polyunsaturation and photodynamic therapy (Kelley *et al.*, 1997) or chemotherapy (Hardman *et al.*, 1999; Hardman *et al.*, 1997).

In summary, these varied therapeutic strategies for using $^{\bullet}NO$ and fatty acids have potential for usefulness in human disease. The examples given are for cancer, but the potential extends to a variety of other human disorders.

ACKNOWLEDGMENTS

This investigation was supported by Grant P01 CA66081 awarded by the National Cancer Institute, Department of Health and Human Services, The Iowa Leukemia and Cancer Research Fund, The Dr. Richard O. Emmons Memorial Fund, and The Mamie C. Hopkins Fund.

REFERENCES

Abu-Soud, H. M., and Hazen, S. L., 2000, Nitric oxide is a physiological substrate for mammalian peroxidases, *J. Biol. Chem.* **275**: 37524-37532.

Awad, A. B., and Spector, A. A., 1976, Modification of the fatty acid composition of Ehrlich ascites tumor cell plasma membranes, *Biochim. Biophys. Acta* **426**: 723-731.

Barclay, L. R. C., 1993, Model biomembranes - Quantitative studies of peroxidation, antioxidant action, partitioning, and oxidative stress, *Canadian Journal of Chemistry* **71**: 1-16

Beck, S. A., Smith, K. L., and Tisdale, M. J., 1991, Anticachectic and antitumor effect of eicosapentaenoic acid and its effect on protein turnover, *Cancer Res.* **51**: 6089-6093.

Beckman, J. S., Beckman, T. W., Chen, J., Marshall, P. A., and Freeman, B. A., 1990, Apparent hydroxyl radical production by peroxynitrite: implications for endothelial injury from nitric oxide and superoxide, *Proc. Natl. Acad. Sci. USA* **87**: 1620-1624.

Beckman, J. S., and Koppenol, W. H., 1996, Nitric oxide, superoxide, and peroxynitrite: the good, the bad, and ugly, *Am. J. Physiol.* **271**: C1424-1437.

Bertling, C. J., Lin, F., and Girotti, A. W., 1996, Role of hydrogen peroxide in the cytotoxic effects of UVA/B radiation on mammalian cells, *Photochem. Photobiol.* **64**: 137-142.

Bogdan, C., 2001, Nitric oxide and the regulation of gene expression, *Trends Cell Biol.* **11**: 66-75.

Borek, C., Ong, A., Mason, H., Donahue, L., and Biaglow, J. E., 1986, Selenium and vitamin E inhibit radiogenic and chemically induced transformation *in vitro* via different mechanisms, *Proc. Natl. Acad. Sci. U S A* **83**: 1490-1494.

Bovaris, A., 2000, The mitochondrial production of free radicals, in: *Sunrise Free Radical School* (G.R. Buettner, ed.), Oxygen Society, San Diego, pp. 1-15.

Bredt, D. S., Hwang, P. M., and Snyder, S. H., 1990, Localization of nitric oxide synthase indicating a neural role for nitric oxide, *Nature* **347**: 768-770.

Broekman, M. J., Eiroa, A. M., and Marcus, A. J., 1991, Inhibition of human platelet reactivity by endothelium-derived relaxing factor from human umbilical vein endothelial cells in suspension: blockade of aggregation and secretion by an aspirin-insensitive mechanism, *Blood* **78**: 1033-1040.

Buettner, G. R., 1993, The pecking order of free radicals and antioxidants: lipid peroxidation, alpha-tocopherol, and ascorbate, *Arch. Biochem. Biophys.* **300**: 535-543.

Buettner, G. R., Kelley, E. E., and Burns, C. P., 1993, Membrane lipid free radicals produced from L1210 murine leukemia cells by photofrin photosensitization: an electron paramagnetic resonance spin trapping study, *Cancer Res.* **53**: 3670-3673.

Buettner, G. R., and Need, M. J., 1985, Hydrogen peroxide and hydroxyl free radical production by hematoporphyrin derivative, ascorbate and light, *Cancer Lett.* **25**: 297-304.

Burns, C. P., 1988, Membranes and cancer chemotherapy, *Cancer Invest.* **6**: 439-451

Burns, C. P., and Dudley, D. T., 1982, Temperature dependence and effect of membrane lipid alteration on melphalan transport in L1210 murine leukemia cells, *Biochem. Pharmacol.* **31**: 2116-2119.

Burns, C. P., Halabi, S., Clamon, G. H., Hars, V., Wagner, B. A., Hohl, R. J., Lester, E., Kirshner, J. J., Vinciguerra, V., and Paskett, E., 1999, Phase I clinical study of fish oil fatty acid capsules for patients with cancer cachexia: cancer and leukemia group B study 9473, *Clin. Cancer Res.* **5**: 3942-3947.

Burns, C. P., Haugstad, B. N., Mossman, C. J., North, J. A., and Ingraham, L. M., 1988, Membrane lipid alteration: effect on cellular uptake of mitoxantrone, *Lipids* **23**: 393-397.

Burns, C. P., Haugstad, B. N., and North, J. A., 1987, Membrane transport of mitoxantrone by L1210 leukemia cells, *Biochem. Pharmacol.* **36**: 857-860.

Burns, C. P., Luttenegger, D. G., and Dudley, D. T., 1980, Fatty acid alteration of L1210 murine leukemia cells. Growth rate and stability of lipid changes in culture, *J. Natl. Cancer Inst.* **65**: 987-991.

Burns, C. P., Luttenegger, D. G., Dudley, D. T., Buettner, G. R., and Spector, A. A., 1979, Effect of modification of plasma membrane fatty acid composition on fluidity and methotrexate transport in L1210 murine leukemia cells, *Cancer Res.* **39**: 1726-1732.

Burns, C. P., Luttenegger, D. G., and Spector, A. A., 1978, Effect of dietary fat saturation on survival of mice with L1210 leukemia, *J. Natl. Cancer Inst.* **61**: 513-515.

Burns, C. P., Luttenegger, D. G., Wei, S. P., and Spector, A. A., 1977, Modification of the fatty acid composition of L1210 murine leukemia cells, *Lipids* **12**: 747-752.

Burns, C. P., and North, J. A., 1986, Adriamycin transport and sensitivity in fatty acid-modified leukemia cells, *Biochim. Biophys. Acta* **888**: 10-17.

Burns, C. P., North, J. A., Petersen, E. S., and Ingraham, L. M., 1988, Subcellular distribution of doxorubicin: comparison of fatty acid-modified and unmodified cells, *Proc. Soc. Exp. Biol. Med.* **188**: 455-460.

Burns, C. P., Petersen, E. S., North, J. A., and Ingraham, L. M., 1989, Effect of docosahexaenoic acid on rate of differentiation of HL-60 human leukemia, *Cancer Res.* **49**: 3252-3258.

Burns, C. P., Rosenberger, J. A., and Luttenegger, D. G., 1983, Selectivity in modification of the fatty acid composition of normal mouse tissues and membranes *in vivo*, *Ann. Nutr. Metab.* **27**: 268-277

Burns, C. P., and Spector, A. A., 1987, Membrane fatty acid modification in tumor cells: a potential therapeutic adjunct, *Lipids* **22**: 178-184.

Burns, C. P., and Wagner, B. A., 1993, Effects of exogenous lipids on cancer and cancer chemotherapy. Implications for treatment, *Drug Saf.* **8**: 57-68.

Burns, C. P., and Wagner, B. A., 1991, Heightened susceptibility of fish oil polyunsaturate-enriched neoplastic cells to ethane generation during lipid peroxidation, *J. Lipid Res.* **32**: 79-87.

Chamulitrat, W., 1998, Nitric oxide inhibited peroxyl and alkoxyl radical formation with concomitant protection against oxidant injury in intestinal epithelial cells, *Arch. Biochem. Biophys.* **355**: 206-214.

Cook, J. A., Krishna, M. C., Pacelli, R., DeGraff, W., Liebmann, J., Mitchell, J. B., Russo, A., and Wink, D. A., 1997, Nitric oxide enhancement of melphalan-induced cytotoxicity, *Br. J. Cancer* **76**: 325-334

Cosgrove, J. P., Church, D. F., and Pryor, W. A., 1987, The kinetics of the autoxidation of polyunsaturated fatty acids, *Lipids* **22**: 299-304.

Croute, F., Soleilhavoup, J. P., Vidal, S., Dupouy, D., and Planel, H., 1982, Extracellular hydrogen peroxide produced under irradiation as the most important factor in the lethality of gamma-irradiated Paramecium tetraurelia, *Int. J. Radiat. Biol. Relat. Stud. Phys. Chem. Med.* **41**: 209-214.

Dong, Z., Staroselsky, A. H., Qi, X., Xie, K., and Fidler, I. J., 1994, Inverse correlation between expression of inducible nitric oxide synthase activity and production of metastasis in K-1735 murine melanoma cells, *Cancer Res.* **54**: 789-793.

Eiserich, J. P., Hristova, M., Cross, C. E., Jones, A. D., Freeman, B. A., Halliwell, B., and van der Vliet, A., 1998, Formation of nitric oxide-derived inflammatory oxidants by myeloperoxidase in neutrophils, *Nature* **391**: 393-397.

Formigli, L., Papucci, L., Tani, A., Schiavone, N., Tempestini, A., Orlandini, G. E., Capaccioli, S., and Orlandini, S. Z., 2000, Aponecrosis: morphological and biochemical exploration of a syncretic process of cell death sharing apoptosis and necrosis, *J. Cell Physiol.* **182**: 41-49.

Gabor, H., and Abraham, S., 1986, Effect of dietary menhaden oil on tumor cell loss and the accumulation of mass of a transplantable mammary adenocarcinoma in BALB/c mice, *J. Natl. Cancer Inst.* **76**: 1223-1229.
Gallo, O., Masini, E., Morbidelli, L., Franchi, A., Fini-Storchi, I., Vergari, W. A., and Ziche, M., 1998, Role of nitric oxide in angiogenesis and tumor progression in head and neck cancer, *J. Natl. Cancer Inst.* **90**: 587-596.
Garcia-Cardena, G., and Folkman, J., 1998, Is there a role for nitric oxide in tumor angiogenesis?, *J Natl. Cancer Inst.* **90**: 560-561.
Garg, U. C., and Hassid, A., 1989, Inhibition of rat mesangial cell mitogenesis by nitric oxide-generating vasodilators, *Am. J. Physiol.* **257**: F60-66.
Giloni, L., Takeshita, M., Johnson, F., Iden, C., and Grollman, A. P., 1981, Bleomycin-induced strand-scission of DNA. Mechanism of deoxyribose cleavage, *J. Biol. Chem.* **256**: 8608-8615.
Gonzalez, M. J., Schemmel, R. A., Gray, J. I., Dugan, L., Jr., Sheffield, L. G., and Welsch, C. W., 1991, Effect of dietary fat on growth of MCF-7 and MDA-MB231 human breast carcinomas in athymic nude mice: relationship between carcinoma growth and lipid peroxidation product levels, *Carcinogenesis* **12**: 1231-1235.
Gorbunov, N. V., Tyurina, Y. Y., Salama, G., Day, B. W., Claycamp, H. G., Argyros, G., Elsayed, N. M., and Kagan, V. E., 1998, Nitric oxide protects cardiomyocytes against tert-butyl hydroperoxide-induced formation of alkoxyl and peroxyl radicals and peroxidation of phosphatidylserine, *Biochem. Biophys. Res. Commun.* **244**: 647-651.
Guffy, M. M., North, J. A., and Burns, C. P., 1984, Effect of cellular fatty acid alteration on adriamycin sensitivity in cultured L1210 murine leukemia cells, *Cancer Res.* **44**: 1863-1866.
Guffy, M. M., Rosenberger, J. A., Simon, I., and Burns, C. P., 1982, Effect of cellular fatty acid alteration on hyperthermic sensitivity in cultured L1210 murine leukemia cells, *Cancer Res.* **42**: 3625-3630.
Gupta, M. P., Evanoff, V., and Hart, C. M., 1997, Nitric oxide attenuates hydrogen peroxide-mediated injury to porcine pulmonary artery endothelial cells, *Am. J. Physiol.* **272**: L1133-1141.
Hardman, W. E., Barnes, C. J., Knight, C. W., and Cameron, I. L., 1997, Effects of iron supplementation and ET-18-OCH3 on MDA-MB 231 breast carcinomas in nude mice consuming a fish oil diet, *Br. J. Cancer* **76**: 347-354
Hardman, W. E., Moyer, M. P., and Cameron, I. L., 1999, Fish oil supplementation enhanced CPT-11 (irinotecan) efficacy against MCF7 breast carcinoma xenografts and ameliorated intestinal side-effects, *Br. J. Cancer* **81**: 440-448.
Hart, C. M., Tolson, J. K., and Block, E. R., 1991, Supplemental fatty acids alter lipid peroxidation and oxidant injury in endothelial cells, *Am. J. Physiol.* **260**: L481-488.
Hogg, N., Kalyanaraman, B., Joseph, J., Struck, A., and Parthasarathy, S., 1993, Inhibition of low-density lipoprotein oxidation by nitric oxide. Potential role in atherogenesis, *FEBS Lett.* **334**: 170-174.
Huie, R. E., and Padmaja, S., 1993, The reaction of NO with superoxide, *Free Radic. Res. Commun.* **18**: 195-199
Ignarro, L. J., Bush, P. A., Buga, G. M., Wood, K. S., Fukuto, J. M., and Rajfer, J., 1990, Nitric oxide and cyclic GMP formation upon electrical field stimulation cause relaxation of corpus cavernosum smooth muscle, *Biochem. Biophys. Res. Commun.* **170**: 843-850.
Ignarro, L. J., Byrns, R. E., Buga, G. M., and Wood, K. S., 1987, Endothelium-derived relaxing factor from pulmonary artery and vein possesses pharmacologic and chemical properties identical to those of nitric oxide radical, *Circ. Res.* **61**: 866-879.

Ischiropoulos, H., Zhu, L., and Beckman, J. S., 1992, Peroxynitrite formation from macrophage-derived nitric oxide, *Arch. Biochem. Biophys.* **298**: 446-451.

Jenkins, D. C., Charles, I. G., Thomsen, L. L., Moss, D. W., Holmes, L. S., Baylis, S. A., Rhodes, P., Westmore, K., Emson, P. C., and Moncada, S., 1995, Roles of nitric oxide in tumor growth, *Proc. Natl. Acad. Sci. U S A* **92**: 4392-4396.

Kanner, J., Harel, S., and Granit, R., 1991, Nitric oxide as an antioxidant, *Arch. Biochem. Biophys.* **289**: 130-136.

Karmali, R. A., Marsh, J., and Fuchs, C., 1984, Effect of omega-3 fatty acids on growth of a rat mammary tumor, *J. Natl. Cancer Inst.* **73**: 457-461.

Keizer, H. G., Pinedo, H. M., Schuurhuis, G. J., and Joenje, H., 1990, Doxorubicin (adriamycin): a critical review of free radical-dependent mechanisms of cytotoxicity, *Pharmacol. Ther.* **47**: 219-231

Kelley, E. E., Buettner, G. R., and Burns, C. P., 1997, Production of lipid-derived free radicals in L1210 murine leukemia cells is an early oxidative event in the photodynamic action of Photofrin, *Photochem. Photobiol.* **65**: 576-580.

Kelley, E. E., Domann, F. E., Buettner, G. R., Oberley, L. W., and Burns, C. P., 1997, Increased efficacy of *in vitro* Photofrin photosensitization of human oral squamous cell carcinoma by iron and ascorbate, *J. Photochem. Photobiol. B* **40**: 273-277.

Kelley, E. E., Wagner, B. A., Buettner, G. R., and Burns, C. P., 1999, Nitric oxide inhibits iron-induced lipid peroxidation in HL-60 cells, *Arch. Biochem. Biophys.* **370**: 97-104.

Komiyama, T., Kikuchi, T., and Sugiura, Y., 1986, Interactions of anticancer quinone drugs, aclacinomycin A, adriamycin, carbazilquinone, and mitomycin C, with NADPH-cytochrome P-450 reductase, xanthine oxidase and oxygen, *J. Pharmacobiodyn.* **9**: 651-664.

Kuzin, B., Roberts, I., Peunova, N., and Enikolopov, G., 1996, Nitric oxide regulates cell proliferation during Drosophila development, *Cell* **87**: 639-649.

Liepkalns, V. A., and Spector, A. A., 1975, Alteration of the fatty acid composition of Ehrlich ascites tumor cell lipids, *Biochem. Biophys. Res. Commun.* **63**: 1043-1047.

Marnett, L. J., and Wilcox, A. L., 1995, The chemistry of lipid alkoxyl radicals and their role in metal-amplified lipid peroxidation, *Biochem. Soc. Symp.* **61**: 65-72

Mitchell, J. B., Wink, D. A., DeGraff, W., Gamson, J., Keefer, L. K., and Krishna, M. C., 1993, Hypoxic mammalian cell radiosensitization by nitric oxide, *Cancer Res.* **53**: 5845-5848.

Moncada, S., and Higgs, A., 1993, The L-arginine-nitric oxide pathway, *N. Engl. J. Med.* **329**: 2002-2012.

Niki, E., Saito, T., Kawakami, A., and Kamiya, Y., 1984, Inhibition of oxidation of methyl linoleate in solution by vitamin E and vitamin C, *J. Biol. Chem.* **259**: 4177-4182.

O'Donnell, V. B., Chumley, P. H., Hogg, N., Bloodsworth, A., Darley-Usmar, V. M., and Freeman, B. A., 1997, Nitric oxide inhibition of lipid peroxidation: kinetics of reaction with lipid peroxyl radicals and comparison with alpha-tocopherol, *Biochemistry* **36**: 15216-15223.

Ozgen, U., Savasan, S., Stout, M., Buck, S., and Ravindranath, Y., 2000, Further elucidation of mechanism of resistance to vincristine in myeloid cells: role of hypochlorous acid in degradation of vincristine by myeloperoxidase, *Leukemia* **14**: 47-51.

Petersen, E. S., Kelley, E. E., Modest, E. J., and Burns, C. P., 1992, Membrane lipid modification and sensitivity of leukemic cells to the thioether lipid analog BM 41.440, *Cancer Res.* **52**: 6263-6269.

Peus, D., Vasa, R. A., Meves, A., Pott, M., Beyerle, A., Squillace, K., and Pittelkow, M. R., 1998, H2O2 is an important mediator of UVB-induced EGF-receptor phosphorylation in cultured keratinocytes, *J. Invest. Dermatol.* **110**: 966-971.

Porter, N. A., 1984, Chemistry of lipid peroxidation, *Methods Enzymol.* **105**: 273-282

Qian, S. Y., Wang, H. P., Schafer, F. Q., and Buettner, G. R., 2000, EPR detection of lipid-derived free radicals from PUFA, LDL, and cell oxidations, *Free Radic. Biol. Med.* **29**: 568-579.

Reed, J. C., 1999, Dysregulation of apoptosis in cancer, *J. Clin. Oncol.* **17**: 2941-2953.

Reed, J. C., 2000, Mechanisms of apoptosis, *Am. J. Pathol.* **157**: 1415-1430.

Riley, P. A., 1994, Free radicals in biology: oxidative stress and the effects of ionizing radiation, *Int. J. Radiat. Biol.* **65**: 27-33.

Rose, D. P., Connolly, J. M., and Meschter, C. L., 1991, Effect of dietary fat on human breast cancer growth and lung metastasis in nude mice, *J. Natl. Cancer Inst.* **83**: 1491-1495.

Rubbo, H., Parthasarathy, S., Barnes, S., Kirk, M., Kalyanaraman, B., and Freeman, B. A., 1995, Nitric oxide inhibition of lipoxygenase-dependent liposome and low-density lipoprotein oxidation: termination of radical chain propagation reactions and formation of nitrogen-containing oxidized lipid derivatives, *Arch. Biochem. Biophys.* **324**: 15-25.

Rubbo, H., Radi, R., Anselmi, D., Kirk, M., Barnes, S., Butler, J., Eiserich, J. P., and Freeman, B. A., 2000, Nitric oxide reaction with lipid peroxyl radicals spares alpha-tocopherol during lipid peroxidation. Greater oxidant protection from the pair nitric oxide/alpha-tocopherol than alpha-tocopherol/ascorbate, *J. Biol. Chem.* **275**: 10812-10818.

Rubbo, H., Radi, R., Trujillo, M., Telleri, R., Kalyanaraman, B., Barnes, S., Kirk, M., and Freeman, B. A., 1994, Nitric oxide regulation of superoxide and peroxynitrite-dependent lipid peroxidation. Formation of novel nitrogen-containing oxidized lipid derivatives, *J. Biol. Chem.* **269**: 26066-26075.

Saran, M., and Bors, W., 1997, Radiation chemistry of physiological saline reinvestigated: evidence that chloride-derived intermediates play a key role in cytotoxicity, *Radiat. Res.* **147**: 70-77.

Sergent, O., Griffon, B., Morel, I., Chevanne, M., Dubos, M. P., Cillard, P., and Cillard, J., 1997, Effect of nitric oxide on iron-mediated oxidative stress in primary rat hepatocyte culture, *Hepatology* **25**: 122-127.

Spector, A. A., and Burns, C. P., 1987, Biological and therapeutic potential of membrane lipid modification in tumors, *Cancer Res.* **47**: 4529-4537.

Sperandio, S., de Belle, I., and Bredesen, D. E., 2000, An alternative, nonapoptotic form of programmed cell death, *Proc. Natl. Acad. Sci. USA* **97**: 14376-14381.

Spitz, D. R., Kinter, M. T., Kehrer, J. P., and Roberts, R. J., 1992, The effect of monosaturated and polyunsaturated fatty acids on oxygen toxicity in cultured cells, *Pediatr. Res.* **32**: 366-372.

Struck, A. T., Hogg, N., Thomas, J. P., and Kalyanaraman, B., 1995, Nitric oxide donor compounds inhibit the toxicity of oxidized low-density lipoprotein to endothelial cells, *FEBS Lett.* **361**: 291-294.

Tisdale, M. J., and Dhesi, J. K., 1990, Inhibition of weight loss by omega-3 fatty acids in an experimental cachexia model, *Cancer Res.* **50**: 5022-5026.

Wagner, B. A., Buettner, G. R., and Burns, C. P., 1994, Free radical-mediated lipid peroxidation in cells: oxidizability is a function of cell lipid bis-allylic hydrogen content, *Biochemistry* **33**: 4449-4453.

Wagner, B. A., Buettner, G. R., and Burns, C. P., 1993, Increased generation of lipid-derived and ascorbate free radicals by L1210 cells exposed to the ether lipid edelfosine, *Cancer Res.* **53**: 711-713.

Wagner, B. A., Buettner, G. R., and Burns, C. P., 1992, Membrane peroxidative damage enhancement by the ether lipid class of antineoplastic agents, *Cancer Res.* **52**: 6045-6051.

Wagner, B. A., Buettner, G. R., and Burns, C. P., 1996, Vitamin E slows the rate of free radical-mediated lipid peroxidation in cells, *Arch. Biochem. Biophys.* **334**: 261-267.

Wagner, B. A., Buettner, G. R., Oberley, L. W., and Burns, C. P., 1998, Sensitivity of K562 and HL-60 cells to edelfosine, an ether lipid drug, correlates with production of reactive oxygen species, *Cancer Res.* **58**: 2809-2816.

Waldman, S. A., and Murad, F., 1988, Biochemical mechanisms underlying vascular smooth muscle relaxation: the guanylate cyclase-cyclic GMP system, *J. Cardiovasc. Pharmacol.* **12**: S115-118.

Wigmore, S. J., Barber, M. D., Ross, J. A., Tisdale, M. J., and Fearon, K. C., 2000, Effect of oral eicosapentaenoic acid on weight loss in patients with pancreatic cancer, *Nutr. Cancer* **36**: 177-184

Wilcox, A. L., and Marnett, L. J., 1993, Polyunsaturated fatty acid alkoxyl radicals exist as carbon-centered epoxyallylic radicals: a key step in hydroperoxide-amplified lipid peroxidation, *Chem. Res. Toxicol.* **6**: 413-416.

Wink, D. A., Cook, J. A., Christodoulou, D., Krishna, M. C., Pacelli, R., Kim, S., DeGraff, W., Gamson, J., Vodovotz, Y., Russo, A., and Mitchell, J. B., 1997, Nitric oxide and some nitric oxide donor compounds enhance the cytotoxicity of cisplatin, *Nitric Oxide* **1**: 88-94.

Wink, D. A., Cook, J. A., Pacelli, R., Liebmann, J., Krishna, M. C., and Mitchell, J. B., 1995, Nitric oxide (NO) protects against cellular damage by reactive oxygen species, *Toxicol. Lett.* **82-83**: 221-226.

Wink, D. A., Feelisch, M., Fukuto, J., Chistodoulou, D., Jourd'heuil, D., Grisham, M. B., Vodovotz, Y., Cook, J. A., Krishna, M., DeGraff, W. G., Kim, S., Gamson, J., and Mitchell, J. B., 1998, The cytotoxicity of nitroxyl: possible implications for the pathophysiological role of NO, *Arch. Biochem. Biophys.* **351**: 66-74.

Wink, D. A., Hanbauer, I., Krishna, M. C., DeGraff, W., Gamson, J., and Mitchell, J. B., 1993, Nitric oxide protects against cellular damage and cytotoxicity from reactive oxygen species, *Proc. Natl. Acad. Sci. USA* **90**: 9813-9817.

Wink, D. A., Vodovotz, Y., Laval, J., Laval, F., Dewhirst, M. W., and Mitchell, J. B., 1998, The multifaceted roles of nitric oxide in cancer, *Carcinogenesis* **19**: 711-721.

Wood, R., Falch, J., and Wiegand, R. D., 1975, Hepatoma, host liver, and normal rat liver neutral lipids as affected by diet, *Lipids* **10**: 202-207.

Yamanaka, N., Oda, O., and Nagao, S., 1996, Nitric oxide released from zwitterionic polyamine/NO adducts inhibits Cu^{2+}-induced low density lipoprotein oxidation, *FEBS Lett.* **398**: 53-56.

Yang, M., Nazhat, N. B., Jiang, X., Kelsey, S. M., Blake, D. R., Newland, A. C., and Morris, C. J., 1996, Adriamycin stimulates proliferation of human lymphoblastic leukaemic cells via a mechanism of hydrogen peroxide (H_2O_2) production, *Br. J. Hemeatol.* **95**: 339-344.

Chapter 7

Oxidized LDL-Induced Apoptosis

Hervé Benoist, Robert Salvayre, and Anne Nègre-Salvayre
Inserm U466, Institut Louis Bugnard, CHU Rangueil, 31403 Toulouse Cedex 4, France

1. INTRODUCTION

Cell death plays an essential role in normal development, homeostasis and pathology. Eukaryotic cells need molecular oxygen and the aerobic metabolic processes produce a variety of reactive oxygen species (ROS) with potential deleterious oxidative effects when they are produced in excess. Lipid oxidation is an example of ROS-mediated cellular damage. Lipoproteins, particularly low-density lipoproteins (LDL), are one of the major putative targets of lipid oxidation processes. *In vivo*, LDL are the main carrier of cholesterol and provide the cells with cholesterol and other nutrients. They can overload cells with oxidized lipids, either through LDL receptor-mediated uptake, when the apolipoprotein is preserved (in mildly oxidized LDL), or through scavenger receptors when more extensively oxidized. Oxidized molecules introduced in the cells may irreversibly alter cellular functions, such as mitosis or protein expression, and finally lead to cell death. In tissues capable of self-renewal there is a balance between cell division and cell death, and oxidized LDL may potentially influence the two processes. Numerous studies have demonstrated that oxidized LDL are toxic, inducing necrosis or apoptosis *in vitro* and contributing to inflammation *in vivo* (Mallat and Tedgui, 2000; Mitchinson *et al.*, 1996). The investigation of mechanisms and pathophysiological consequences of oxidized LDL-mediated cell death may contribute to our understanding of the pathogenesis of atherosclerosis and related diseases.

Subcellular Biochemistry, Volume 36: Phospholipid Metabolism in Apoptosis.
Edited by Quinn and Kagan. Kluwer Academic/Plenum Publishers, New York, 2002

2. OXIDIZED LDL-MEDIATED CELL DEATH AND PATHOLOGY

2.1 Involvement in pathology

Apoptosis is an essential feature of various diseases including atherosclerosis. Indeed, by using the TUNEL technique numerous histological studies have evidenced apoptosis in atherosclerotic lesions, at levels depending upon the lesion stage (Kockx *et al.*, 1998*a,b;* Mallat and Tedgui, 1998). The percentage of TUNEL-positive cells is very low in fatty streaks and high in advanced atherosclerotic plaques (up to 30-60%) (Mallat *et al.*, 1997; Björkerud and Björkerud, 1996*a*; Geng and Libby, 1995). However, little is known about the *in vivo* apoptosis inducers. Oxidized LDL is probably one of the factors implicated, because of the continual entrapment and trafficking of LDL in vascular cells and because of occurrence of lipid peroxidation during this physiological process (Hajjar and Haberland, 1997). In addition, oxidized LDL, oxidized lipids and apoptotic cells are clearly present within the atherosclerotic plaque (Okura *et al.*, 2000). For instance, oxysterols are found in atherosclerotic lesions, especially in the lipid core, and in macrophage-rich regions (Brown and Jessup, 1999; Brown *et al.*, 1997) and observations in animals and human have suggested the concomitant presence of lipid oxidation and apoptosis (Rader and Dugi, 2000). *In vivo*, apoptosis has been detected in vascular endothelial cells (EC), macrophages, smooth muscle cells (SMC) and in foam cells (Pirillo *et al.*, 2000; Rader and Dugi, 2000; Cai *et al.*, 1997; Björkerud and Björkerud, 1996*a*; Geng and Libby, 1995; Han *et al.*, 1995). The proportion of apoptotic cells in the vascular wall appears to be related to the blood LDL concentration. In addition, it probably depends on the conditions of interaction between vascular cells and oxLDL (duration of exposure, concentration, oxidation level, presence of cytokines and of infectious agents...) (Hajjar, 2000; Hajjar and Haberland, 1997; Hansson, 1997). Finally, numerous *in vitro* experiments have demonstrated that oxLDL and oxidized lipids can induce apoptosis in the different cells of the vascular wall (Heermeier *et al.*, 2001; Hsieh *et al.*, 2001; Wintergerst *et al.*, 2000; Li *et al.*, 2000a, 1998a; Alcouffe *et al.*, 1999; Bachem *et al.*, 1999; Farber *et al.*, 1999; Meilhac *et al.*, 1999; Dimmeler *et al.*, 1997; Clare *et al.*, 1995; Fossel *et al.*, 1994; Hessler *et al.*, 1979)

The pathophysiological role of oxidized LDL-induced apoptosis in vessel wall homeostasis is complex and probably depends on the cell type that oxLDL interacts with. For instance, EC are potential early targets, oxidized LDL may alter their functions and lead to apoptosis, contributing to early pro-atherosclerotic events. Apoptosis of macrophages or SMC could have more lasting consequences on vascular homeostasis. Death of SMC could be involved in plaque destabilization. Conversely, macrophage apoptosis could lead to an inhibition of collagen breakdown and consequently to plaque stabilization (Davies, 1996; Libby *et al.*, 1996). In addition, it may be expected that the cell phenotype and the physiological state (activation, proliferation or quiescence) modulate the cell susceptibility to oxLDL-mediated apoptosis. For example, proliferative EC are susceptible to oxLDL-mediated apoptosis but confluent cells are not (Escargueil-Blanc *et al.*, 1998, 1997). On the other hand, oxidized LDL may have indirect effects on cell death in the vascular wall. Indeed, oxLDL alter multiple signaling pathways and gene expressions (Napoli *et al.*, 2000; Maziere *et al.*, 2000, 1999; Auge *et al.*, 1999), that might modulate the molecular microenvironment of vascular cells by growth factor and cytokine production, and finally modulate cell death. Thus, ox-LDL induces the release of TNFα, a cytokine with potential apoptotic effect, by resting macrophages (Janabi *et al.*, 2000; Jovinge *et al.*, 1996), but suppress its expression in stimulated macrophages (Ohlsson *et al.*, 1996). Conversely, the cytokines can modulate cellular functions and cell sensitivity to oxLDL-induced apoptosis. For instance, TNFα may potentiate oxysterol-induced apoptosis in vascular smooth muscle cells (Ares *et al.*, 2000), whereas FGF, the secretion of which can be triggered by oxLDL (Ananyeva *et al.*, 1997), could prevent it (Mitchinson *et al.*, 1996).

In addition to the question of cardio-vascular diseases, hyperlipidemia and high LDL levels accelerate arteriosclerosis in chronic allograft rejection (Dietrich *et al.*, 2000; Shi *et al.*, 1997), perhaps by potentiating apoptosis in the donor tissue. Besides, oxidized LDL may be involved in other organ diseases. Apoptosis induction by oxidized lipids has been observed in neuronal cells suggesting that oxidized LDL may be involved in brain neurotoxicity (Keller *et al.*, 1999; Sugawa *et al.*, 1997). A role for lipids and oxLDL has been proposed in the progression of renal disease (Galle *et al.*, 1999). Lipid injury in the glomerulus might, at least partly, involve oxLDL-mediated apoptosis (Tashiro *et al.*, 1999). Finally, recent *in vitro* data suggest that oxLDL can inhibit the T lymphocyte response and induce death in activated lymphocytes (Caspar-Bauguil *et al.*, 1999, 1998; Alcouffe *et al.*, 1999). If that is confirmed *in vivo,* hypercholesterolemia, oxLDL and oxLDL-mediated apoptosis might weaken the protective role of immune system towards infectious agents.

2.2 Oxidized LDL and infectious agents

Numerous experiments have suggested that LDL oxidation generates determinants that lead to a specific immune response that, together with the innate immune response, acts as an oxLDL clearance mechanism (Figure 1A) (Horkko *et al.*, 2000; Shaw *et al.*, 2000).

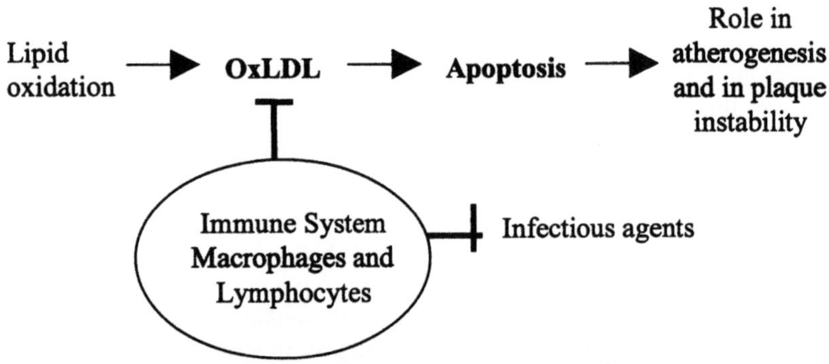

A- The immune system participates in tissue homeostasis by the elimination of oxLDL and infectious agents, and consequently by the decrease of vascular cell death.
Clearance: —┼ Induction: —▶

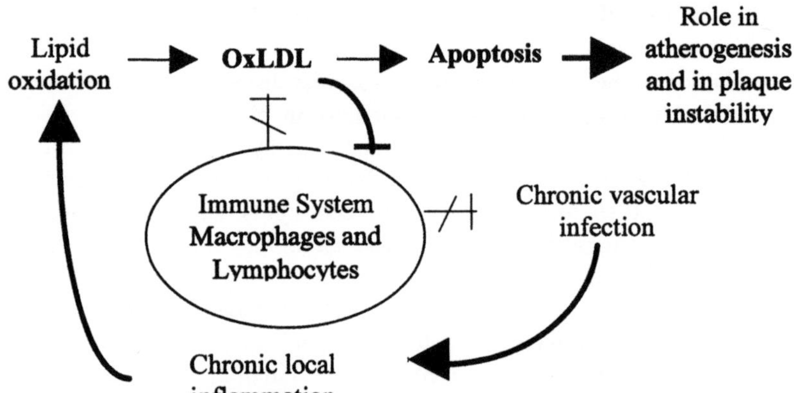

B - Chronic hyperlipidemia and chronic infection potentiate inflammation and lipid oxidation. An increase of lipid oxidation might inhibit the immune system and thus facilitate a chronic infection and a rise in oxLDL-mediated apoptosis.
Inhibition: —┤ Decrease of the clearance: —/┤

Figure 1. Hypothetical interactions between infectious agents, oxLDL and the immune system in atherogenesis.

However, hypercholesterolemia and conditions of chronic inflammation (such as a chronic and asymptomatic infection of areas of vascular wall) can locally produce a rise in oxidized lipids that may alter the immune system by unspecifically activating multiple signaling pathways in macrophages and lymphocytes (Janabi *et al.*, 2000; Meilhac *et al.*, 1999; Hansson, 1997; Malorni *et al.*, 1997; Jovinge *et al.*, 1996; Ohlsson *et al.*, 1996; Wu *et al.*, 1996), dysregulating lymphocyte responses (Caspar-Bauguil *et al.*, 1999, 1998) or inducing apoptosis in activated lymphocytes or macrophages (Wintergerst *et al.*, 2000; Alcouffe *et al.*, 1999, Aupeix *et al.*, 1995). This immune dysfunction might decrease the removal of oxidized lipoprotein particles and also the immune response against infectious agents. The interaction between infectious agent and hyperlipidemia could install a vicious cycle participating, at least through oxidized LDL and cell death induction, in the development and instability of the atherosclerotic plaque (Hajjar, 2000; Davies, 1996; Libby *et al.*, 1996) (Figure 1-B). Experimental and epidemiological studies have involved cytomegalovirus, *Chlamydia pneumoniae* and herpes viruses in atherosclerosis and particularly in the coronary heart disease. These agents have common characteristics: they can be detected in atherosclerotic lesions, are obligate intracellular pathogens and can lead to chronic infections, inflammatory disorders and persistent life-long immune responses (Khovidhunkit *et al.*, 2000). Association between these infectious agents and hyperlipidemia could accelerate the atherogenic process. For instance, *C. pneumoniae* induces foam cell formation in presence of LDL, mainly through chlamydial LPS (Kalayoglu and Byrne, 1998*a,b*), and accelerates atherogenesis in hypercholesterolemic apoE-deficient mice (Moazed *et al.*, 1999). More recently, it was demonstrated *in vivo* that the host response to infection and inflammation induces LDL oxidation (Memon *et al.*, 2000). One infectious proatherogenic mechanism could be the potentiation of lipid oxidation and a possible subsequent increase of oxLDL-mediated apoptosis in the vascular wall (Netea *et al.*, 2000; Kalayoglu *et al.*, 1999). Finally, recent data have shown that LPS Binding Protein (LBP) binds to LDL and VLDL and circulates with these lipoproteins in healthy persons. The circulating LBP-LDL/VLDL complexes bind predominantly LPS in serum from septic patients. These results demonstrate a new link between bacteria and lipoproteins, suggesting a role for LDL in the defense against bacteria and the scavenging of LPS (Vreugdenhil *et al.*, 2001). The putative effect of oxidation on this carrier system and on vascular cell death remains to be investigated.

3. MECHANISMS OF OXIDIZED-LDL INDUCED APOPTOSIS

Mechanisms of oxLDL-induced apoptosis have been studied in numerous *in vitro* models, such as SMC, EC, macrophages and lymphocytes, for which cytotoxic and apoptotic effect of oxidized lipids were demonstrated. *In vitro* experiments have shown that slightly or strongly oxidized LDL may have dual effect on vascular cells. For instance slightly oxidized LDL have a proliferative effect on SMC and macrophages (Augé *et al.*, 1999; Heery *et al.*, 1995), whereas strongly oxidized LDL promote apoptosis (Hsieh *et al.*; 2001, Siow *et al.*, 1999; Björkerud and Björkerud, 1996*b*). These opposite effects are probably related to the concentration and the quality of oxidized lipids internalized by the cells, but could also result from biochemical differences between cellular pathways of oxLDL uptake.

3.1 Role of receptors

The level of LDL in plasma is considered as the major risk factor for atherosclerosis. It depends on the diet and on the hepatic clearance by the LDL receptors. In the body, the cellular uptake of cholesterol is generally mediated through the high affinity LDL receptor. To avoid an excess of intracellular free cholesterol induced by a rise in extracellular concentration, cells can inhibit cholesterol and LDL receptor synthesis and increase cholesterol ester synthesis (Sniderman *et al.*, 2000). LDL entrapped in the vascular wall can be altered by various enzymatic and chemical molecules (Girotti, 1998), many of them mediate lipid oxidation and consequently the production of putative apoptotic molecules. These lipophilic invaders can be taken up through the apoB/E receptor or the scavenger receptor pathways. The ubiquitous apoB/E receptor allows cell loading with toxic lipids when the apolipoprotein is not or only slightly modified, i.e. in early stages of lipoprotein oxidation such as in minimally oxidized LDL (Siess *et al.*, 1999; Meilhac *et al.*, 1999).

The scavenger (SC) receptors are cell-surface molecules strongly interacting with modified-LDL such as highly oxidized LDL. These receptors are predominantly expressed on macrophages, but are also found on activated SMC and endothelial cells (Steinbrecher, 1999). Three classes of scavenger receptors are now described: the class A receptors, the class B receptors and the third class including CD68. A common feature of these molecules is their ability to bind to various ligands, such as polyanions, acetyl LDL, oxLDL, apoptotic cells and bacteria. The class A receptors include SR-AI/II receptors and the scavenger receptor MARCO (macrophage receptor with collagenous structure). These molecules are transmembrane glycoproteins with an amino-terminal cytoplasmic domain. The extracellular region is characterized: i) by a collagenous domain that is thought to be responsible for ligand binding, ii) by a cysteine-rich globular C terminal domain (that allows attachment of these molecules to the SC receptor cysteine-rich domain superfamily). SR-AI/II receptors clearly bind oxLDL whereas MARCO probably also bind other oxidized lipids. In addition these receptors can bind different bacteria, suggesting a putative role in innate immunity (Memon *et al.*, 2000; Steinbrecher, 1999; Pearson, 1996). The class B receptors include CD36, SR-B1 in murine tissues, and CLA-1 in humans. CLA-1 is strongly expressed in the adrenal, and moderately expressed in the liver, testis and monocytes. The main function for SR-B1/CLA-1 is probably a role in cholesterol transport, however it has been recently shown that CLA-1 at the monocyte surface may serve in the clearance of senescent/apoptotic cells (Murao *et al.*, 1997). In the same way, CD36 on monocyte/macrophages may be involved in phagocytosis of apoptotic cells. This receptor was demonstrated to interact with numerous different ligands such as thrombospondin-1, oxLDL, fatty acids and anionic phospholipids (Steinbrecher, 1999). In addition, a recent report showed that CD36 is significantly expressed on foam cells in atherosclerotic lesions (Nakata *et al.*, 1999). The third class of SC receptors, including SR-C1 in Drosophilia and Macrosialin/CD68 (CD68 in human), are endosomal membrane glycoproteins. The Macrosialin expression appears limited to macrophages and dendritic cells suggesting a role in immunity induction (Martinez-Pomares *et al.,* 1996; Pearson, 1996). These molecules might be receptors for oxLDL and apoptotic cells *in vivo* (Ramprasad *et al.*, 1996). Finally, several other molecules might be receptors for oxLDL, alone or in cooperation with the previous receptors: β2-integrins, FcγRIIb2 on macrophages (Steinbrecher, 1999), or SREC and LOX-1 on endothelial cells (Li *et al.*, 2000a).

In summary, the binding of oxLDL and/or lipids to several SC receptors (SRAI/II, CD36, CD68) has been clearly demonstrated. Most of other receptors may potentially bind oxLDL or lipids. Consequently, the cellular internalization of oxidized lipids by the SC receptor pathways may be involved in the biological effect of oxLDL, such as apoptosis induction. As a likely major ligand and because of their oxidized lipid contents, oxLDL may have many putative effects on different cells, such as alteration of cell signaling and gene expression.

3.2 Toxic oxidized molecules

Whatever the pathway for oxLDL uptake, numerous observations have demonstrated that their oxidized lipid contents can be directly toxic for the cells or can trigger oxidative mechanisms. Some of them are described below.

Oxysterols: Oxysterols are formed during cholesterol catabolism (Breuer and Björkem, 1995) occurring *in vivo* through enzymatic oxidation (27-hydroxycholesterol) (Björkem, 1992) or non-enzymatically (7α- and 7β-hydroxycholesterol, 7-ketocholesterol, 25-hydroxycholesterol). They are commonly found in human atherosclerotic plaques (Brown *et al.*, 1997) and are severely cytotoxic for vascular cells (Brown and Jessup, 1999). Oxysterols are present in native LDL and are abundantly formed during LDL oxidation (Chang *et al.*, 1997). Generally, 7-oxygenated sterols, and particularly 7-ketocholesterol, are mostly generated during LDL oxidation, whereas 24-, 25-, and 27-hydroxycholesterols are less abundant (Brown *et al.*, 1996, 1997; Carpenter *et al.*, 1994).

The toxic effect of 7-oxysterols, 25- and 27-hydroxycholesterols and their involvement in LDL cytotoxicity have been extensively studied on the different vascular cell types (Lizard *et al.*, 1999; Aupeix *et al.*, 1995; Clare *et al.*, 1995; Ramasamy *et al.*, 1992). 7α and 7β-hydroxycholesterols, 7-ketocholesterol, 25 and 27-hydroxycholesterol induce apoptosis (Brown and Jessup, 1999; Lizard *et al.*, 1999, 1998; Zhang *et al.*, 1997; Hughes *et al.*, 1994). 7β-hydroperoxycholesterol is one of the most toxic oxysterols present in oxidized LDL (Brown and Jessup, 1999; Colles *et al.*, 1996). 25-hydroxycholesterol, though less active (Aupeix *et al*, 1995), is able to trigger a cytochrome c release and subsequent caspase activation in CHO cells, but also calcium increase in relation with apoptosis (Rusinol *et al.*, 2000).

Oxysterol-mediated cytoxicity occurs by necrosis or apoptosis (Lizard *et al.*, 1999; Yin *et al.*, 2000), and can be partially prevented by Bcl2 overexpression (which inhibits caspase-3 activation and subsequent apoptosis), and by glutathione or antioxidants (Lizard *et al.*, 1998). The mechanism of cytotoxicity is unclear but could involve an oxidative stress (Lizard *et al.*, 1998), a direct stimulation of apoptotic cascade (Harada *et al.*, 1997; Christ *et al.*, 1993), a calcium rise (Spyridopoulos *et al.*, 2001), or metabolic dysfunctions such as inhibition of HMGCoA reductase or incorporation of oxysterols into membranes (Brown and Jessup, 1999).

The contribution of oxysterols to the cytotoxicity of oxidized LDL is likely, but it is to be noted that oxysterol concentrations necessary to trigger apoptosis in cultured cells are much higher than that reported in toxic concentrations of oxidized LDL and in atherosclerotic plaques (Brown and Jessup, 1999). This could be due to a synergistic effect of the different toxic molecules brought by the oxLDL.

Lipid hydroperoxides and phospholipid oxidation products: Lipid hydroperoxides are the major primary products of LDL oxidation. Lipid hydroperoxides (LOOH) are relatively polar, non-radical intermediates of lipid peroxidation that can exert deleterious effects on cells through modification of membrane structure (Girotti, 1998). LOOH are more active than lysophosphatidylcholine (LPC) to inhibit the production of nitric oxide (NO) (Huang *et al.*, 1999*a*). Moreover, LOOH in particular 15-HPETE and 13-HPODE, alter rapidly the redox balance with a significant decrease in the cellular glutathione, and initiate apoptosis through an activation of caspase 3 and cleavage of PARP (Wang *et al.*, 2000), these events being inhibited by pre-incubation with vitamin C (Siow *et al.*, 1999).

Lysophospholipids: Native and oxidized LDL contain appreciable amounts of lysophosphatidylcholine (LPC) (Steinberg, 1997) and lysophosphatidic acid (LPA) (Siess *et al.*, 1999), which could originate from the presence in LDL of a lipoprotein-associated phospholipase A(2) (Hakkinen *et al.*, 1999). In addition, oxidized LDL activate cellular phospholipases, PLA2 (Anthonsen *et al.*, 2000), and PLD (Gomez-Munoz *et al.*, 2000). The effect of LPA and LPC is different, since LPA is not involved in apoptosis but rather in cell proliferation and contraction (Siess *et al.*, 2000; Cunnick *et al.*, 1998). LPC is a very bioactive lysophospholipid able to induce various biological effects in vascular cells, such as vasoconstriction and inhibition of endothelium-dependent relaxation (Murohara *et al.*, 1994), inhibition of cell signaling through receptors linked to G-proteins, secretion of pro-inflammatory cytokines (Huang *et al.*, 1999*b*) and apoptotic or non-apoptotic cell death to smooth muscle cells (Hsieh *et al.*, 2001; Nilsson *et al.*, 1998). LPC possesses an ischemia-like effect on the heart (Hashizume *et al.*, 1997), and could be involved in the suppression of antiapoptotic signaling (suppression of NF-kappaB activation) induced by oxidized LDL in endothelial cells activated by TNF-alpha, or lipopolysaccharide (Heermeier *et al.*, 2001). LPC (in particular LPC-C16 and –C18, but not -C12) may induce cell death by sensitizing endothelial cells to Fas-mediated apoptosis (Sata and Walsh, 1998*a,b*). It is to be noted, however, that antioxidants, such as probucol, or metal chelators, such desferrioxamine, are unable to prevent LPC-induced cell death, whereas these agents protect against oxidized LDL-mediated cytotoxicity (Naito *et al.*, 1994).

Aldehydes: The oxidative breakdown of polyunsaturated fatty acids generates a number of aldehydes, which are highly bioactive and well characterized molecules. Aldehydes are longer-lived than free radicals, and are able to attack targets extra or intracellularly (for review see Esterbauer *et al.*, 1991). Among the different aldehydes formed during lipid and LDL oxidation, malondialdehyde and 4-hydroxyl alkenals, in particular 4-hydroxynonenal (4-HNE), have been intensively studied. 4-HNE results from the oxidation of arachidonic acid and, to a lesser extent, linoleic acid (Esterbauer *et al.*, 1991). 4-HNE and related hydroxyalkenals react rapidly with a number of amino acids such as histidine, serine or tyrosine, with thiol compounds, causing a rapid loss of SH-groups, and modify the free NH2 amino groups of lysine, (Esterbauer *et al.*, 1991). 4-HNE and other aldehydes are involved in the modification of apoB of LDL, that leads to a progressive lack of recognition by the apoB/E receptor and a metabolic deviation of LDL towards the scavenger receptor of macrophages (Steinberg, 1997).

Hydroxyalkenals are highly cytotoxic for cultured cells through multiple mechanisms, namely a rapid depletion of glutathione (Romero *et al.*, 1998), a decrease in thiol proteins (Van der Vliet *et al.*, 1991), disturbance of calcium homeostasis (Griffin and Segall, 1987), alteration in DNA, RNA and protein synthesis, induction of mitochondrial permeability transition and apoptosis signaling (Kristal *et al.*, 1996). Stress-activated protein kinase kinase (SEK1), and c-Jun N-terminal protein kinase (JNK) may be involved in 4-HNE-mediated apoptosis in PC12 cells (Soh *et al.*, 2000), which requires also an activation of the caspase-8, caspase-9 and caspase-3 (Liu *et al.*, 2000). One of the most marked effects of 4-HNE is the inhibition of NFkappaB activation, by inhibiting the phosphorylation and subsequent proteolysis of IkB (Page *et al.*, 1999). This could be due, at least in part, to an inhibition of the ubiquitin/proteasome pathway, through the 4-HNE-proteasome subunit formation (Vieira *et al.*, 2000).

The apoptotic effect of 4-HNE and related aldehydes occurs at concentrations over 10 to 20 μM and depends on the cell type, endothelial cells and fibroblasts being very sensitive to low 4-HNE concentrations (Esterbauer *et al.*, 1991). Oxidized LDL exhibit certain amounts of 4-HNE (range between 10 to 50 μM depending on the oxidation stage) which are compatible with an involvement of this aldehyde in the cytotoxicity (Jürgens *et al.*, 1987). Moreover, local tissue concentrations of 4-HNE may reach concentrations higher than 100 μM, in particular in membranes because of its high hydrophobicity (Benedetti *et al.*, 1986).

Besides 4-HNE, a number of aldehydes are formed during LDL oxidation, which are termed generically as TBARS or thiobarbituric reactive substances, among them malondialdehyde (MDA). Though less bioactive than 4-HNE, these aldehydes may participate in the toxicity of oxidized LDL, by interfering with proteins (modification of thiol and free reactive-amino groups), and with nucleosides (Esterbauer *et al.*, 1991). Like 4-HNE, MDA and related compounds inhibit the DNA repair systems, increase the rate of mutations, and are genotoxic (although no clear evidence exists for a direct modification or crosslink-generation of DNA.

3.3 Involvement of the redox balance

In higher eukaryotes, O_2 can be considered as a terminal electron acceptor required for energy production during respiratory mechanism. Several enzymatic systems, such as oxidases and superoxide dismutases, produce reactive oxygen species (ROS) ($O_2^{\cdot-}$, H_2O_2, $OH^{\cdot}$) that can also accept electrons and play an essential role in many of the metabolic processes associated with an aerobic existence (Buttke and Sanstrom, 1994). In addition, ROS can interact with NO to generate reactive nitrogen species. All of these oxygen derivatives have various biological effects, notably through their signaling properties. Indeed, numerous observations have shown that ROS may interact with signaling mechanisms and modulate gene and protein expression, particularly in vascular cells (Wolin, 2000; Girotti, 1998).

On the other hand, several ROS are highly cytotoxic. Consequently, eukaryotic cells have developed an elaborate arsenal of antioxidant mechanisms to neutralize their deleterious effects (enzymes such as superoxide dismutases, catalases, glutathione peroxidases, thioredoxin; inhibitors of free-radical chain reaction such as tocopherol, carotenoids, ascorbic acid; Fe^{2+} chelating proteins such as lactoferrin and transferrin). It can be postulated that ROS may induce an oxidative stress leading to cell death when the level of intracellular ROS exceeds an undefined threshold. Indeed, numerous observations have shown that ROS are mediators of cell death, particularly apoptosis (Maziere *et al.*, 2000; Girotti, 1998; Kinscherf *et al.*, 1998; Suzuki *et al.*, 1997; Buttke and Sanstrom, 1994; Albina *et al.*, 1993).

The possible involvement of ROS in oxidized LDL-mediated apoptosis has been reported, in particular through the activation of the caspase cascade. Oxidized LDL induce the generation of ROS in different vascular cell types (Hsieh *et al.*, 2001; Mazière *et al.*, 1999), possibly via oxLDL uptake using SC receptors, for instance through a mechanism involving the binding to a lectin-like oxLDL receptor (LOX-1) (Cominacini *et al.*, 2000). However the cellular source of intracellular ROS production is still unknown. H_2O_2 is produced, possibly through an activation of peroxisomes involved in the degradation of oxidized fatty acids, and the overexpression of catalase protects against oxidized lipid-induced cytotoxicity (Santanam *et al.*, 1999). A role for the NAD(P)H oxidase system is possible, since Cominacini reports a concomitant increase in extracellular H_2O_2 and $O_2^{\cdot-}$ production (Cominacini *et al.*, 2000). Mitochondrial respiratory chain and lipoxygenase may also be involved (Hsieh *et al.*, 2001).

Their role in oxidized-mediated cell death is also assessed (i) by the protective effect of antioxidants which reduce or inhibit oxidized LDL-induced ROS generation, caspase activation and subsequent apoptosis, whereas agents unable to inhibit the ROS increase do not protect against cell death (Hsieh *et al.*, 2001), and (ii) by the apoptotic effect of oxidants such as H_2O_2. Oxidized LDL decrease the activity of antioxidant enzymes, Cu/Zn superoxide dismutase (Li *et al.*, 1998b), and glutathione peroxidase (Rosenblat and Aviram, 1998). On the contrary, GPX overexpression protects against apoptosis (Brigelius-Flohe *et al.*, 2000). Moreover, besides (or consequently to) the ROS increase, oxLDL trigger a rise of intracellular TBARS content (Mabile *et al.*, 1995), and a strong derivatization of cellular proteins by 4-HNE, this being correlated with apoptosis (Vieira *et al.*, 2000). Finally, it is likely that intracellular ROS are involved in oxidized LDL-mediated cell death, but this depends probably of several parameters including the oxidant species and the source and amplitude of ROS production, which could determine the activation or inhibition of survival or stress-activated pathways and transcription factors.

3.4 Evidence for the implication of the death receptor pathway

Apoptosis is triggered by various factors such as DNA damaging agents, withdrawal of growth factors, UV and γ radiations, ROS, proinflammatory cytokines (e.g. TNFα) and immune mediators (e.g. Fas Ligand (FasL)). There are at least two main signaling pathways for apoptosis: a mitochondrial pathway and a death receptor pathway. The latter is physiologically activated by the binding of death factors (TNFα, FasL, DR-3, -4, -5, and -6) to plasma membrane receptors. Some of these mechanisms can be modulated by oxLDL.

TNF receptors, with or without death function, are ubiquitously expressed on cells. Contradictory data indicate that oxLDL modulate TNFα production, they may induce (Janabi *et al.*, 2000; Jovinge *et al.*, 1996) or inhibit (Niemann-Jonsson *et al.*, 2000) the release of TNFα by monocytes/macrophages, its putative main source in vascular wall. More interesting, recent results demonstrate that oxLDL can induce apoptosis by activating Fas and TNF receptor signaling pathways, and FasL-Fas interaction system in human endothelial cells. For instance, oxLDL down-regulate Bcl-2 expression (Napoli *et al.*, 2000) and the cellular caspase inhibitor FLIP (Sata and Walsh, 1998*a*) that interact with Fas-associated death domain protein, whereas blocking of FasL and TNF receptors reduced oxLDL-mediated apoptosis. In addition oxLDL can activate PKC, PTK, MAP and Jun kinases, whereas the kinase inhibitors reduce oxLDL-induced apoptosis (Napoli *et al.*, 2000; Li *et al.*, 1998a). Thus, oxLDL might induce apoptosis by direct modulation of intracellular pathways of apoptosis. But oxLDL could also promote Fas-mediated endothelial cell suicide by increasing Fas and FasL expression on the cell surface and by decreasing cellular inhibitors of apoptosis. Thus oxLDL-mediated apoptosis is reduced in *gld* or *lpr* mice that lack functional FasL or Fas respectively (Sata and Walsh, 1998*b*), suggesting that the Fas pathway plays a critical role in oxLDL-induced apoptosis. Moreover, a positive correlation has been observed *in vivo* between acceleration of atherosclerotic lesion development, apoptosis and Fas/FasL expression in rabbit and humans (Schneider *et al.*, 2000; Esaki *et al.*, 2000; Geng *et al.*, 1997; Cai *et al.*, 1997). OxLDL is clearly a putative candidate as inducer of cell death by the way of Fas/Fas ligand system.

3.5 Intracellular death signals

Oxidized LDL trigger a large array of signaling pathways potentially cytotoxic and depending on the relative level of LDL oxidation, the concentration of oxidized LDL in the culture medium, and the cell type. Among the various events relative to oxLDL-induced cytotoxicity, one can cite NFkappaB dysregulation (Heermeier *et al.*, 2001; Cominacini *et al.*, 2000; Janabi *et al.*, 2000; Caspar-Bauguil *et al.*, 1999; Ohlsson *et al.*, 1996), ceramide release (Escargueil-Blanc *et al*, 1998), caspase activation (Dimmeler *et al.*, 1997), proteasome dysfunction (Vieira *et al.*, 2000) and deregulation of cytosolic calcium homeostasis (Escargueil-Blanc *et al.*, 1997) (Figure 2). To date, the primary molecular targets leading to these responses, their mechanisms of activation and their links remain largely unknown.

Caspase activation: In human endothelial cells and macrophages, oxLDL induce the activation of caspases involved in the apoptotic process, namely the major executioner CPP-32/caspase-3 (Wintergerst *et al.*, 2000; Dimmeler *et al.*, 1997). Mitochondria are probably involved in caspase activation because toxic concentrations of oxLDL elicited a release of mitochondrial cytochrome C to the cytosol, and inhibition of this process by cyclosporin A blocked the apoptosis triggered by oxidized LDL (Walter *et al.*, 1998). Activation of caspase-3 in macrophages involves CD36 (Wintergest *et al.*, 2000), whereas the class A scavenger receptor (MSR-A) protects against apoptosis in another macrophagic cell type THP-1. Caspase inhibitors with large specificity (Z-VAD.fmk) or more specific for caspase-3 (Z-DVED.fmk) reduce significantly oxLDL-mediated apoptosis (Farber *et al.*, 1999; Wintergest *et al.*, 2000). Until now, the molecular mechanism of caspase activation by oxidized LDL remains unknown. Antisense studies and overexpression of Bcl2 in lymphocytes and HL60 demonstrated that Bcl2 protects against apoptosis, probably by inhibiting caspase activation, but does not protect against oxLDL-mediated necrosis. This suggests that oxLDL trigger death signaling pathway leading to both apoptosis and necrosis. A sustained and intense calcium signal may constitute such a death signal (Meilhac *et al.*, 1999).

Calcium: During toxic cell injury, a disruption of calcium homeostasis plays often a critical role by triggering activation of calcium-dependent degradative enzymes, resulting in irreversible damage to cellular components and leads to cell death. As documented by Kass and Orrenius (1999), alterations in calcium homeostasis are commonly observed during apoptosis, through various mechanisms involving an emptying of intracellular calcium stores, followed by an influx of calcium across the plasma membrane, or a disruption of mitochondrial function and subsequent oxidative stress that leads to inhibition of Ca^{2+} transport systems located in the plasma membrane, endoplasmic reticulum or mitochondria. Oxidized LDL trigger an intense and sustained cytosolic calcium rise in smooth muscle cells, endothelial cells, and lymphocytes, that in turn activates calcium-dependent enzymes involved in cellular events leading to apoptosis (or necrosis) (Meilhac *et al.*, 1999; Escargueil-Blanc *et al.*, 1997). For instance, it was reported that the calcium rise activates calcium-dependent endonucleases which induce chromatin cleavage and apoptosis, but also calcium-dependent proteases which are involved in cytoskeletal alterations, blebbing of the plasma membrane, and necrosis (Meilhac *et al.*, 1999; Escargueil-Blanc *et al.*, 1997).

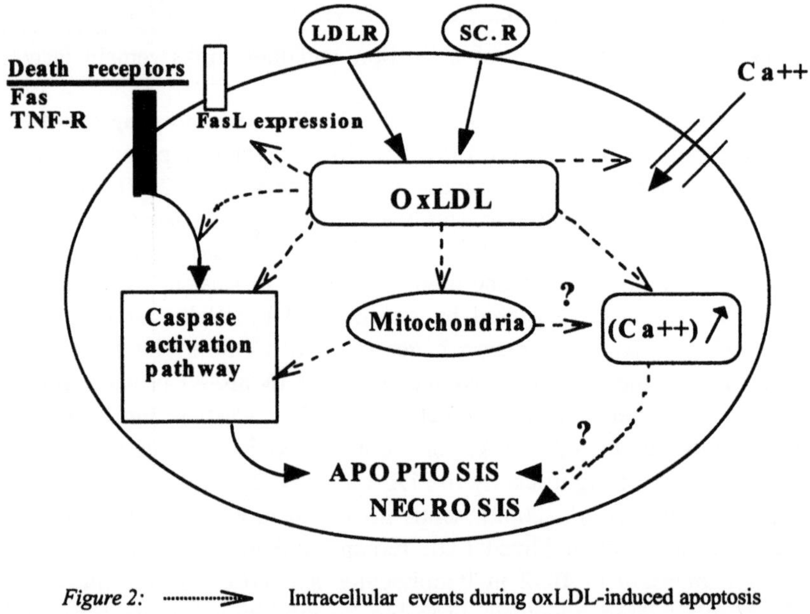

Figure 2: ⋯⋯> Intracellular events during oxLDL-induced apoptosis

The deregulation of cytosolic calcium homeostasis is due, at least partly, to an influx in extracellular calcium because EGTA inhibits the calcium rise and cell death (apoptosis and necrosis) (Escargueil-Blanc *et al.*, 1997). Dihydropyridine calcium channel antagonists block the intracellular free calcium increase and cytotoxicity (Sevanian *et al.*, 2000; Negre-Salvayre *et al.*, 1992). Oxidized LDL generate intracellular ROS that may inhibit plasma membrane Ca^{2+}/ATPases pumps and contribute to the rise of cytosolic calcium (Zhao *et al.*, 1996), and cell death, like that observed with TNF-alpha in fibroblasts L929 (Ko *et al.*, 2000). The mechanisms leading to the calcium rise and to caspase activation might not be mutually exclusive since cyclosporin A which inhibits oxidized LDL-mediated apoptosis, by blocking cytochrome C release and caspase3 activation (Wintergest *et al.*, 2000), can also block calcium-dependent apoptosis by inhibition of the Ca^{2+}-calmodulin-dependent calcineurin (Liu *et al.*, 1991). However, the primary targets responsible for the toxic cell signaling and the mechanisms of oxidized LDL-mediated calcium rise and/or caspase activation, and their links or respective involvement in cell death remain to be elucidated.

Ceramide: Ceramide is generated through activation of the sphingomyelin-ceramide pathway, i.e. sphingomyelin hydrolysis by signal-regulated sphingomyelinase, or through ceramide biosynthesis (Bose *et al.*, 1995). The precise role of ceramide in apoptosis is largely hypothetical (Hannun and Luberto, 2000). Harada-Shiba *et al.* (1998) reported a role for an early generation of ceramide, in oxidized LDL-mediated apoptosis of endothelial cells. Furthermore, the sphingomyelin/ceramide pathway could play a role in the enhanced expression of manganese superoxide dismutase (MnSOD) and p53 observed during oxidized LDL-mediated apoptosis of human macrophages (Kinscherf *et al.*, 1999). But the involvement of ceramide in cell death is discussed because serpins (TPCK and DCIC) which inhibit the sphingomyelin hydrolysis and ceramide generation, do not protect against cell death triggered by oxidized LDL. Furthermore, the permeant C2- and C6-ceramides induced-apoptosis is not a calcium-dependent process (in contrast to that triggered by oxidized LDL) (Escargueil-Blanc *et al.*, 1998).

Oxidized LDL alter cellular functions: role in cell death: Oxidized LDL seem to be poorly degraded by lysosomal enzymes and accumulate in lysosomes altering in turn their functionality (Dean *et al.*, 1997). It has been proposed that inhibition of oxidized LDL degradation and subsequent lipid accumulation may induce a destabilization of the acidic compartment, and lysosomal rupture with a relocation of lysosomal enzymes in the cytosol (Li W *et al.*, 1998). This process, also called 'endopepsis', occurs early and could precede mitochondrial dysfunction and cell death (Fossel *et al.*, 1994). Moreover, oxidized LDL trigger a dysfunction of the intracellular proteolytic ubiquitin/proteasome pathway (early activation followed by inhibition) which leads to an accumulation of oxidatively modified proteins, and participate in several cellular events such as dysfunction of NFkappaB activation (Heermeier *et al.*, 2001; Cominacini *et al.*, 2000; Janabi *et al.*, 2000; Caspar-Bauguil *et al.*, 1999; Ohlsson *et al.*, 1996), accumulation of p53 (Maziere *et al.*, 2000) and apoptosis (Vieira *et al.*, 2000). Protein tyrosine kinases (PTK) and PKC are activated by oxidized LDL (Li D *et al.*, 1998), and are supposed to participate in cell death because PTK inhibitors reduce apoptosis (Kapiotis *et al.*, 1997; Li *et al.*, 1998a). But these data need further clarification, and they do not support unequivocally a pro-apoptotic role of PTK and PKC in oxLDL-mediated apoptosis, because the inhibitors used exhibit also a potent antioxidant activity (Kapiotis *et al.*, 1997).

In summary, apoptosis results from uptake of oxidized LDL and subsequent intracellular release of oxidation products able to promote cell signaling leading to cell death. When apoptosis is inhibited by Bcl2 overexpression, oxidized LDL trigger necrosis (Meilhac *et al.*, 1999). Finally the cell death evoked by oxidized LDL may be involved in the formation of the necrotic core of atherosclerotic plaques, and in plaque rupture (Mallat *et al.*, 2000, 1997). Apoptotic lipidic bodies and cell debris resulting from cell death exhibit a potent thrombogenic potential and may play a role in thrombotic events accompanying plaque rupture (Mallat and Tedgui, 2000; Mallat *et al.*, 1997). Therefore, the protective (antiapoptotic) effect of physiological (HDL, estrogens), dietary (antioxidants) or pharmacological agents, may be of importance in preventing the formation of instable plaques and subsequent atherothrombotic events.

4. PROTECTIVE MECHANISMS

HDL: HDL promote the cholesterol efflux from arterial. In addition, HDL protect LDL from oxidation and induction of subsequent cytotoxicity (Von Eckardstein *et al.*, 2001; Kwiterovich, 1998; Berliner and Heinecke, 1996). This effect is apparently due to the presence of 2 enzymes (paraoxonase and PAF-acetylhydrolase bound to HDL) which may hydrolyse the oxidized lipids of LDL, thus blocking the subsequent biological responses (Hajjar and Haberland, 1997). Moreover, HDL could reduce hydroperoxides to their corresponding hydroxides, or extract oxidized lipids from oxidized LDL (Bonnefont-Rousselot *et al.*, 1999). HDL inhibit oxidized LDL cytotoxicity (Hessler *et al.*, 1979), through inhibition of oxLDL-induced stress and calcium rise. HDL act as an anti-oxidant lowering the oxidative modification of LDL, and preventing the cells of arterial wall from the deleterious effect of oxidized LDL, but their precise cellular mechanism of protection still remains unknown.

Estrogens: Estrogens are considered as beneficial for the cardiovascular system since they inhibit inflammatory and proatherogenic responses such as smooth muscle cell proliferation and migration, cytokine activation and expression of adhesion molecules (Haynes *et al.*, 2000; Zhu *et al.*, 1999). Estrogens, in particular 17beta-estradiol, have antioxidant effects on LDL (Zhu *et al.*, 1999) and inhibit apoptosis of cultured endothelial cells (Alvarez *et al.*, 1997). The protective mechanism could involve an inhibition of the oxidative stress triggered by oxidized LDL, as reported in endothelial cells exposed to H2O2 (Sudoh *et al.*, 2001).

Antioxidants: The rationale for the use of antioxidants as antiatherogenic agents is very simple: if LDL oxidation contributes to atherogenesis through multiple biological effects including apoptosis, antioxidants should prevent or delay atherogenesis. Some discrepancies exist anyway: *in vitro* studies show that lipid- and water-soluble antioxidants prevent LDL oxidation, and have a very efficient cytoprotective effect against the various biological effects triggered by oxLDL such as apoptosis; in contrast, *in vivo* the beneficial effect of antioxidant intake is more discussed (Berliner and Heinecke, 1996).

Vitamin E or α-tocopherol (α-TOH) (a lipid-soluble antioxidant present in lipoproteins) is very efficient in preventing LDL oxidation (Esterbauer *et al.*, 1991, Thomas and Stocker, 2000). α-TOH and analogs provide significant cytoprotection against apoptosis triggered by oxidized LDL. This protective effect results from inhibition of the oxLDL-induced calcium rise (Mabile *et al.*, 1995), and early events such as the degradation of IkappaB and subsequent NF-kappaB activation (Li *et al.*, 2000b, 1999). Ascorbic acid (a hydrophilic water-soluble antioxidant) is also able to block LDL oxidation (Heller *et al.*, 1998) and dehydroascorbic acid prevents the oxidized LDL-mediated macrophage apoptosis (Asmis and Wintergest, 1998). It may be noted that α-TOH could exhibit anti- or pro-oxidant activity depending on the environment, and the presence of co-antioxidants such as ascorbic acid, is necessary for an efficient anti-oxidant activity of α-TOH (Thomas and Stocker, 2000).

Polyphenols and flavonoids present in red wine and grape juice, fruits and vegetables, have potent antioxidant activity, which may slow down oxidative modification of LDL, and their subsequent toxicity (Wedworth and Lynch, 1995). Phenolic compounds exert cytoprotection on vascular cultured cells by inhibiting the calcium rise and subsequent oxidized LDL-mediated cell death (Vieira *et al.*, 1998). These compounds may play a role in the relatively low level of coronary heart disease in France ("French Paradox") and other Mediterranean countries (Renaud and Ruf, 1994).

Several lines of evidences suggest that all these antioxidants may have some beneficial effect against the development of atherosclerosis. However, the final effects of antioxidants observed in experiments on animal models and in human cardiovascular diseases are still largely discussed (Berliner and Heinecke, 1996). Some studies suggest that antioxidants slow down the LDL oxidation process (possibly involved in the development of primary stages of atherosclerosis), and also promote the regression of established advanced lesions (Berliner and Heinecke, 1996). But additional *in vivo* studies and new strategies utilizing combinations of antioxidants, are needed to determine conclusively their efficiency in atherogenesis prevention.

5. CONCLUSION

There is now substantial evidence establishing that oxLDL and lipid peroxidation products generated in the vascular wall, can elicit proinflammatory and pathological events involved in atherosclerosis. As reported by D. Hajjar and M. Haberland, the oxLDL should be considered as true molecular " Trojan Horses " or " cellular saboteurs " bearing a lot of bioreactive products of lipid peroxidation, that modify cell functions, induce intracellular oxidative stress, trigger multiple and contradictatory cell signalings and transcription factors, and finally induce cell death by apoptosis or necrosis. To date, the signaling mechanisms triggered by oxLDL and leading to physiopathological events, are only partly understood. The oxLDL-mediated apoptosis may participate in the progression of atherosclerosis, and in subsequent plaque rupture and thrombotic events which lead finally to acute cardiovascular diseases. In cultured cells antioxidants inhibit LDL oxidation and biological effect of oxLDL, such as apoptosis induction, but their ability to prevent efficiently vascular diseases and to induce regression of established lesions remains to be established. In conclusion, it is very important to clarify the mechanisms of cell-mediated LDL oxidation, as well as the cellular dysfunctions induced by lipid peroxidation products, in order to define new therapeutic strategies efficient in blocking and preventing the atherosclerotic process which always represent a major cause of death in western countries.

ACKNOWLEDGMENTS

Dr. Nelly Blaes and Dr. Mogens Thomsen are thanked for helpful comments and discussions.

REFERENCES

Albina, J.E., Cui, S., Mateo, R.B., and Reichner, J.S., 1993, Nitric oxide-mediated apoptosis in murine peritoneal macrophages, *J Immunol.* **150**: 5080-5085.

Alcouffe, J., Caspar-Bauguil, S., Garcia , V., Salvayre , R ., Thomsen, M., and Benoist, H., 1999, Oxidized low density lipoproteins induce apoptosis in PHA-activated peripheral blood mononuclear cells and in the Jurkat T-cell line, *J Lipid Res.***40**: 1200-1210.

Alvarez, R.J., Gips, S.J., Moldovan, N., Wilhide, C.C., Milliken, E.E., Hoang, A.T., Hruban, R.H., Silverman, H.S., Dang, C.V., Goldschmidt-Clermont, P.J., 1997, 17beta-estradiol inhibits apoptosis of endothelial cells, *Biochem. Biophys. Res. Commun.,* **237**: 372-81

Ananyeva, N.M., Tjurmin, A.V., Berliner, J.A., Chisolm, G.M., Liau, G, Winkles, J.A., and Haudenschild, C.C., 1997, Oxidized LDL mediates the release of fibroblast growth factor-1, *Arterioscler Thromb Vasc Biol.* **17**: 445-453.

Anthonsen, M.W., Stengel, D., Hourton, D., Ninio, E., Johansen, B., 2000, Mildly oxidized LDL induces expression of group IIa secretory phospholipase A(2) in human monocyte-derived macrophages, *Arterioscler Thromb Vasc Biol,* **20**: 1276-1282.

Ares, M.P., Porn-Ares, MI, Moses, S, Thyberg, J, Juntti-Berggren, L, Berggren, P, Hultgardh-Nilsson, A, Kallin, B, Nilsson, J., 2000, 7beta-hydroxycholesterol induces Ca(2+) oscillations, MAPK activation and apoptosis in human aortic smooth muscle cells, *Atherosclerosis* **153**: 2 3-35.

Asmis, R., Wintergerst, E.S., 1998, Dehydroascorbic acid prevents apoptosis induced by oxidized low-density lipoprotein in human monocyte-derived macrophages, *Eur. J. Biochem.* **255**: 147-155.

Augé, N., Nikolova-Karakashian, M., Carpentier, S., Parthasarathy, S., Negre-Salvayre, A., Salvayre, R., Merrill, A.H. Jr, Levade, T., 1999, Role of sphingosine 1-phosphate in the mitogenesis induced by oxidized LDL in smooth muscle cells via activation of sphingomyelinase, ceramidase, and sphingosine kinase, *J. Biol. Chem.* **274:** 21533-21538.

Aupeix, K, Weltin, D, Mejia, JE, Christ, M, Marchal, J, Freyssinet, JM, Bischoff, P., 1995, Oxysterol-induced apoptosis in human monocytic cell lines, *Immunobiology* **194**: 15-28.

Bachem, M.G., Wendelin, D., Schneiderhan, W., Haug, C., Zorn, U., Gross, H.J., Schmid-Kotsas, A., and Grunert, A., 1999, Depending on their concentration oxidized low density lipoproteins stimulate extracellular matrix synthesis or induce apoptosis in human coronary artery smooth muscle cells, *Clin Chem Lab Med.* **37:** 319-326.

Benedetti, A., Pompella, A., Fulceri, R., Romani, A., Comporti, M., 1986, 4-Hydroxynonenal and other aldehydes produced in the liver *in vivo* after bromobenzene intoxication, *Toxicol. Pathol.* **14**: 457-461.

Berliner, J.A., and Heinecke, J.W., 1996, The role of oxidized lipoproteins in atherogenesis, Free Radic. *Biol. Med.* **20**: 707-27.

Björkerud, S., and Bjorkerud, B., 1996*a*, Apoptosis is abundant in human atherosclerotic lesions, especially in inflammatory cells (macrophages and T cells), and may contribute to the accumulation of gruel and plaque instability, *Am J Pathol.* **149**: 367-380.

Björkerud, B., and Bjorkerud, S., 1996*b*, Contrary effects of lightly and strongly oxidized LDL with potent promotion of growth versus apoptosis on arterial smooth muscle cells, macrophages, and fibroblasts, *Arterioscler Thromb Vasc Biol.* **16**: 416-424.

Björkerud, I., 1992, Mechanism of degradation of the steroid side chain in the formation of bile acids, *J. Lipid Res.* **33**: 455-471.

Bonnefont-Rousselot, D., Therond, P., Beaudeux, J.L., Peynet, J., Legrand, A., and Delattre, J., 1999, High density lipoproteins (HDL) and the oxidative hypothesis of atherosclerosis, Clin. *Chem. Lab. Med.* **37**: 939-948.

Bose, R., Verheij, M., Haimovitz-Friedman, A., Scotto, K., Fuks, Z., and Kolesnick, R., 1995, Ceramide synthase mediates daunorubicin-induced apoptosis: an alternative mechanism for generating death signals, *Cell* **82**: 405-414.

Breuer, O., and Bjorkhem, I., 1995, Use of an 18O2 inhalation technique and mass isotopomer distribution analysis to study oxygenation of cholesterol in rat. Evidence for *in vivo* formation of 7-oxo-, 7beta-hydroxy-, 24-hydroxy-, and 25-hydroxycholesterol, *J. Biol. Chem.* **270**: 20278-20284.

Brigelius-Flohe,R., Maurer, S., Lotzer, K., Bol, G., Kallionpaa, H., Lehtolainen, P., Viita, H., and Yla-Herttuala, S., 2000, Overexpression of PHGPx inhibits hydroperoxide-induced oxidation, NFkappaB activation and apoptosis and affects oxLDL-mediated proliferation of rabbit aortic smooth muscle cells, *Atherosclerosis* **152**: 307-316.

Brown, A.J., Dean, R.T., and Jessup, W., 1996, Free and esterified oxysterol: formation during copper-oxidation of low density lipoprotein and uptake by macrophages, *J. Lipid Res.* **37**: 320-335.

Brown, A.J., and Leong, S.L., Dean, R.T., Jessup, W., 1997, 7-Hydroperoxycholesterol and its products in oxidized low density lipoprotein and human atherosclerotic plaque, *J. Lipid Res.* **38:** 1730-1745.

Brown, A.J. and Jessup, W., 1999, Oxysterols and atherosclerosis. *Atherosclerosis* **142**: 1-28.

Buttke, T.M., and Sanstrom, P.A., 1994, Oxidative stress as a mediator of apoptosis, *Immunol Today*, **15**: 7-10.

Cai, W., Devaux, B., Schaper, W., Schaper, J., 1997, The role of Fas/APO 1 and apoptosis in the development of human atherosclerotic lesions, *Atherosclerosis* **131**: 177-186.

Carpenter, K.L., Wilkins, G. M., Fussell, B., Ballantine, J.A., Taylor, S.E., Mitchinson, M.J., and Leake, D.S., 1994, Production of oxidized lipids during modification of low-density lipoprotein by macrophages or copper, *Biochem. J.* **304**: 625- 633.

Caspar-Bauguil, S., Tkaczuk, J., Haure, M.J., Durand, M., Alcouffe, J., Thomsen, M., Salvayre, R., and Benoist, H., 1999, Mildly oxidized low-density lipoproteins decrease early production of interleukin 2 and nuclear factor kappaB binding to DNA in activated T-lymphocytes, *Biochem. J.* **337**: 269-274.

Caspar-Bauguil, S., Saadawi, M., Negre-Salvayre, A., Thomsen, M., Salvayre, R., and Benoist, H., 1998, Mildly oxidized low-density lipoproteins suppress the proliferation of activated CD4+ T-lymphocytes and their interleukin 2 receptor expression *in vitro*, *Biochem. J.* **330**: 659-666.

Chang, Y.H., Abdalla, D.S., and Sevanian, A., 1997, Characterization of cholesterol oxidation products formed by oxidative modification of low density lipoprotein, *Free Radic. Biol. Med.* **23**: 202-214.

Christ, M., Luu, B., Mejia, J.E., Moosbrugger, I., and Bischoff, P., 1993, Apoptosis induced by oxysterols in murine lymphoma cells and in normal thymocytes, *Immunology* **78**: 455-460.

Clare, K., Hardwick, S.J., Carpenter, K.L., Weeratunge, N., and Mitchinson, M.J., 1995, Toxicity of oxysterols to human monocyte-macrophages, *Atherosclerosis* **118**: 67-75.

Colles, S.M., Irwin, K.C., Chisolm, G.M., 1996, Roles of multiple oxidized LDL lipids in cellular injury: dominance of 7 beta-hydroperoxycholesterol, *J. Lipid Res.* **37**: 2018-2028.

Cominacini, L., Pasini, A.F., Garbin, U., Davoli, A., Tosetti, M.L., Campagnola, M., Rigoni, A., Pastorino, A.M., Lo Cascio, V., and Sawamura, T., 2000, Oxidized low density lipoprotein (ox-LDL) binding to ox-LDL receptor-1 in endothelial cells induces the activation of NF-kappaB through an increased production of intracellular reactive oxygen species, *J. Biol. Chem.* **275**: 12633-12638.

Cunnick, J.M., Dorsey, J.F., Standley, T., Turkson, J., Kraker, A.J., Fry, D.W., Jove, R., and Wu, J., 1998, Role of tyrosine kinase activity of epidermal growth factor receptor in the lysophosphatidic acid-stimulated mitogen-activated protein kinase pathway, *J. Biol. Chem.* **273**: 14468-14475.

Davies, M.J., 1996, Stability and instability: two faces of coronary atherosclerosis, *Circulation* **94**: 2013-2020.

Dean, R.T., Fu, S., Stocker, R., Davies, M.J., 1997, Biochemistry and pathology of radical-mediated protein oxidation, *Biochem. J.* **324**: 1-18.

Dietrich, H, Hu, Y, Zou, Y, Huemer, U, Metzler, B, Li, C, Mayr, M, and Xu, Q., 2000, Rapid development of vein graft atheroma in ApoE-deficient mice, *Am. J. Pathol.* **157**: 659-669.

Dimmeler, S., Haendeler, J., Galle, J., and Zeiher, A.M., 1997, Oxidized low-density lipoprotein induces apoptosis of human endothelial cells by activation of CPP32-like proteases. A mechanistic clue to the 'response to injury' hypothesis, *Circulation* **95**: 1760-1763.

Esaki, T., Hayashi, T., Muto, E., Kano, H., Kumar, T.N., Asai, Y., Sumi D., and Iguchi, A., 2000, Expression of nitric oxide synthase and Fas/Fas ligand correlates with the incidence of apoptosis cell death in atheromatous plaques of human coronary arteries, *Nitric Oxide* **4**: 561-571.

Escargueil-Blanc, I., Meilhac, O., Pieraggi, M.T., Arnal, J.F., Salvayre, R., and Negre-Salvayre, A., 1997, Oxidized LDLs induce massive apoptosis of cultured human endothelial cells through a calcium-dependent pathway. Prevention by aurintricarboxylic acid, *Arterioscler. Thromb. Vasc. Biol.* **17**: 331-339.

Escargueil-Blanc, I., Andrieu-Abadie, N., Caspar-Bauguil, S., Brossmer, R., Levade, T., Negre-Salvayre, A., and Salvayre, R., 1998, Apoptosis and activation of the sphingomyelin-ceramide pathway induced by oxidizedLDL are not causally related in ECV-304 endothelial cells, *J. Biol. Chem.* **273**: 27389-27395.

Esterbauer, H.J., Schaur, R.J. and Zollner, H., 1991, Chemistry and biochemistry of 4-hydroxynonenal, malonaldehyde and related aldehydes, *Free Radic. Biol.Med.* **11**: 81-128.

Farber, A., Kitzmiller, T, Morganelli, PM, Pfeiffer, J, Groveman, D, Wagner, RJ, Cronenwett, J.L., and Powell, R.J., 1999, A caspase inhibitor decreases oxidized low-density lipoprotein-induced apoptosis in bovine endothelial cells, *J. Surg. Res.* **85**: 323-330.

Fossel, E.T., Zanella, C.L., Fletcher, J.G., and Hui, K.K., 1994, Cell death induced by peroxidized low-density lipoprotein: endopepsis, *Cancer Res.* **54**: 1240- 1248.

Galle, J., Schneider, R., Heinloth, A., Dimmeler, S., and Heermeier, K., 1999, Lp(a) and LDL induce apoptosis in human endothelial cells and in rabbit aorta: role of oxidative stress, *Kidney Int.* **55**: 1450-1461.

Geng, Y.J., and Libby, P., 1995, Evidence for apoptosis in advanced human atheroma. Colocalization with interleukin-1 beta-converting enzyme, *Am. J. Pathol.* **147**: 251-266.

Geng, Y.J., Henderson, L.E., levesque, E.B., Muszynski, M., and Libby, P., 1997, Fas is expressed in human atherosclerotic intima and promotes apoptosis of cytokine-primed human vascular smooth muscle cells, *Arterioscler. Thromb. Vasc. Biol.* **17**: 2200-2208.

Girotti, A.W., 1998, Lipid hydroperoxide generation, turnover, and effector action in biological systems, *J. Lipid Res.* **39**: 1529-1542.

Gomez-Munoz, A., Martens, J.S., and Steinbrecher, U.P., 2000, Stimulation of phospholipase D activity by oxidized LDL in mouse peritoneal macrophages, *Arterioscler. Thromb. Vasc. Biol.* **20**: 135-143.

Griffin, D.S., and Segall, H.J., 1987, Role of cellular calcium homeostasis in toxic liver injury induced by the pyrrolizidine alkaloid senecionine and the alkenal trans-4-OH-2-hexenal, *J. Biochem. Toxicol. Fall-Winter* **2**: 155-167.

Hajjar, D.P., and Haberland, M.E., 1997, Lipoprotein trafficking in vascular cells. Molecular Trojan horses and cellular saboteurs, *J Biol Chem.* **272**: 22975-229778.

Hajjar, D.P., 2000, Oxidized lipoproteins and infectious agents. Are they in collusion to accelerate atherogenesis?, *Arterioscler. Thromb. Vasc. Biol.* **20**: 1421-1422.

Hakkinen, T, Luoma, JS, Hiltunen, MO, Macphee, CH, Milliner, KJ, Patel, L, Rice, SQ, Tew, DG, Karkola, K, and Yla-Herttuala, 1999, Lipoprotein-associated phospholipase A(2), platelet-activating factor acetylhydrolase, is expressed by macrophages in human and rabbit atherosclerotic lesions, *Arterioscler. Thromb. Vasc. Biol.* **19**: 2909-2917.

Han, DK, Haudenschild, CC, Hong, MK, Tinkle, BT, Leon, MB, and Liau, G., 1995 , Evidence for apoptosis in human atherogenesis and in a rat vascular injury model, *Am. J. Pathol.* **147**: 267-77.

Hannun, YA, and Luberto, C, 2000, Ceramide in the eukaryotic stress response, *Trends Cell. Biol.* **10**: 73-80.

Hansson, G.K., 1997, Cell-mediated immunity in atherosclerosis. *Curr Opin Lipidol.* **8**: 301-311.

Harada, K., Ishibashi, S., Miyashita, T., Osuga, J., Yagyu, H., Ohashi, K., Yazaki, Y., and Yamada, N., 1997, Bcl-2 inhibits oxysterol-induced apoptosis through suppressing CPP32-mediated pathway, *FEBS Lett.* **411**: 63-66.

Harada-Shiba, M., Kinoshita, M., Kamido, H., and Shimokado, K., 1998, Oxidized low density lipoprotein induces apoptosis in cultured human umbilical vein endothelial cells by common and unique mechanisms, *J. Biol. Chem.* **273**: 9681-9687.

Hashizume, H., Hoque, A.N., Magishi, K., Hara, A., and Abiko, Y., 1997, A new approach to the development of anti-ischemic drugs. Substances that counteract the deleterious effect of lysophosphatidylcholine on the heart, *Jpn. Heart J.* **38**: 11-25.

Haynes, M.P., Russell, K.S., and Bender, J.R., 2000, Molecular mechanisms of estrogen actions on the vasculature, *J. Nucl. Cardiol.* **7**: 500-508.

Heermeier, K., Leicht, W., Palmetshofer, A., Ullrich, M., Wanner, C., Galle, J., 2001, Oxidized LDL Suppresses NF-kappaB and Overcomes Protection from Apoptosis in Activated Endothelial Cells. *J. Am. Soc. Nephrol.* **12**: 456-463.

Heery, J.M., Kozak, M., Stafforini, D.M., Jones, D.A., Zimmerman, G.A., McIntyre, T.M., and Prescott, S.M., 1995, Oxidatively modified LDL contains phospholipids with platelet-activating factor-like activity and stimulates the growth of smooth muscle cells, *J. Clin. Invest.* **96**: 2322-2330.

Heller, F.R., Descamps, O., and Hondekijn, J.C., 1998, LDL oxidation: therapeutic perspectives, *Atherosclerosis* **137** Suppl: S25-31.

Hessler, J.R., Robertson, A.L. Jr, and Chisolm, G.M., 1979, LDL-induced cytotoxicity and its inhibition by HDL in human vascular smooth muscle and endothelial cells in culture, *Atherosclerosis* **32**: 213-229.

Horkko, S., Binder, C.J., Shaw, P.X., Chang, M.K., Silverman, G., Palinski, W., and Witztum, J.L., 2000, Immunological responses to oxidized LDL, *Free Radic. Biol. Med.* **28**: 1771-1779.

Hsieh, C.C., Yen, M.H., Yen, C.H., Lau, Y.T., 2001, Oxidized low density lipoprotein induces apoptosis via generation of reactive oxygen species in vascular smooth muscle cells, *Cardiovasc. Res.* **49**: 135-145.

Huang, A., Li, C., Kao, R.L., and Stone, W.L., 1999*a*, Lipid hydroperoxides inhibit nitric oxide production in RAW264.7 macrophages, *Free Radic. Biol. Med.* **26**: 526-537.

Huang, Y.H., Schafer-Elinder, L., Wu, R., Claesson, H.E., and Frostegard J., 1999*b*, Lysophosphatidylcholine (LPC) induces proinflammatory cytokines by a platelet-activating factor (PAF) receptor-dependent mechanism, *Clin. Exp. Immunol.* **116**: 326-331.

Hughes, H., Mathews, B., Lenz, M.L., and Guyton, J.R., 1994, Cytotoxicity of oxidized LDL to porcine aortic smooth muscle cells is associated with the oxysterols 7-ketocholesterol and 7-hydroxycholesterol, *Arterioscler. Thromb.* **14**: 1177- 1185.

Janabi, M., Yamashita, S., Hirano, K., Sakai, N., Hiraoka, H., Matsumoto, K., Zhang, Z., Nozaki, S., and Matsuzawa, Y., 2000, Oxidized LDL-induced NF-kappa B activation and subsequent expression of proinflammatory genes are defective in monocyte-derived macrophages from CD36-deficient patients, *Arterioscler. Thromb. Vasc. Biol.* **20**: 1953-1960.

Jovinge, S., Ares, M.P., Kallin, B., and Nilsson, J., 1996, Human monocytes/macrophages release TNF-alpha in response to Ox-LDL, *Arterioscler. Thromb. Vasc. Biol.* **16**: 1537-1539.

Jurgens, G., Hoff, H.F., Chisolm, G.M., and Esterbauer, H., 1987, Modification of human serum low density lipoprotein by oxidation--characterization and pathophysiological implications, *Chem. Phys. Lipids* **45**: 315-336.

Kalayoglu, M.V., and Byrne, G.I., 1998*a*, A Chlamydia pneumoniae component that induces macrophage foam cell formation is chlamydial lipopolysaccharide, *Infect. Immun.* **66**: 5067-5072.

Kalayoglu, M.V., Byrne, G.I., 1998*b*, Induction of macrophage foam cell formation by Chlamydia pneumoniae, *J. Infect. Dis.* **177**: 725-729.

Kalayoglu, M.V., Hoerneman, B., LaVerda, D., Morrison, S.G., Morrison, R.P., Byrne, G.I., 1999, Cellular oxidation of low-density lipoprotein by Chlamydia pneumoniae, *J. Infect. Dis.* **180**: 780-790.

Kapiotis, S., Hermann, M., Held, I., Seelos, C., Ehringer, H., and Gmeiner, B.M., 1997, Genistein, the dietary-derived angiogenesis inhibitor, prevents LDL oxidation and protects endothelial cells from damage by atherogenic LDL, *Arterioscler. Thromb. Vasc. Biol.* **17**: 2868-2874.

Kass, G.E., and Orrenius, S., 1999, Calcium signaling and cytotoxicity, *Environ. Health Perspect.* **107** Suppl 1: 25-35.

Keller, J.N., Hanni, K.B., Markesbery, W.R., 1999, Oxidized low-density lipoprotein induces neuronal death: implications for calcium, reactive oxygen species, and caspases, *Neurochem.* **72**: 2601-2609.

Khovidhunkit, W., Memon, R.A., Feingold, K.R., Grunfeld, C., 2000, Infection and inflammation-induced proatherogenic changes of lipoproteins. *J. Infect. Dis.* **181** Suppl 3: S462-472.

Kinscherf, R., Claus, R., Wagner, M., Gehrke, C., Kamencic, H., Hou, D., Nauen, O., Schmiedt, W., Kovacs, G., Pill, J., Metz, J., and Deigner, H.P., 1998, Apoptosis caused by oxidized LDL is manganese superoxide dismutase and p53 dependent, *FASEB J.* **12**: 461-467.

Kinscherf, R., Wagner, M., Kamencic, H., Bonaterra, G.A., Hou, D., Schiele, R.A., Deigner, H.P., Metz, J., 1999, Characterization of apoptotic macrophages in atheromatous tissue of humans and heritable hyperlipidemic rabbits, *Atherosclerosis* **144**: 33-39.

Ko, S., Kwok, T.T., Fung, K.P., Choy, Y.M., Lee, C.Y., and Kong, S.K., 2000, Slow rise of Ca2+ and slow release of reactive oxygen species are two cross-talked events important in tumour necrosis factor-alpha-mediated apoptosis, *Free Radic. Res.* **33**: 295-304.

Kockx, M.M., De Meyer, G.R., Buyssens, N., Knaapen, M.W., Bult, H., Herman, A.G.., 1998*a*, Cell composition, replication, and apoptosis in atherosclerotic plaques after 6 months of cholesterol withdrawal, *Circ. Res.* **83**: 378-387.

Kockx, M.M., De Meyer, G.R., Muhring, J., Jacob, W., Bult, H., and Herman, A.G.., 1998*b*, Apoptosis and related proteins in different stages of human atherosclerotic plaques, *Circulation* **97**: 2307-2315.

Kristal, B.S., Park, B.K., Yu, B.P., 1996, 4-Hydroxyhexenal is a potent inducer of the mitochondrial permeability transition, *J. Biol. Chem*. **271**: 6033-6038.

Kwiterovich, P.O. Jr, 1998, The antiatherogenic role of high-density lipoprotein cholesterol, *Am. J. Cardiol.* **82**: 13-21.

Li, D., Yang, B., and Mehta, J.L., 1998a, OxLDL induces apoptosis in human coronary endothelial cells: role of PKC, PTK, bcl-2, and Fas, *Am. J. Physiol.* **275**: H568-576.

Li, D., Saldeen, T., Mehta, J.L., 1999, gamma-tocopherol decreases ox-LDL-mediated activation of nuclear factor-kappaB and apoptosis in human coronary artery endothelial cells. *Biochem. Biophys. Res. Commun.* **259**: 157-161.

Li, D., and Mehta, J.L., 2000*a*, Upregulation receptor for oxidized LDL (LOX-1) by oxidized LDL and implications in apoptosis of human coronaary artery endothelial cells, *Arterioscle. Thromb. Vasc. Biol.* **20**: 1116-1122.

Li, D., Saldeen, T., and Mehta, J.L., 2000*b*, Effects of alpha-tocopherol on ox-LDL-mediated degradation of IkappaB and apoptosis in cultured human coronary artery endothelial cells, *J. Cardiovasc. Pharmacol.* **36**: 297-301.

Li, W., Yuan, X.M., Brunk, U.T., 1998b, OxLDL-induced macrophage cytotoxicity is mediated by lysosomal rupture and modified by intralysosomal redox-active iron, *Free Radic. Res.* **29**: 389-398.

Libby, P., Geng, Y.J., Aikawa, M., Schoenbeck, U., Mach, F., Clinton, S.K., Sukhova, G.K., and Lee, R.T., 1996, Macrophages and atherosclerotic plaque stability. *Curr. Opin. Lipidol.*. **7**: 330-335.

Liu, J., Farmer, J.D. Jr, Lane, W.S., Friedman, J., Weissman, I., and Schreiber, S.L, 1991, Calcineurin is a common target of cyclophilin-cyclosporin A and FKBP- FK506 complexes, *Cell* **66**: 807-815.

Liu, W., Kato, M., Akhand, A.A., Hayakawa, A., Suzuki, H., Miyata T., Kurokawa, K., Hotta, Y., Ishikawa, N., and Nakashima, I., 2000, 4-hydroxynonenal induces a cellular redox status-related activation of the caspase cascade for apoptotic cell death, *J. Cell. Sci.* **113**: 635-461.

Lizard, G., Gueldry, S., Sordet, O., Monier, S., Athias, A., Miguet, C., Bessede, G., Lemaire, S., Solary, E., and Gambert, P., 1998, Glutathione is implied in the control of 7-ketocholesterol-induced apoptosis, which is associated with radical oxygen species production, *FASEB J.* **15**: 1651-1663.

Lizard, G., Monier, S., Cordelet, C., Gesquiere, L., Deckert, V., Gueldry, S., Lagrost, L., and Gambert, P., 1999, Characterization and comparison of the mode of cell death, apoptosis versus necrosis, induced by 7beta-hydroxycholesterol and 7-ketocholesterol in the cells of the vascular wall, *Arterioscler. Thromb. Vasc. Biol.* **19**: 1190-1200.

Mabile, L., Fitoussi, G., Periquet, B., Schmitt, A., Salvayre, R., and Negre-Salvayre, A., 1995, alpha-Tocopherol and trolox block the early intracellular events (TBARS and calcium rises) elicited by oxidized low density lipoproteins in cultured endothelial cells, *Free Radic. Biol. Med.* **19**: 177-187.

Mallat, Z., Ohan, J., Leseche, G., and Tedgui, A., 1997, Colocalization of CPP-32 with apoptotic cells in human atherosclerotic plaques, *Circulation* **96**: 424-428.

Mallat, Z., and Tedgui, A., 1998, Rate of apoptosis in human atherosclerosis, *Am. J. Pathol.* **153:** 1320-1321.

Mallat, Z., Hugel, B., Ohan, J., Leseche, G., Freyssinet, J.M., and Tedgui, A., 1999, Shed membrane microparticles with procoagulant potential in human atherosclerotic plaques: a role for apoptosis in plaque thrombogenicity, *Circulation* **99**: 348-353.

Mallat, Z., and Tedgui, A., 2000, Apoptosis in the vasculature: mechanisms and functional importance, *Br. J. Pharmacol.* **130**: 947-962.

Malorni, W., Straface, E., Di Genova, G., Fattorossi, A., Rivabene, R., Camponeschi, B., Masella, R., and Viora, M., 1997, Oxidized low-density lipoproteins affect natural killer cell activity by impairing cytoskeleton function and altering the cytokine network, *Exp. Cell Res.* **236**: 436-445.

Martinez-Pomares, L., Platt, N., McKnight, A.J., da Silva, R.P., and Gordon, S., 1996, Macrophage membrane molecules: markers of tissue differentiation and heterogeneity, *Immunobiology* **195**: 407-416.

Maziere, C., Alimardani, G., Dantin, F., Dubois, F., Conte, M.A., and Maziere, J.C., 1999, Oxidized LDL activates STAT1 and STAT3 transcription factors: possible involvement of reactive oxygen species, *FEBS Lett.* **448**: 49-52.

Maziere, C., Meignotte, A., Dantin, F., Conte, M.A., and Maziere, J.C., 2000, Oxidized LDL induces an oxidative stress and activates the tumor suppressor p53 in MRC5 human fibroblasts, *Biochem. Biophys. Res. Commun.* **276**: 718-723.

Meilhac, O., Escargueil-Blanc, I., Thiers, J.C., Salvayre, R., and Negre-Salvayre, A., 1999, Bcl-2 alters the balance between apoptosis and necrosis, but does not prevent cell death induced by oxidized low density lipoproteins, *FASEB J.* **13**: 485-494.

Memon, R.A., Staprans, I., Noor, M., Holleran, W.M., Uchida, Y., Moser, A.H., Grunfeld, and C. Feingold, K.R., 2000, Infection and inflammation induce LDL oxidation *in vivo*, *Arterioscler. Thromb. Vasc. Biol.* **20**: 1536-1542.

Mitchinson, M.J., Hardwick, S.J., and Bennett, M.R., 1996, Cell death in atherosclerosis, *Curr. opin. lipidol.* **7**: 324-329.

Moazed, T.C., Campbell, L.A., Rosenfeld, M.E., Grayston, J.T., and Kuo, C.C.., 1999, Chlamydia pneumoniae infection accelerates the progression of atherosclerosis in apolipoprotein E-deficient mice, *J. Infect. Dis.* **180**: 238-241.

Murao, K., Terpstra, V., Green, S.R., Kondratenko, N., Steinberg, D., and Quehenberger, O., 1997, Characterization of CLA-1, a human homologue of rodent scavenger receptor BI, as a receptor for high density lipoprotein and apoptotic thymocytes, *J. Biol. Chem.* **272**: 17551-17557.

Murohara, T., Kugiyama, K., Ohgushi, M., Sugiyama, S., Ohta, Y., and Yasue, H., 1994, LPC in oxidized LDL elicits vasocontraction and inhibits endothelium-dependent relaxation, *Am. J. Physiol.* **267**: H2441-2449.

Naito, M., Yamada, K., Hayashi, T., Asai, K., Yoshimine, N., Iguchi, A., 1994, Comparative toxicity of oxidatively modified low-density lipoprotein and lysophosphatidylcholine in cultured vascular endothelial cells, *Heart Vessels* **9**: 183-187.

Nakata, A., Nakagawa, Y., Nishida, M., Nozaki , S., Miyagawa, J., Nakagawa, T., Tamura, R., Matsumoto, K., Kameda-Takemura, K., Yamashita, S., Matsuzawa, Y., 1999, CD36, a novel receptor for oxidized low-density lipoproteins, is highly expressed on lipid-laden macrophages in human atherosclerotic aorta, *Arterioscler. Thromb. Vasc. Biol.* **19**: 1333-1339.

Napoli, C., Quehenberger, O., De Nigris, F., Abete, P., Glass, C.K., and Palinski, W., 2000, Mildly oxidized low density lipoprotein activates multiple signaling pathways in human coronary cells, *FASEB J.* **14**: 1996-2007.

Negre-Salvayre, A., and Salvayre, R., 1992, Protection by Ca^{2+} channel blockers (nifedipine, diltiazem and verapamil) against the toxicity of oxidized low density lipoprotein to cultured lymphoid cells, *Br. J. Pharmacol.* **107**: 738-744.

Netea, M.G., Dinarello, C.A., Kullberg, B.J., Jansen, T., Jacobs, L., Stalenhoef, A.F., Van Der Meer, J.W., 2000, Oxidation of low-density lipoproteins by acellular components of Chlamydia pneumoniae, *J. Infect. Dis.* **181**: 1868-1870.

Niemann-Jonsson, A., Dimayuga, P., Jovinge, S., Calara, F., Ares, M.P., Fredrikson, G.N., and Nilsson, J., 2000, Accumulation of LDL in rat arteries is associated with activation of tumor necrosis factor-alpha expression, *Arterioscler. Thromb. Vasc. Biol.* **20**: 2205-2211.

Nilsson, J., Dahlgren, B., Ares, M., Westman, J., Hultgardh Nilsson, A., Cercek, B., and Shah, P.K., 1998, Lipoprotein-like phospholipid particles inhibit the smooth muscle cell cytotoxicity of lysophosphatidylcholine and platelet-activating facto, *Arterioscler. Thromb. Vasc. Biol.* **18**: 13-19.

Ohlsson, B.G., Englund, M.C., Karlsson, A.L., Knutsen, E., Erixon, C., Skribeck, H., Liu, Y., Bondjers, G., and Wiklund, O., 1996, Oxidized low density lipoprotein inhibits lipopolysaccharide-induced binding of nuclear factor-kappaB to DNA and the subsequent expression of tumor necrosis factor-alpha and interleukin-1beta in macrophages, *J. Clin. Invest.* **98**: 78-89.

Okura, Y., Brink, M., Itabe, H., Scheidegger, K.J., Kalangos, A., and Delafontaine, P., 2000, Oxidized low-density lipoprotein is associated with apoptosis of vascular smooth muscle cells in human atherosclerotic plaques, *Circulation* **102**: 2680-2686.

Page, S., Fischer, C., Baumgartner, B., Haas, M., Kreusel, U., Loidl, G., Hayn, M., Ziegler-Heitbrock, H.W., Neumeier, D., and Brand, K., 1999, 4-Hydroxynonenal prevents NF-kappaB activation and tumor necrosis factor expression by inhibiting IkappaB phosphorylation and subsequent proteolysis, *J. Biol. Chem.* **274**: 11611-11618.

Pearson, A.M., 1996, Scavenger receptors in innate immunity. *Curr. Opin. Immunol.* **8**: 20-28.

Pirillo, A., Zhu, W., Norata, G.D., Zanelli, T., Barberi, L., Roma, P., and Catapano, A.L., 2000, Oxidized lipoproteins and endothelium, *Clin. Chem. Lab. Med.* **38**: 155-160.

Rader, D.J., and Dugi, K.A., 2000, The endothelium and lipoproteins: insights from recent cell biology and animal studies, *Semin. Thromb. Hemost.***26**: 521-528.

Ramasamy, S., Boissonneault, G.A., and Hennig, B., 1992, Oxysterol-induced endothelial cell dysfunction in culture, *J. Am. Coll. Nutr.* **11**: 532-538.

Ramprasad, M.P., Terpstra, V., Kondratenko, N., Quehenberger, O., Steinberg, D., 1996, Cell surface expression of mouse macrosialin and human CD68 and their role as macrophage receptors for oxidized low density lipoprotein, *Proc. Natl. Acad. Sci. U S A* **93**: 14833-14838.

Renaud, S., and Ruf, J.C., 1994, The French paradox: vegetables or wine, *Circulation*, **90**: 3118-3119.

Romero, F.J., Bosch-Morell, F., Romero, M.J., Jareno, E.J., Romero, B., Marin, N., and Roma, J., 1998, Lipid peroxidation products and antioxidants in human disease, *Environ. Health Perspect.* **106** Suppl 5: 1229-1234

Rosenblat, M., and Aviram, M., 1998, Macrophage glutathione content and glutathione peroxidase activity are inversely related to cell-mediated oxidation of LDL: in vitro and *in vivo* studies, *Free Radic. Biol. Med.* **24**: 305-317.

Rusinol, A.E., Yang, L., Thewke, D., Panini, S.R., Kramer, M.F., and Sinensky, M.S., 2000, Isolation of a somatic cell mutant resistant to the induction of apoptosis by oxidized low density lipoprotein, *J. Biol. Chem.* **275**: 7296-7303.

Santanam, N., Augé, N., Zhou, M., Keshava, C., and Parthasarathy, S., 1999, Overexpression of human catalase gene decreases oxidized lipid-induced cytotoxicity in vascular smooth muscle cells, *Arterioscler. Thromb. Vasc. Biol.* **19**: 1912-1917.

Sata, M., and Walsh, K., 1998*a*, Endothelial cell apoptosis induced by oxidized LDL is associated with down-regulation of the cellular caspase inhibitor FLIP, *J. Biol. Chem.* **273**: 33103-33106.

Sata, M. , and Walsh, K., 1998*b*, Oxidized LDL activates Fas-mediated endothelial apoptosis, *J. Clin. Invest.* **102**: 1682-1689.

Schneider, D.B., Vassali, G., Wen, S., Driscoll, R.M., Sassani, A.B., DeYoung, M.B., Linnemann, R., Virmani, R., and Dichek, D.A., 2000, Expression of Fas ligand in arteries of hyper cholesterolemis rabbits accelerates athesclerotic lesion formation. *Arterioscler. Thromb. Vasc. Biol.* **20**: 298-308.

Sevanian, A., Shen, L., and Ursini, F., 2000, Inhibition of LDL oxidation and oxidized LDL-induced cytotoxicity by dihydropyridine calcium antagonists, *Pharm. Res.* **17**: 999-1006.

Shaw, P.X., Horkko, S., Chang, M.K., Curtiss, L.K., Palinski, W., Silverman, G.J., and Witztum, J.L., 2000, Natural antibodies with the T15 idiotype may act in atherosclerosis, apoptotic clearance, and protective immunity, *J. Clin. Invest.* **105**: 1731-1740.

Shi, C., Lee, W.S., Russell, M.E., Zhang, D., Fletcher, D.L, Newell, J.B., and Haber, E., 1997, Hypercholesterolemia exacerbates transplant arteriosclerosis via increased neointimal smooth muscle cell accumulation: studies in apolipoprotein E knockout mice, *Circulation* **96**: 2722-2728.

Siess, W., Zangl, K.J., Essler, M., Bauer, M., Brandl, R., Corrinth, C., Bittman, R., Tigyi, G.,and Aepfelbacher, M., 1999, Lysophosphatidic acid mediates the rapid activation of platelets and endothelial cells by mildly oxidized low density lipoprotein and accumulates in human atherosclerotic lesions, *Proc. Natl. Acad. Sci. U S A* **96**: 6931-6936.

Siess, W., Essler, M., and Brandl, R., 2000, Lysophosphatidic acid and sphingosine 1-phosphate: two lipid villains provoking cardiovascular diseases?, *IUBMB Life* **49**: 167-171.

Siow, R.C., Richards, J.P., Pedley, K.C., Leake, D.S., and Mann, G.E., 1999, Vitamin C protects human vascular smooth muscle cells against apoptosis induced by moderately oxidized LDL containing high levels of lipid hydroperoxides, *Arterioscler. Thromb. Vasc. Biol.* **19**: 2387-2394.

Sniderman, A.D., Zhang, X.J., and Cianflone, K., 2000, Governace of the concentration of plasma LDL: a reevaluation of the LDL receptor paragdim, *Atherosclerosis* **148:** 215-229.

Soh, Y., Jeong, K.S., Lee, I.J., Bae, M.A., Kim, Y.C., and Song, B.J., 2000, Selective activation of the c-Jun N-terminal protein kinase pathway during 4-hydroxynonenal-induced apoptosis of PC12 cells, *Mol. Pharmacol.* **58**: 535- 541.

Spyridopoulos, I., Wischhusen, J., Rabenstein, B., Mayer, P., Axel, D.I., Frohlich, K.U., and Karsch, K.R., 2001, Alcohol Enhances Oxysterol-Induced Apoptosis in Human Endothelial Cells by a Calcium-Dependent Mechanism. Arterioscler. *Thromb. Vasc. Biol.* **21**: 439-444.

Steinberg, D., 1997, Oxidative modification of LDL and atherogenesis, *Circulation* **95**: 1062-1071.

Steinbrecher, U.P., 1999, Receptors for oxidized low density lipoprotein, *Biochim. Biophys. Acta.* **1436:** 279-298.

Sudoh, N., Toba, K., Akishita, M., Ako, J., Hashimoto, M., Iijima, K., Kim, S., Liang, .Q., Ohike, Y., Watanabe, T., Yamazaki, I., Yoshizumi, M., Eto, M., and Ouchi, Y., 2001, Estrogen prevents oxidative stress-induced endothelial cell apoptosis in rats, *Circulation* **103**: 724-729.

Sugawa, M., Ikeda, S., Kushima, Y., Takashima, Y., and Cynshi, O., 1997, Oxidized low density lipoprotein caused CNS neuron cell death, *Brain Res.* **761**: 165-172.

Suzuki, Y.J., Forman H.J., and Sevanian, A., 1997, Oxidants as stimulators of signal transduction, *Free Radic. Biol. Med.* **22**: 269-285.

Tashiro, K., Makita, Y., Shike, T., Shirato, I., Sato, T., Cynshi, O., and Tomino, Y., 1999, Detection of cell death of cultured mouse mesangial cells induced by oxidized low-density lipoprotein, *Nephron* **82**: 51-58.

Thomas, S.R., and Stocker, R., 2000, Molecular action of vitamin E in lipoprotein oxidation: implications for atherosclerosis, *Free Radic. Biol. Med.* **28**: 1795-1805.

Van der Vliet, A., Van der Aar, E.M., and Bast, A., 1991, The lipid peroxidation product 4-hydroxy-2,3-trans-1 nonenal decreases rat intestinal smooth muscle function in-vitro by alkylation of sulphydryl groups, J. *Pharm. Pharmacol.* **43**: 515-517.

Vieira, O., Escargueil-Blanc, I., Meilhac, O., Basile, J.P., Laranjinha, J., Almeida, L., Salvayre, R., and Negre-Salvayre, A., 1998, Effect of dietary phenolic compounds on apoptosis of human cultured endothelial cells induced by oxidized LDL, *Br. J. Pharmacol.* **123**: 565-573.

Vieira, O., Escargueil-Blanc, I., Jurgens, G., Borner, C., Almeida, L., Salvayre, R., and Negre-Salvayre, A., 2000, Oxidized LDLs alter the activity of the ubiquitin- proteasome pathway: potential role in oxidized LDL-induced apoptosis, *FASEB J.* **14**: 532-542.
Von Eckardstein, A., Nofer, J.R., and Assmann, G., 2001, High density lipoproteins and arteriosclerosis. Role of cholesterol efflux and reverse cholesterol transport, *Arterioscler. Thromb. Vasc. Biol.* **21**: 13-27.
Vreugdenhil, A.C., Snoek, A.M., van 't Veer, C., Greve, J.W., and Buurman, W.A., 2001, LPS-binding protein circulates in association with apoB-containing lipoproteins and enhances endotoxin-LDL/VLDL interaction, *J. Clin. Invest.* **107**: 225-234.
Walter, J., Grunberg, J., and Schindzielorz, A., and Haass, C., 1998, Proteolytic fragments of the Alzheimer's disease associated presenilins-1 and –2 are phosphorylated *in vivo* by distinct cellular mechanisms, *Biochemistry* **37**: 5961-5967.
Wang, T.G., Gotoh, Y., Jennings, M.H., Rhoads, C.A., and Aw T.Y., 2000, Lipid hydroperoxide-induced apoptosis in human colonic CaCo-2 cells is associated with an early loss of cellular redox balance, *FASEB J* .**14**: 1567- 1576.
Wedworth, S.M., and Lynch, S., 1995, Dietary flavonoids in atherosclerosis prevention, *Pharmacother.* **29**: 627-628.
Wintergerst, E.S., Jelk, J., Rahner, C., Asmis, R., 2000, Apoptosis induced by oxidized low density lipoprotein in human monocyte-derived macrophages involves CD36 and activation of caspase-3, *Eur. J. Biochem.* **267**: 6050-6059.
Wolin, M.S., 2000, Interactions of oxidants with vascular signaling systems, *Arterioscler. Thromb. Vasc. Biol.* **20**: 1430-1442.
Wu, R., Giscombe, R., Holm, G., Lefvert, A.K., 1996, Induction of human cytotoxic T lymphocytes by oxidized low density lipoproteins, *Scand. J. Immunol.* **43**: 381-384.
Yin, J., Chaufour, X., McLachlan, C., McGuire, M., White, G., King, N., and Hambly, B., 2000, Apoptosis of vascular smooth muscle cells induced by cholesterol and its oxides *in vitro* and *in vivo*, *Atherosclerosis* **148**: 365-374.
Zhang, J., Xue, Y., Jondal, M., and Sjovall, J., 1997, 7alpha-Hydroxylation and 3-dehydrogenation abolish the ability of 25-hydroxycholesterol and 27- hydroxycholesterol to induce apoptosis in thymocytes, *Eur. J. Biochem.* **247**: 129-135.
Zhao, B., Dierichs, R., Miller, F.N., and Dean, W.L., 1996, Oxidized low density lipoprotein inhibits platelet plasma membrane Ca(2+)-ATPase, *Cell Calcium* **19**: 453-458.
Zhu, X., Bonet, B., Gillenwater, H., and Knopp, R.H., 1999, Opposing effects of estrogen and progestins on LDL oxidation and vascular wall cytotoxicity: implications for atherogenesis, *Proc. Soc. Exp. Biol.* Med. **222**: 214-221.

Chapter 8

Induction of Apoptosis by Redox-Cycling Quinones

Karin Öllinger and Katarina Kågedal
Division of Pathology II, Faculty of Health Sciences, Linköping University, Linköping, Sweden

1. INTRODUCTION

Studies have shown that apoptosis can be induced in many different ways, whereas the morphological features associated with such cell death are highly conserved. This implies the coexistence of multiple signaling pathways that converge upstream of a common sequence of events and culminate in apoptosis. There is growing evidence that the final outcome of an apoptosis-triggering signal is dictated by the ability of a cell to maintain an appropriate oxidant-antioxidant balance and metabolism of redox cycling compounds, such as quinones, will alter the intracellular redox balance. Quinones comprise a large group of naturally occurring and synthetic compounds that have functional, toxicological, mutagenic, and antitumor actions. In this chapter, we discuss the mechanisms of quinone-induced cell damage and death, and we also summarize results in the literature indicating that lysosomal enzymes may take part in quinone-elicited apoptosis.

2. QUINONES

Quinones are widespread in nature (Thomson, 1971) and have a variety of functions in the life cycles of most kinds of living organisms. These diketones are found in higher plants, fungi, bacteria, and throughout the animal kingdom, and they play a central role in many biosynthetic processes that involve electron transport, such as cellular respiration (ubiquinone) and photosynthesis (plastoquinone). Vitamin K is an important factor in blood

Subcellular Biochemistry, Volume 36: Phospholipid Metabolism in Apoptosis.
Edited by Quinn and Kagan. Kluwer Academic/Plenum Publishers, New York, 2002

coagulation, and topaquinone has been identified as a cofactor of amine oxidases (Dooley, 1999). Both plants and insects use quinones as defensive compounds, examples of which are juglone in the walnut tree and hydroquinone in bombardier beetles. Figure 1 exemplifies different quinone structures.

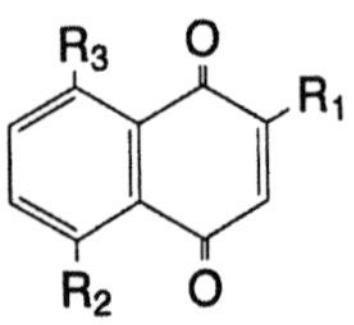

Menadion R_1=CH_3, R_2=H, R_3=H
Naphthazarin R_1=H, R_2=OH, R_3=OH
Juglone R_1=H, R_2=OH, R_3=H

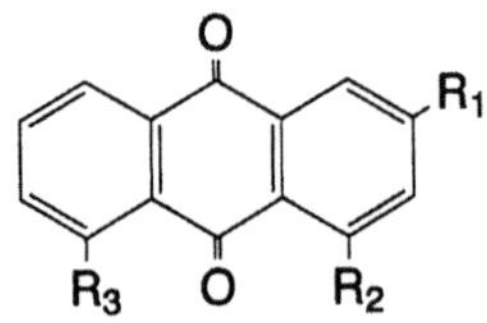

Danthron R_1=H, R_2=OH, R_3=OH
Rhein R_1=COOH, R_2=OH, R_3=OH
Chrysophanol R_1=CH_3, R_2=OH, R_3=OH

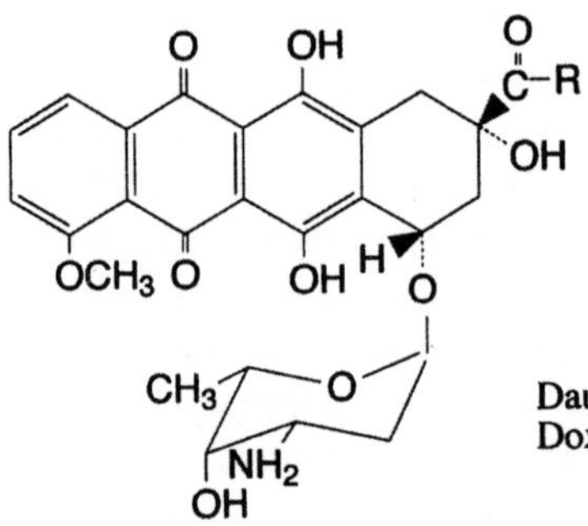

Daunorubicin R=CH_3
Doxorubicin R=CH_2OH

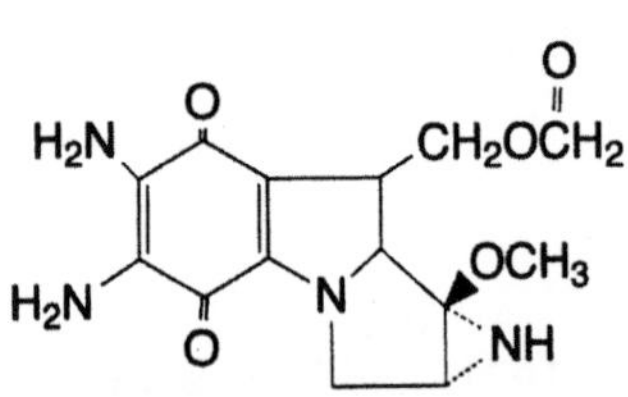

Mitomycin C

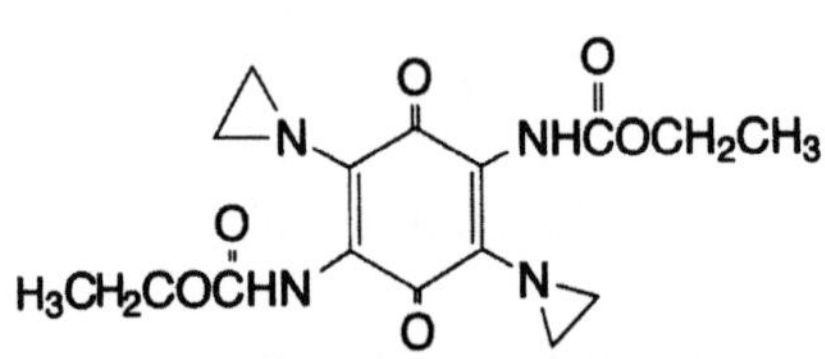

Diaziquinone

Figure 1. Examples of quinoid structures.

Different strains of *Streptomyces* produce several substances that exhibit quinoid structure and have been shown to exert antibiotic and antineoplastic effects in pre-clinical tests. The largest group of these compounds is the anthracyclines, which are used extensively in the treatment of several different forms of cancer. The first anthracycline to be identified was daunorubicin, which was isolated in 1963, followed by aclacinomycin A and doxorubicin (Adriamycin®), and the antitumor action of these drugs was subsequently demonstrated in animal systems. Other quinoid antineoplastics generated by *Streptomyces* are mitomycin C and streptonigrin. Clinical use of anthracyclines is highly limited, because these substances induce cumulative heart toxicity that leads to cardiomyopathy and arrhythmia. Accordingly, the goal of chemists has been to synthesize quinoids that have substantial antineoplastic activity but cause fewer side effects. The synthetic anticancer drugs that contain quinone moieties and are presently used as pharmaceuticals include epirubicin, idarubicin, mitoxantron, and aziridinyl benzoquinones.

2.1 Intracellular metabolism of quinones

To understand the overall biological activity of quinoid compounds, it is necessary to study the chemical properties of these substances. Inasmuch as the cellular damage that is induced by quinones resembles that seen after radiolysis, the most prominent reactions involving quinones are probably DNA damage and generation of oxygen free radicals.

Quinones and other xenobiotics are metabolized primarily in the liver, and the end products are water-soluble compounds that are excreted from the body. The first step in metabolism is usually an oxidative process (phase I detoxification), after which further oxidation and conjugation can occur (phase II detoxification). Although those reactions normally yield easily excreted products, some steps in the metabolic pathway may generate metabolites that are more toxic than the parent compound.

2.1.1 Redox cycling

Quinones can accept one or two electrons to form the semiquinone anion ($Q^{\bullet -}$) and the hydroquinone dianion (Q^{2-}). Single-electron reduction of a quinone is catalyzed by flavoenzymes with relatively low substrate selectivity (Kappus, 1986), for instance NADPH cytochrome P-450 reductase (E.C. 1.6.2.3), NADPH cytochrome b_5 reductase (E.C. 1.6.2.2), and NADPH ubiquinone oxidoreductase (E.C. 1.6.5.3). The rate of reduction depends on several interrelated chemical properties of a quinone, including the single-electron reduction potential, as well as the number, position, and chemical characteristics of the substituent(s). The flavoenzyme DT-diphorase (NAD(P)H: quinone acceptor oxidoreductase; E.C. 1.6.99.2) catalyzes the two-electron reduction of a quinone to a hydroquinone.

A semiquinone can be readily oxidized to the parent compound by molecular oxygen and can then re-enter the reductase-catalyzed reaction. The enzymatic reduction and autoxidation of quinones under aerobic conditions generates superoxide anion radicals, and this process is known as redox cycling (Figure 2). Hydroquinones are less prone to transfer electrons to oxygen, because the second-electron potential is often too high.

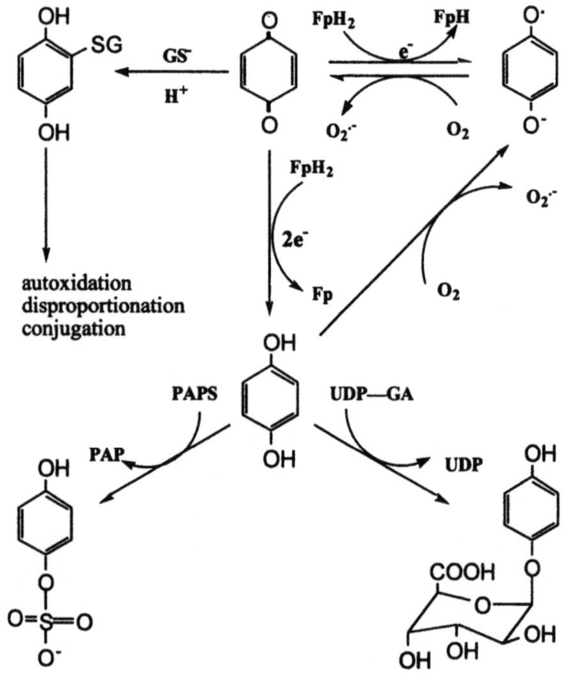

Figure 2. Possible reactions during intracellular metabolism of a quinone.

The superoxide anion radical ($O_2^{\bullet -}$) is produced when oxygen accepts one electron. This radical has a short lifetime in aqueous solutions, where it mainly undergoes spontaneous dismutation to hydrogen peroxide and oxygen (Reaction 1). The superoxide radical is in equilibrium with its conjugated acid, the hydroperoxyl radical (HO_2), which is a stronger oxidant and generally more reactive than $O_2{}^{-}$.

$$HO_2^{\bullet} + O_2^{\bullet -} + H^+ \longrightarrow H_2O_2 + O_2 \qquad (1)$$

$$H_2O_2 + Fe^{2+} \longrightarrow OH^- + HO^{\bullet}\, Fe^{3+} \qquad (2)$$

$$QH^{\bullet} + Fe^{3+} \longrightarrow Fe^{2+} + Q + H^+ \qquad (3)$$

The superoxide anion radical and hydrogen peroxide are not particularly harmful to cells. It is the product of hydrogen peroxide decomposition, the hydroxyl radical ($HO^{\bullet}$), that is responsible for most of the cytotoxicity of oxygen radicals. The reaction can be catalyzed by several transition metals, including copper, manganese, cobalt, and iron, of which iron is the most abundant in the human body (Reaction 2; also called the Fenton reaction). To avoid iron-catalyzed reactions, iron is transported and stored chiefly as Fe(III), although redox active iron can be formed in oxidative reactions, and Fe(III) can be reduced by semiquinone radicals (Reaction 3).

2.1.2 Conjugation- and addition-reactions

Addition of sulfur to quinones is particularly important due to the wide distribution and high intracellular concentration of such compounds in the form of high- and low-molecular-weight thiols, such as glutathione. The addition reaction is classified as 1,4-reductive additions of Michael type (Finley, 1974). Glutathione can react with electrophile targets in a cell, spontaneously or in interactions catalyzed by glutathione S-transferases, and this generates a glutathionyl conjugate (Jakoby and Ziegler, 1980; Figure 2). Conjugation of glutathione to an electrophile increases the hydrophilic character of this compound, hence it has been proposed to be a detoxification reaction during the biotransformation of electrophilic compounds. In contrast, we have found that two-electron reduction of glutathionyl conjugates of several substituted 1,4-naphthoquinones and 2-methyl-1,4-naphthoquinones was followed by redox cycling, which in some cases proceeded more rapidly for the glutathionyl derivative than for the parent hydroquinone (Buffinton *et al.*, 1989).

Glucuronidation is one of the major conjugation reactions involved in metabolic conversion of xenobiotics and endogenous substances to polar, water-soluble metabolites. UDP-glucuronosyl-transferases catalyze the transfer of glucuronic acid from UDP-α-D-glucuronic acid (UDP-GA) to a nucleophilic acceptor, such as the hydroxyl group of a hydroquinone (Kasper and Henton, 1980). Sulfontransferases catalyze the transfer of the sulfuryl group (SO_3^-) from adenosine 3'-phosphate-5'-phosphosulfate (PAPS) to an appropriate acceptor, either a hydroxyl or an amino group (Jakoby and Ziegler, 1980), which results in formation of sulfate monoesters (Figure 2). When conjugated to either glucuronic acid or sulfate, hydroquinones lose both their ability to redox cycle and their electrophilic character and can therefore no longer participate in reactions that are potentially dangerous to the cell.

3. MECHANISMS FOR QUINONE-MEDIATED CELL DAMAGE

Although all quinones have the same functional group, their physicochemical behavior and mechanisms of toxicity vary due to the presence of different substituents. Thus, the cellular aspects of quinone metabolism are diverse, and a single mechanism explaining these actions has not yet been identified. Furthermore, it is noteworthy that the cytotoxicity of some xenobiotic compounds, such as benzene, benzo[a]pyrene, and 1-naphthol, may partly be caused by metabolic conversion of these compounds to quinones (Snyder *et al.*, 1987; Zheng *et al.*, 1997).

All aerobic organisms contain substances that help prevent injury mediated by free radicals, and these include antioxidants such as α-tocopherol and the enzymes superoxide dismutase and glutathione peroxidase. When the protective effect of the antioxidants is overwhelmed by the production of reactive oxygen species, the intracellular milieu becomes oxidative, leading to a state known as oxidative stress (Halliwell and Gutteridge, 1999). Thus the balance between the generated free radicals and the efficiency of the protective antioxidant system determines the extent of cellular damage.

3.1 Redox cycling

The amounts of NAD(P)H and oxygen used during redox cycling will be disproportionately larger than the amount of quinone present (Buffinton *et al.,* 1989; Cohen and d'Arcy Doherty, 1987). Research in our laboratory has clearly shown the cytotoxic impact of redox cycling under aerobic and anaerobic conditions: the toxic effect of naphthazarin (5,8-dihydroxy-1,4-naphthoquinone) on primary rat hepatocytes increased 10-fold when the culture conditions were changed from anaerobiosis to normal oxygen tension (Öllinger and Brunmark, 1991), and O_2-dependent toxicity was observed in rat cardiomyocytes exposed to doxorubicin (Öllinger and Brunmark, 1994).

It is widely assumed that DT-diphorase-catalyzed two-electron reduction of quinones to hydroquinones is a detoxification reaction, because the hydroquinone is generally less redox active and may also be readily conjugated to glucuronic acid or sulfate and thereafter easily excreted. In support of this theory, the effective DT-diaphorase inhibitor dicumarol has been found to potentiate the toxic effect of quinones on hepatocytes presumably by blocking two-electron reduction and thereby increasing the amount of quinone available for one-electron reduction (Miller *et al.*, 1986; Öllinger and Brunmark 1991). Nonetheless, for several antineoplastic drugs, the reduced metabolite is biologically the most active form and is responsible for alkylation and cross-linking of DNA and/or oxidative stress. Mitomycin C, streptonigrin, diaziquinon, and dinitrophenyl-azirinin are examples of therapeutic antitumor quinones that depend on DT-diaphorase activity for bioactivation (Ngo *et al.*, 1998; Riley and Workman, 1992; Sun and Ross, 1996).

3.2 DNA damage

Exposing cells to quinoid compounds results in cell cycle arrest and stabilization of p53, leading to initiation of either repair mechanisms or the cell death program. Moreover, several substances have been documented to cause malignant transformation, for example menadione and benzo[a]-pyrene-3,6-quinone as detected in the Salmonella TA 104 tester strain (Chesis *et al.*, 1984). This implies that damage to DNA is an essential aspect of quinone cytotoxicity. DNA damage caused by doxorubicin may occur through intercalation and formation of covalent adducts. In a recent review Gewirtz (1999) suggests that interaction with topoisomerase II is the most likely reaction at clinically used concentrations of the drug.

Intercalation entails non-covalent binding to DNA, in which an agent is inserted between the base pairs of DNA. This increases the melting point by stabilizing the DNA double helix and may contribute to inhibition of formation of the single-stranded templates that are needed by DNA and RNA polymerases for replication and transcription (O'Brien, 1985; Zunino *et al.,* 1975). Although antitumor anthracyclines are able to intercalate, a more important process that occurs during cell death is bioreductive activation, which causes covalent binding to DNA. Quinones can act as electrophiles and bind to DNA, as has been shown for menadione (Morrison *et al.*, 1984) and *p*-benzoquinone (Snyder *et al.,* 1987). Mitomycin C may cross-link with DNA by one-electron reduction reactions, followed by formation of a quinone methide, and finally a nucleophilic attack leading to new covalent bonds (Pan *et al.*, 1984). Zeman and colleagues (1998) demonstrated that doxorubicin forms covalent adducts exclusively at GpC sequences and thereby blocks RNA polymerase. The adducts are, however, formed only in the presence of molecular oxygen and formaldehyde. In addition to the quinone-DNA adducts that are formed through intercalation and covalent binding, redox active quinones may generate radicals or activated oxygen species. These often attack DNA in either the sugar or the base region, giving rise to a large number of effects, such as DNA fragmentation, base loss, and strand breaks (Imlay and Linn, 1988; Morrison *et al.*, 1984; Robles *et al.*, 1999). The hydroxyl radical, which can be generated during quinone metabolism, effectively induces DNA strand scission via addition to a DNA base forming e.g. 8-hydroxy-guanosine (Serrano *et al.*, 1999).

3.2.1 Inhibition of topoisomerase

Topoisomerase II is an essential, sulfhydryl (SH)-dependent endonuclease that is required for replication, recombination, chromosome segregation, and chromosome structure. Several antineoplastic drugs, for instance doxorubicin (Tarr and van Helden, 1990) and ß-lapachone (Frydman *et al.*, 1997), stabilize the topoisomerase II-DNA cleavable complex, which leads to scission of double-stranded DNA. ß-lapachone is a naturally occurring plant quinone and a potent inhibitor of DNA repair that is used to sensitize tumor cells to DNA-damaging agents. Inasmuch as ß-lapachone possesses thiol reactivity, a likely mechanism of topoisomerase inhibition is protein oxidation of the topoisomerase-DNA enzyme complex (Shiah *et al.*, 1999). Furthermore, N-benzyladriamycin, which is a semi-synthetic congener of doxorubicin, has been found to inhibit the topoisomerase II activity on DNA (Lothstein *et al.*, 2000). Also, substituted derivatives of naphthazarin have been shown to exert a toxic effect on L1210 cells through inhibition of topoisomerase I (Song *et al.*, 1999).

3.3 Protein damage and thiol balance

Covalent protein adducts of quinones are formed through Michael-type addition reaction with protein sulfhydryl groups or glutathione. Metabolic activation of several toxins (e.g., naphthalene, pentachlorophenol, and benzene) into quinones has been shown to result in protein quinone adducts (Lin *et al.*, 1997; Rappaport *et al.*, 1996; Zheng *et al.*, 1997). Conversion of substituted hydroquinones such as *p*-aminophenol-hydroquinone and 2-bromo-hydroquinone to their respective glutathione S-conjugates must occur to allow bioactivation into nephrotoxic metabolites (Dekant, 1993). Western blot analysis of proteins from the kidneys of rats treated with 2-bromo-hydroquinone has revealed three distinct protein adducts conjugated to quinone-thioethers (Kleiner *et al.*, 1998).

Thiol groups are easily oxidized by reactive oxygen species, and intracellular reduced glutathione is depleted by formation of glutathione disulfides and mixed protein thiol disulfides (Cohen and d'Arcy Doherty, 1987). Our research team has found that hepatocytes exposed to naphthoquinones exhibited rapid depletion of reduced glutathione and an increase in the oxidized form, and also that curbing glutathione synthesis by pretreating the cells with buthionin sulfoximin increased the toxicity of the redox-cycling quinones juglone and naphthazarin (Öllinger and Brunmark, 1991). Furthermore, glutathione depletion may be due to conjugation reactions that occur in the absence of oxidative stress (Rossi *et al.*, 1986). Anthracycline cytotoxicity is also accompanied by some degree of oxidation of glutathione, whereas depletion of GSH has little effect on either the toxicity of doxorubicin to rat cardiomyocytes (Öllinger and Brunmark, 1994) or mutagenicity and DNA damage (Capranico *et al.*, 1986).

3.4 Lipid peroxidation

It seems that cellular damage induced by quinone metabolism and increased generation of $O_2^{\bullet -}$ and H_2O_2 is seldom caused primarily by lipid peroxidation, but instead occurs due to depletion of the antioxidant defense system. We have found that primary rat hepatocytes exposed to naphthazarin exhibited lipid peroxidation, measured as accumulation of malondialdehyde and 4-hydroxyalkenal, and this resulted in destabilization of the lysosomes, detected as loss of lysosomal proton gradient and subsequent cell death (Öllinger and Brunk 1995). Pretreatment of the hepatocytes with the lipid-soluble radical scavenger DPPD (N,N'-diphenyl-*p*-phenylenediamine), which mainly removes carbon-centered radicals, completely inhibited lipid peroxidation and protected against quinone-induced loss of the lysosomal destabilization and viability loss. In our experiments (Öllinger and Brunmark, 1994), death of doxorubicin-treated heart myocytes was delayed by DPPD addition. In contrast, Cummings and coworkers (1992) injected subcutaneous carcinomas with doxorubicin, in either the free form or incorporated in albumin microspheres, and observed that the compound did not cause an increase, but instead a statistically significant decrease, in lipid peroxidation (Cummings *et al.*, 1992).

Doxorubicin binds readily to cell membranes, changing their structure and function. The targets of doxorubicin binding are compounds with a negative charge, of which the most extensively studied is the phospholipid cardiolipin (Pollakis *et al.*, 1983). Cardiolipin occurs in high concentrations in the inner mitochondrial membrane, where it is required for full activity of cytochrome c oxidase. In a recent study (Das and Mazumdar, 2000), the interaction between cytochrome c oxidase and cardiolipin in the presence of doxorubicin was analysed, and the results of pico-second time-resolved fluorescence depolarization showed that the cardiolipin layer was depleted due to complexation with the drug.

3.5 Effects on iron metabolism

Doxorubicin has a strong metal-binding capacity and is especially prone to interact with iron and copper, which leads to formation of iron(III) and copper(II) complexes at ligand ratios of 1: 3 and 1: 2, respectively. Considering iron, the binding constant is among the largest known, therefore the presence of Fe(III)- doxorubicin complexes formed *in vivo* should not be overlooked (Hasinoff and Davey, 1988). Furthermore, complexes between iron and daunorubicin, doxorubicin, and epidoxorubicin can bind to DNA and mediate covalent attachment to G-bases of DNA (Taatjes and Koch, 2001).

In addition, several redox-cycling quinones, including paraquat, diaquat, doxorubicin, and daunomycin, have been found to mediate the release of iron from iron stores (e.g., ferritin) that is induced by organic radicals or the superoxide anion radical (Aust *et al.*, 1985)

4. APOPTOSIS

4.1 General overview

Apoptosis is a form of physiological death in which an individual cell undergoes an internally controlled transition from being in an intact metabolically active state to consisting of a number of shrunken remnants with maintained membrane integrity. It is presumed that apoptosis does not entail lysis of internal organelles, although studies have indicated that there is extensive two-way translocation of proteins between organelles and the cytosol. Release of cytochrome c from the mitochondria leads to what may be the ultimate mechanisms of apoptosis, namely, formation of the apoptosome, activation of procaspase-9, and initiation of the caspase cascade. Reactive oxygen species are involved in various biological processes, including induction of gene expression, regulation of proliferative events, and cellular responses to cytokines. Buttke and Sandstrom (1994) summarized evidence that oxidative stress is a common mediator of apoptosis in eukaryotic cells. Apoptosis can be induced by intracellularly or extracellularly generated free radicals, and, not surprisingly, free radical scavengers can suppress apoptosis induction. Moreover, such scavengers can prevent apoptosis caused by factors that are not in themselves oxidants, as shown by the following examples: apoptosis caused by deprivation of follicle-stimulating hormones in ovarian follicles is inhibited by superoxide dismutase and ascorbic acid (Tilly and Tilly, 1995); nitrone spin traps restrain apoptosis induced by glucocorticoids in thymocytes (Slater *et al.*, 1995); and free radical scavengers blocks TNF-α-induced apoptosis (Matthews *et al.*, 1987)

In light of the results we discussed in previous sections, which indicate that the cytotoxicity of quinones is due to generation of free radicals and DNA damage, it seems that these compounds would be potent triggers of apoptosis.

4.2 Quinone-induced apoptosis

It is now generally accepted that anthracyclines can induce apoptosis in tumor cells. Treatment with doxorubicin has been found to cause apoptosis in HeLa cells and murine thymocytes, as well as P388, HL-60, and U937 human leukemic cells (Ling *et al.*, 1993; Skladanowski and Konopa, 1993; Zaleskis *et al.*, 1994). The outcome of anthracycline anticancer therapy is highly dependent on two-electron reduction leading to bioactivation of the quinone in use (Riley and Workman, 1992). Diaziquone and its derivatives diaziridinequinone and methyl-diaziridinequinone induced p53 in a dose- and time-dependent manner, analyzed by the electrophoretic mobility shift assay. Expression of wild-type p53 elicited by diaziquone was suppressed when DT-diaphorase activity was inhibited by pretreating the cells with dicumarol. Furthermore, inhibition of oxidative stress by the antioxidant enzyme catalase reduced induction of p53 by about 45% (Ngo *et al.*, 1998). Two other substrates of DT-diaphorase, streptonigrin and mitomycin C, caused apoptosis in HT29 cells that was prevented by dicumarol (Sun and Ross, 1996), thus cancer cells expressing high levels of DT-diaphorase can be treated more successfully than those expressing little or no DT-diaphorase.

There is limited evidence that quinone-induced apoptosis is mediated by lipid peroxidation. It has been reported that doxorubicin was cytotoxic to ovarian carcinomas, and that apoptosis was mediated through lipid peroxidation together with activation of p53, Apaf-1, and caspase-9 (Minko *et al.*, 1999). In our laboratory (Kågedal *et al.*, 1999), the three anthraquinones rhein, dantron, and chrysophanol were compared regarding their toxicity and their ability to induce apoptosis in primary rat hepatocytes, and only rhein was found to have redox-cycling capacity and induce apoptosis. These results are in accordance with previous observations, in that they show that the toxicity of the quinone is substantially altered by the chemical properties of the substituents. Since DPPD and the iron chelator desferal completely inhibited apoptosis, we concluded that the cytotoxic effect of rhein was due to formation of free radicals and lipid peroxidation. In another study (Roberg and Öllinger, 1998), naphthazarin-treated rat heart myocytes underwent apoptotis but no lipid peroxidation was detected; the apoptotic death of those cells was, however, inhibited by preincubation with the lipid-soluble antioxidant α-tocopherol succinate, which is an effective free radical scavenger.

Studies of the oxidation products of catecholamines (i.e., seretonin, dopamine, dopa, adrenaline, and noradrenaline) have indicated that protein oxidation by quinones may lead to apoptosis. Oxidation results in formation of *ortho*-quinones, which contribute to cytotoxicity and have been suggested to underlie disorders such as Parkinson's disease and schizophrenia. In soluble brain extracts, *o*-quinones have been observed to undergo redox cycling and covalent binding to the free sulfhydryl groups of proteins. Experiments *in vitro* have shown that this binding is enhanced by Fe(II), which results in the formation of high-molecular protein aggregates, lipid peroxidation, and apoptosis (Velez-Pardo *et al.*, 1997). Other investigators (Baez *et al.*, 1997) have proposed that these effects can be prevented by conjugation of *o*-quinones to glutathione and that they are involved in protecting the cells of the nervous system from degenerative processes.

5. THE ROLE OF LYSOSOMES IN APOPTOSIS

5.1 Lysosomal function

Lysosomes are responsible for degradation and recycling of macromolecules through hetero- and autophagocytosis. Lysosomes are acidic and contain a variety of hydrolytic enzymes, most of which are active at pH values below 7. Upon rupturing, a lysosome releases its contents of hydrolytic enzymes, which inundate the cytosol. Disruption of lysosomes and weakening of the lysosomal membrane have been implicated in a number of pathological conditions, such as myocardial ischemia, Alzheimer's disease, cancer, and inflammation (Agha and Gad, 1995; Cataldo *et al.*, 1996; Shibata *et al.*, 1997; Tedone *et al.*, 1997). Different studies have demonstrated three distinct lysosomal mechanisms by which lysosomes participate in the pathogenesis of diseases. First, it has been shown that the uptake of toxic substances into a cell is followed by rupture of the lysosomes and release of the enzymes they contain to the cytosol, which results in necrosis. Second, overloading of the lysosomes may occur if some of the lysosomal enzymes are inhibited or absent (e.g., as seen in lipid storage diseases), leading to an excess of stored matter that ultimately results in disruption of the lysosomal membrane. Third, discharge of lysosomal enzymes from the cell into the extracellular space has been observed in some types of cancer cells.

5.2 The importance of lysosomal enzymes during apoptosis

Biochemical and genetic analysis of apoptosis has shown that intracellular proteases are key effectors of pathways that lead to cell death.

In particular, studies have demonstrated the primacy of changes in the mitochondria and activation of caspases. For quite some time, scientists have assumed that lysosomes are rather durable organelles that are not broken down until late in apoptosis. However, recent publications have suggested that lysosomes are instead fairly fragile and therefore release their hydrolytic enzymes to the cytosol even during mild oxidative stress, despite an appearance of being ultrastructurally intact (Öllinger and Brunk, 1995; Roberg and Öllinger, 1998; Zhang *et al.*, 1995). Furthermore, there is growing evidence that the lysosomal proteases called cathepsins are also involved in mediating and promoting apoptosis. Like caspases, the cathepsins are synthesized as inactive zymogens and are activated through proteolytic processing. The most convincing results showing an association between cathepsins and apoptosis have been obtained in studies focused on the cysteine protease cathepsin B and the aspartate protease cathepsin D.

Wu *et al.* (1998) noted that doxorubicin-induced apoptosis in lymphoid cells was blocked by pepstatin A, which is an inhibitor of cathepsin D. These investigators also observed that cathepsin D was induced through p53 DNA-binding sites at the cathepsin D promoter. Moreover, they have found that, compared to fibroblasts from wild-type mice, cathepsin D-/- fibroblasts from gene knock-out mice exhibited increased resistance to death caused by doxorubicin. Also, in serum-deprived rat PC12 cells undergoing apoptosis, the amount of cathepsin B has been observed to decline, while the level of cathepsin D increased (Shibata *et al.*, 1998), and, in our laboratory (Kågedal *et al.*, 2001), the same phenomenon was recently seen in human fibroblasts exposed to naphthazarin.

Further support for participation of lysosomal enzymes in apoptosis has been gained in studies of several systems that do not involve a quinone-triggering signal. For example, apoptosis induced in hepatocytes by bile salts appears to be regulated through a cathepsin-D-mediated increase in cathepsin B activity (Roberts *et al.*, 1999) and Bid was cleaved in the presence of lysosomal extracts (Stoka *et al.*, 2001). Also, a number of investigations have indicated that lysosomes and lysosomal enzymes are involved in regulation of cell death caused by TNF-α receptor ligation. Monney *et al.* (1998) reported that alkalization of acidic vesicles attenuates TNF-α-induced apoptosis, and Guicciardi *et al.* (2000) showed that TNF-α ligation caused release of cathepsin B from lysosomes. The cathepsin D inhibitor pepstatin A has been found to block TNF-α-elicited apoptosis in U937 histiocytic lymphoma cells (Deiss *et al.*, 1996).

5.3 Lysosomal membrane stability during apoptosis

The results of several studies suggest that destabilization of lysosomes, detected as loss of lysosomal proton gradient, plays an important role in apoptosis initiated by a variety of stimuli, such as oxidative stress, growth factor withdrawal, α-tocopheryl succinate exposure, or Fas-activation (Brunk *et al.*, 1997; Brunk and Svensson, 1999; Neuzil *et al.*, 1999). Moreover, treatment of human macrophages or the mouse macrophage cell line J-774 with oxidized LDL has been found to alter immunostaining of cathepsin D from a granular lysosomal pattern in non-apoptotic cells to a diffuse cytoplasmic pattern at the onset of apoptosis (Li *et al.*, 1998). Since the core apoptosis machinery is localized to the cytoplasmic compartment, activation of such cell death depends on multiple post-translational mechanisms, in particular translocation of proteins. To address the question of whether the enzymes in lysosomes must be translocated to the cytosol in order to mediate apoptosis, our research group has studied the subcellular location of cathepsins D, B, and L during naphthazarin-induced apoptosis (Kågedal *et al.*, 2001; Öllinger, 2000; Roberg and Öllinger, 1998). We found that, in both human fibroblasts and rat myocytes, naphthazarin induced an early release of cathepsin D from lysosomes to the cytosol, and this effect preceded the release of cytochrome c from mitochondria (Roberg *et al.*, 1999; Roberg, 2001). Moreover, we noted that fibroblasts exposed to naphthazarin also released cathepsins B, D, and L early in apoptosis (Kågedal *et al.*, 2001). Furthermore, addition of naphthazarin to the cell cultures induced a rapid increase in reactive oxygen species (detected as enhanced luminol-dependent chemiluminescence) that declined within 30 min. We also observed that depletion and subsequent normalization of the level of intracellular reduced glutathione mirrored the increased formation of oxidants, which implies that naphthazarin metabolism instantly alters the redox balance that may be responsible for the signal transduction. Our experiments also revealed an initial and rapid fluctuation in both ATP and mitochondrial membrane potential in fibroblasts exposed to naphthazarin. Pretreatment with pepstatin A blocked the release of cytochrome c and the onset of apoptosis (Roberg *et al.*, 1999; Roberg, 2001). However, exposing the fibroblasts to pepstatin A before naphthazarin did not affect the increase in formation of reactive oxygen species or depletion of glutathione, nor did it block the translocation of cathepsin D from the lysosomes to the cytosol, which strongly suggests that lysosomal destabilization is regulated by the intracellular redox balance (Öllinger, 2000). This is supported by our earlier experiments, in which pretreatment of rat cardiomyocytes with a lipid-soluble antioxidant prevented release of cathepsin D from lysosomes upon exposure to naphthazarin (Roberg and Öllinger, 1998).

The cytosolic targets of cathepsin D have not been ascertained. However, since we found that inhibition of cathepsin D prevented both release of cytochrome c and activation of caspase-3-like caspases, we suggest that cathepsin D exerts its effect upstream of both the release of cytochrome c from mitochondria and the onset of the caspase cascade. Cathepsin B seems to be of minor importance in this system, because we observed that the activity of cathepsin B decreased and the cathepsin B inhibitor CA074-Me had no influence on the rate of apoptosis (Kågedal *et al.*, 2001).

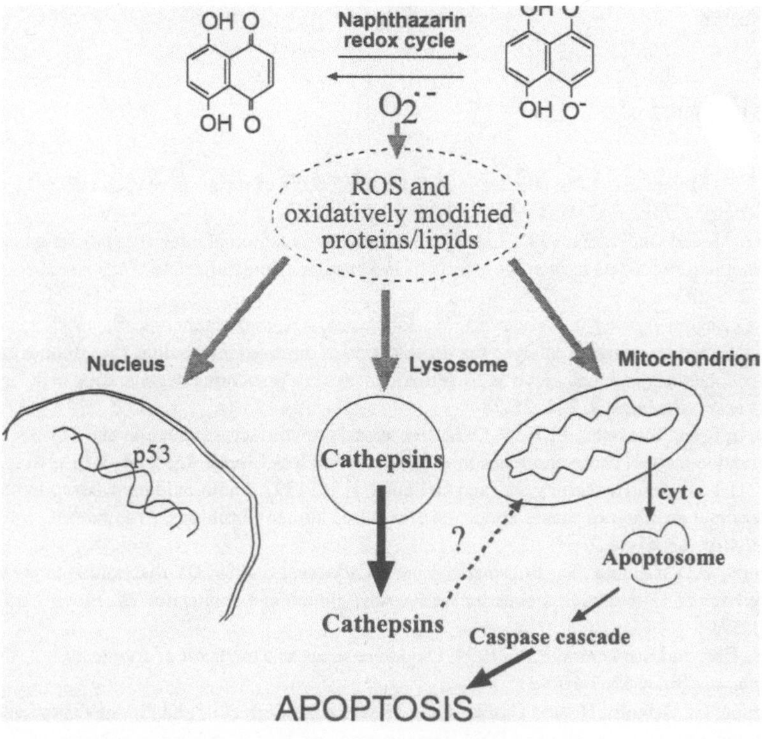

Figure 3. Proposed mechanism for quinone induced apoptosis

5.3.1 Lysosomal damage

The mechanism responsible for releasing the contents of lysosomes is not yet known. According to one hypothesis (Öllinger and Brunk, 1995), the membranes of lysosomes are damaged during apoptosis induced by oxidative stress, possibly due to diffusion of intracellularly produced hydrogen peroxide into these organelles. Inside the lysosomal apparatus, low-molecular-weight iron from degraded metalloproteins may promote iron reduction and Fenton-like chemistry and thereby destabilize the membranes, resulting in leakage of the lysosomal contents. Another plausible mechanism is that free radicals produced in the cytosol functionally impair the lysosomal membrane from the outside. Moreover, the compound atractyloside, which is commonly used to induce the mitochondrial permeability transition and the release of cytochrome c from mitochondria, has been found to liberate cathepsin B from isolated lysosomes (Vancompernolle *et al.*, 1998), suggesting that a pore-opening mechanism exists in both mitochondria and lysosomes.

REFERENCES

Aust, S.D., Morehouse, L.A., and Thomas, C.E., 1985, Role of metals in oxygen radical reactions. *J. Free Rad. Biol. Med.* **1**: 3-25

Agha, A.M. and Gad, M.Z., 1995, Lipid peroxidation and lysosomal integrity in different inflammatory models in rats: the effects of indomethacin and naftazone. *Pharmacol. Res.* **32**: 279-285

Baez, S., Segura-Aguilar, J., Widersten, M., Johansson, AS., and Mannervik, B., 1997, Glutathione transferases catalyse the detoxication of oxidized metabolites (o-quinones) of catecholamines and may serve as an antioxidant system preventing degenerative cellular processes. *Biochem. J.* **324**: 25-28

Brunk, U.T. and Svensson, I., 1999, Oxidative stress, growth-factor starvation and Fas-activation may all cause apoptosis through lysosomal leak. *Redox Report* **4**: 3-11

Brunk, U.T., Dalen, H., Roberg, K., and Hellquist, H.B., 1997, Photo-oxidative disruption of lysosomal membranes causes apoptosis of cultured human fibroblasts. *Free Radic. Biol.Med.* **23**: 616-626

Buffinton, G.D., Öllinger, K., Brunmark, A., and Cadenas, E., 1989, DT-diaphorase catalyzed reduction of 1,4-naphthoquinone derivatives and glutathione-conjugates. *Biochem. J.* **257**: 561-571

Buttke, T.M. and Sandstrom, P.A., 1994, Oxidative stress as a mediator of apoptosis. *Immunol. Today* **15**: 7-10

Capranico, G., Babudri, N., and Casciarri, G., 1986, Lack of effect of glutathione depletion on cytotoxicity, mutagenicity and DNA damage produced by doxorubicin in cultured cells. *Chem.-Biol. Interact.* **57**: 189-201

Cataldo, A.M., Hamilton, D.J., Barnett, J.L., Paskevich, P.A., and Nixon, R.A., 1996, Properties of the endosomal-lysosomal system in the human central nervous system: disturbances mark most neurons in populations at risk to degenerate in Alzheimer's disease. *J. Neurosci.* **16:** 186-199

Chesis, P.L., Levin, D.E., Smith, M.T., Ernster, L., and Ames, B.N., 1984, Mutagenicity of quinones: pathways of metabolic activation and detoxification. *Proc. Natl. Acad. Sci. USA* **81**: 1696-1700

Cohen, G.M. and d'Arcy Doherty, M., 1987, Free radical mediated toxicity by redox cycling chemicals. *Br. J. Cancer* **55**: 46-52

Cummings, J., Willmott, N., Hoey, B.M., Marley, E.S., and Smyth, J.F., 1992, The consequences of doxorubicin quinone reduction *in vivo* in tumor tissue. *Biochem. Pharmacol.* **44**: 2165-2174

Das, T.K. and Mazumdar, S., 2000, Effect of Adriamycin on the boundary lipid structure of cytochrome c oxidase: pico-second time-resolved fluorescence depolarization studies. *Biophys. Chem.* **86**: 15-28

Deiss, L.P., Galinka, H., Barissi, H., Cohen, O., and Kimchi, A., 1996, Cathepsin D protease mediates programmed cell death induced by interferon-gamma, Fas/APO-1 and TNF-α. *EMBO J.* **15**: 3861-3870

Dekant, W., 1993, Bioactivation of nephrotoxins and renal carcinogens by glutathione S-conjugate formation. *Toxicol. Lett.* **67**: 151-160

Dooley, D.M., 1999, Structure and biogenesis of topaquiones and related cofactors. *J. Biol. Inorg. Chem.* **4**: 1-11

Finley, K.T., 1974, The addition and substitution chemistry of quinones. In *The chemistry of the quinoid compounds* (S. Patai, ed.), John Wiley and Sons, London, pp. 877-1144

Frydman, B., Marton, L.J., Sun, J.S., Neder, K., Witiak, D.T., Liu, A.A., Wang, H.M., Mao, Y., Wu, Y., Sanders, M.M., and Liu, L.F., 1997, Induction of DNA topoisomerase II-mediated DNA cleavage by beta-lapachone and related naphthoquinones. *Cancer Res.* **57**: 620-627

Gewirtz, D.A, 1999, A critical evaluation of the mechanisms of action proposed for the antitumor effects of the anthracycline antibiotics adriamycin and daunorubicin. *Biochem. Pharmacol.* **57**: 727-741

Guicciardi, M.E., Deussing, J., Miyoshi, H., Bronk, S.F., Svingen, P.A., Peters, C., Kaufmann, S.H. and Gores, G.J., 2000, Cathepsin B contributes to TNF-α-mediated hepatocyte apoptosis by promoting mitochondrial release of cytochrome c. *J. Clin. Invest.* **106**: 1127-1137

Halliwell, B. and Gutteridge, J.M.C., 1999, *Free Radicals in Biology and Medicine.* Clarendon Press, Oxford

Hasinoff, B.B. and Davey, J.P., 1988, Adriamycin and its iron(III) and copper(II) complexes. *Biochem. Pharmacol.* **37**: 3663-3669.

Imlay, J.A. and Linn, S., 1988, DNA damage and oxygen radical toxicity. *Science* **240**: 1302-1309

Jakoby, W.B. and Ziegler, D.M., 1980, The enzymes of detoxication. *J. Biol. Chem.* **265**: 20715-20718

Kappus, H., 1986, Overview of enzyme systems involved in bioreduction of drugs and in redox cycling. *Biochem. Pharmacol.* **35**: 1-6

Kasper, C.B. and Henton, D., 1980, Glucuronidation. In *Enzymatic basis of detoxication.* (W.B. Jacoby, ed.), Academic Press, New York, pp.4-36

Kleiner, H.E., Rivera, M.I., Pumfird, N.R., Monks, T.J., and Lau, S.S., 1998, Immunochemical detection of quinol-thioether-derived protein adducts. *Chem. Res. Toxicol.* **11**: 1283-1290

Kågedal, K., Bironaite, D., and Öllinger, K., 1999, Anthraquinone cytotoxicity and apoptosis in primary cultures of hepatocytes. *Free Rad. Res.* **31**: 419-428

Kågedal, K., Johansson, U., and Öllinger, K., 2001, The lysosomal protease cathepsin D mediates apoptosis induced by oxidative stress. *FASEB J.* (May 18, 2001) 10.1096/fj.00-0708fje

Li, W., Yuan, X.M., and Brunk, U.T., 1998, OxLDL-induced macrophage cytotoxicity is mediated by lysosomal rupture and modified by intralysosomal redox-active iron. *Free Radic. Res.* **29**: 389-398

Lin, P.H., Waidyanatha, S., Pollack, G.M., and Rappaport, S.M., 1997, Dosimetry of chlorinated quinone metabolites of pentachlorophenol in the livers of rats and mice based upon measurement of protein adducts. *Toxicol. Appl. Pharmacol.* **145**: 399-408

Ling, Y.H., Priebe, W., and Perez-Soler, R., 1993, Apoptosis induced by anthracycline antibiotics in P388 parent and multidrug-resistant cells. *Cancer Res.* **53**: 1845-1852

Lothstein, L., Suttle, D.P., Roaten, J.F., Koseki, Y., Israel, M., and Sweatman, T.W., 2000, Catalytic inhibition of DNA topoisomerase II by N-benzyladriamycin (AD 288). *Biochem. Pharmacol.* **60**: 1621-1628

Matthews, N., Neale, M.L., Jackson, S.K., and Stark, J.M., 1987, Tumor cell killing by tumor necrosis factor: inhibition by anearobiotic condition, free-radical scavengers and inhibitors of arachidonate metabolism. *Immunology* **62**: 153-155

Miller, M.G., Rodgers, A., and Cohen, G.M., 1986, Mechanisms of toxicity of naphthoquinones to isolated hepatocytes. *Biochem. Pharmacol.* **35**: 1177-1184

Minko, T., Kopeckova, P., and Kopecek, J., 1999, Comparison of the anticancer effect of free and HPMA copolymer-bound adriamycin in human ovarian carcinoma cells. *Pharmaceut. Res.* **16**: 986-996

Monney, L., Olivier, R., Otter, I., Jansen, B., Poirier, GG., and Borner, C., 1998, Role of an acidic compartment in tumor-necrosis-factor-α-induced production of ceramide, activation of caspase-3 and apoptosis. *Eur. J. Biochem.* **251**: 295-303

Morrison, H., Jernström, B., Nordenskjöld, M., Thor, H., and Orrenius, S., 1984, Induction of DNA damage by menadione (2-methyl-1,4-naphthoquinone) in primary cultures of rat hepatocytes. *Biochem. Pharmacol.* **33**: 1763-1769

Neuzil, J., Svensson, I., Weber, T., Weber, C., and Brunk, U.T., 1999, alpha-tocopheryl succinate-induced apoptosis in Jurkat T cells involves caspase-3 activation, and both lysosomal and mitochondrial destabilisation. *FEBS Lett.* **445**: 295-300

Ngo, E.O., Nutter, L.M., Sura, T., and Gutierrez, P. L., 1998, Induction of p53 by the concerted actions of aziridine and quinone moieties of diaziquone. *Chem. Res. Toxicol.* **11**: 360-368

O'Brien, P.J., 1985, Free-radical-mediated DNA binding. *Environ. Health Perspect.* **64**: 219-232

Öllinger, K., 2000, Inhibition of cathepsin D prevents free-radical-induced apoptosis in rat cardiomyocytes. *Arch. Biochem. Biophys.* **373**: 346-351

Öllinger, K. and Brunk, U.T., 1995, Oxidative stress-induced cellular injury is mediated through lysosomal damage. *Free Radic. Biol. Med.* **19**: 565-574

Öllinger, K. and Brunmark, A., 1991, Effect of hydroxy substitution on 1,4-naphthoquinone toxicity to rat hepatocytes. *J. Biol. Chem.* **266**: 21496-21503

Öllinger, K. and Brunmark, A., 1994, Effect of different oxygen pressures and DPPD on adriamycin toxicity to cultured neonatal rat heart myocytes. *Biochem. Pharmacol.* **48**: 1707-1715

Pan, S., Andrews, P.A., Glover, C.J., and Bachur, N.R., 1984, Reductive activation of mitomycin C and mitomycin C metabolites catalysed by NADPH-cytochorome P-450 reductase and xantine oxidase. *J. Biol. Chem.* **259**: 959-962

Pollakis, G., Goormaghtigh, E., and Ruysschaert, J.-M., 1983, Role of quinone structure in the mitochondrial damage induced by antitumor anthracyclines. *FEBS Lett.* **155**: 267-272

Rappaport, S.M., McDonald, T.A., and Yeowell-O'Connell, K., 1996, The use of protein adducts to investigate the disposition of reactive metabolites of benzene. *Environ. Health Perspect.* **104** Suppl 6: 1235-1237

Riley, R.J. and Workman, P., 1992, DT-diaphorase and cancer chemotherapy. *Biochem. Pharmacol.* **43**: 1657-1669

Roberg, K., 2001, Relocalization of cathepsin D and cytochrome c early in apoptosis revealed by immunoelectron microscopy. *Lab. Invest.* **81**: 149-158

Roberg, K., Johansson, U., and Öllinger, K., 1999, Lysosomal release of cathepsin D precedes relocation of cytochrome c and loss of mitochondrial transmembrane potential during apoptosis induced by oxidative stress. *Free Radic. Biol. Med.* **27**: 1228-1237

Roberg, K. and Öllinger, K., 1998, Oxidative stress causes relocation of the lysosomal enzyme cathepsin D with ensuing apoptosis in neonatal rat cardiomyocytes. *Am. J. Pathol.* **152**: 1151-1156

Roberts, L.R., Adjei, P.N., and Gores, G.J., 1999, Cathepsins as effector proteases in hepatocyte apoptosis. *Cell Biochem. Biophys.* **30**: 71-88

Robles, S.J., Buehler, P.W., Negrusz, A., and Adami, G.R., 1999, Permanent cell cycle arrest in asynchronously proliferating normal human fibroblasts treated with doxorubicin or etoposide but not camtothecin. *Biochem. Pharmacol.* **58**: 675-685

Rossi, L., Moore, G.A., Orrenius, S., and O'Brien, P.J., 1986, Quinone toxicity in hepatocytes without oxidative stress. *Arch. Biochem. Biophys.* **251**: 25-35

Serrano, J., Palmeira, C.M., Kuehl, D.W., and Wallace, K.B., 1999, Cardioselective and cumulative oxidation of mitochondrial DNA following subchronic doxorubicin administration. *Biochim. Biophys. Acta* **1411**: 201-205

Shiah, S.G., Chuang, S.E., Chau, Y.P., Shen, S.C., and Kuo, ML., 1999, Activation of c-Jun NH_2-terminal kinase and subsequent CPP32/Yama during topoisomerase inhibitor beta-lapachone-induced apoptosis through an oxidation-dependent pathway. *Cancer Res.* **59**: 391-398

Shibata, M., Kanamori, S., Isahara, K., Ohsawa, Y., Konishi, A., Kametaka, S., Watanabe, T.S., Ishido, K., Kominami, E., and Uchiyama, Y., 1998, Participation of cathepsins B and D in apoptosis of PC12 cells following serum deprivation. *Biochem. Biophys. Res. Commun.* **251**: 199-203

Skladanowski, A., and Konopa, J., 1993, Adriamycin and daunomycin induce programmed cell death (apoptosis) in tumor cells. *Biochem. Pharmacol.* **46**: 375-382

Slater, A.F.G., Nobel, C.S.I., Maelaro, E., Bustamante, J., Kimland, M., and Orrenius, S., 1995, Nitrone spin traps and a nitroxide antioxidant inhibit a common pathway in thymocyte apoptosis. *Biochem. J.* **306**: 771-778

Snyder, R., Jowa, L., Witz, G., Kalf, G., and Rushmore, T., 1987, Formation of reactive metabolites from benzene. *Arch. Toxicol.* **60**: 61-64.

Song, G.Y., Zheng, X.G., Kim, Y., Sok, D.E., Ahn, B.Z., 1999, Naphthazarin derivatives (II): formation of glutathione conjugate, inhibition of topoisomerase I and cytotoxicity. *Bioorg. Med. Chem. Lett.* **9**: 2407-2412

Stoka, V., Turk, B., Schendel, S.L., Kim, T.-H., Cirman, T., Snipas, S.J., Ellerby, L.M., Bredesen, L., Freeze, H, Abrahamson, M., Brömme, D., Krajewski, S., Reed, J.C., Yin, X.-M., Turk, V., and Salvesen G.S., 2001, Lysosomal Protease Pathways to Apoptosis. Cleavage of Bid, not pro-caspases, is the most likely route. *J. Biol. Chem.* **276**: 3149-3157

Sun, X. and Ross, D., 1996, Quinone-induced apoptosis in human colon adenocarcinoma cells via DT-diaphorase mediated bioactivation. *Chem. Biol. Interact.* **100**: 267-76

Taatjes, D.J. and Koch, T.H., 2001, Nuclear targeting and retention of anthracycline antitumor drugs in sensitive and resistant tumor cells. *Curr. Med. Chem.* **8**: 15-29

Tarr, M. and van Helden, P.D., 1990, Inhibition of transcription by adriamycin is a consequence of the loss of negative superhelicity in DNA mediated by topoisomerase II. *Mol. Cell Biochem.* **93**: 141-146

Tedone, T., Correale, M., Barbarossa, G., Casavola, V., Paradiso, A., and Reshkin, S.J., 1997, Release of the aspartyl protease cathepsin D is associated with and facilitates human breast cancer cell invasion. *FASEB J.* **11**: 785-792

Tilly, J.L. and Tilly, K.I., 1995, Inhibitors of oxidative stress mimic the ability of follicle-stimulating hormone to supress apoptosis in cultured ovarian follicles. *Endocrinology* **136**: 242-252

Thomson, R.H., 1971, *Naturally occurring quinones.* Academic Press, New York

Vancompernolle, K., van Herreweghe, F., Pynaert, G., van de Craen, M., De Vos, K., Totty, N., Sterling, A., Fiers, W., Vandenabeele, P., and Grooten, J., 1998, Atractyloside-induced release of cathepsin B, a protease with caspase-processing activity. *FEBS Lett.* **438**: 150-158

Velez-Pardo, C., Jimenez Del Rio, M., Verschueren, H., Ebinger, G., and Vauquelin, G., 1997, Dopamine and iron induce apoptosis in PC12 cells. *Pharmacol. Toxicol.* **80**: 76-84

Wu, G.S., Saftig, P., Peters, C., and El-Deiry, W.S., 1998, Potential role for cathepsin D in p53-dependent tumor suppression and chemosensitivity. *Oncogene* **16**: 2177-2183

Zaleskis, G., Berleth, E., Verstovsek, S., Ehrke, and M.J., Mihich, E., 1994, Doxorubicin-induced DNA degradation in murine thymocytes. *Mol. Pharmacol.* **46**: 901-908

Zeman, S.M., Phillips, D.R., and Crothers, D.M., 1998, Characterization of covalent Adriamycin-DNA adducts. *Proc. Natl. Acad. Sci. USA* **95**: 11561-11565

Zhang, H., Dahlen H., Öllinger, K., and Brunk, U.T., 1995, Exposure of cells to non lethal concentrations of hydrogen peroxide induces degeneration-repair mechanisms involving lysosomal destabilization. *Free Radic. Biol. Med.* **19**: 813-822

Zheng, J., Cho, M., Jones, A.D., and Hammock, B.D., 1997, Evidence of quinone metabolites of naphthalene covalently bound to sulfur nucleophiles of proteins of murine Clara cells after exposure to naphthalene. *Chem. Res. Toxicol.* **10**: 1008-1014

Zunino, F., Gambetta, R., and Di Marco, A., 1975, The inhibition *in vitro* of DNA polymerase and RNA polymerases by daunomycin and adriamycin. *Biochem. Pharmacol.* **24**: 309-311

Chapter 9

Apoptosis Induced by Ionizing Radiation

The biological basis of radiosensitivity

Cristiano Ferlini[*], Raffaele D' Amelio[#], and Giovanni Scambia[*]
[*]*Laboratory of Antineoplastic Pharmacology, Dept. of Obstetrics and Gynaecology, Università Cattolica Sacro Cuore, L.go A. Gemelli 1, 00168 Rome, Italy;*[#] *Seconda Facoltà di Medicina e Chirurgia, Università degli studi "La Sapienza", S. Andrea Hospital, Rome, Italy*

1. INTRODUCTION

Ionizing radiation consists of high frequency electromagnetic waves with a sufficient energy to ionize biologic matter and produce biological effects also within living organisms. Moreover, it can be focused at the pathologic target, thereby minimising side effects for the normal tissues. Taking advantage of these properties, radiotherapy is a mainstay of cancer therapy. The outcome of a cell hit by ionizing radiation is either cell death or repair of the lesions. In this last decade much research effort has been spent in the investigation of the mechanisms underlying cell death induced by ionizing radiation. If some years ago the common thinking was that the fate of irradiated cells was dependent on nuclear events, now recent discoveries have drawn our attention to mitochondria as the actual leading actors of the cell death (repair) drama. These mechanisms are essential to explain the inherent radiosensitivity or radioresistance of certain tumors as well as to understand the basis of acquired radioresistance.

The sensitivity to ionizing radiation is maximal in those cells able to activate a co-ordinate program of cell death (apoptosis) primed by the radiation-induced oxidative stress. Albeit apoptosis is a nuclear event and radiation-induced DNA damage is probably the most relevant mechanism of initiation of apoptosis, the control of the execution phase (and sometime also the initiation) takes place at the mitochondrial level. Radioresistance occurs

Subcellular Biochemistry, Volume 36: Phospholipid Metabolism in Apoptosis.
Edited by Quinn and Kagan. Kluwer Academic/Plenum Publishers, New York, 2002

when cancer cells are able to escape from the radiation-induced cell death. Among the inherent mechanisms of radioresistance, mutations at the level of the genes involved in the control of apoptosis (p53, bcl-2, bax) often take place in solid tumors, where hypoxia-selected clones resistant to oxidative stress and apoptosis are found. Adaptive response to ionizing radiation is responsible for the acquired radioresistance observed in patients relapsing from previous treatments. Adaptive response is mediated by raised levels of the DNA repair systems and by the increase of the intracellular thiols. The investigation of the pathways underlying radioresistance has prompted the development of drugs able to either reduce the lethal effects in normal tissues or to potentiate the biological activity of ionizing radiation in cancer cells. Until now, this second approach has been more fruitful and we expect that new methods of DNA screening in cancer cells will allow more effective combinations of radiotherapy/radiosensitizers to any given set of DNA mutations. This chapter discusses some of these mechanisms in the context of their relevance to the clinical effects of radiation on cancer and normal cells.

2. BIOLOGICAL EFFECTS OF IONIZING RADIATIONS

2.1 Direct and indirect effects of ionizing radiation

Although radiation can directly damage nucleic acids, proteins and lipids, the biological effects are mostly dependent on the indirect effects (Roots and Okada, 1975). Since biological matter mainly consists of water, the commonest chemicals responsible for such effects are the reactive oxygen species (ROS), generated by the acceptor oxygen upon the ionizing activity of radiation. Lesions induced by ROS depend on the degree of protection that intracellular structure has. The most "protected" compartment is the nucleus, where DNA is tightly bound with an equal mass of positively charged proteins (histones), forming a repeating array of DNA-protein particles called nucleosomes. The complex between nucleosomes and non-histones chromosomal proteins generates chromatin. Non-histone proteins hold nucleosomes and contribute to protection of the genome. Only in those regions where DNA is actively transcribed, this shell is turned off, and this explains because radiosensitivity as well as sensitivity to DNAases is maximal in the proliferative phases of the cell cycle. The shell is also absent during mitosis, when there is no DNA transcription, but chromosome segregation, where chromatin structure is very different from that occurring in the other cell cycle phases.

Another "protected" site is the mitochondrion where intracellular ROS scavengers are concentrated. Despite this protection, when the radiation dose overcomes a given threshold, both cell nucleus and mitochondria are damaged, thereby stimulating a cellular response. In normal cells, if the damage cannot be repaired, the program of cell death is loaded. The commonest program of cell death is apoptosis. This kind of cell death is very efficient and is aimed to eliminate damaged cells without activating inflammation.

2.2 DNA damage and genetic control of DNA repair

A variety of DNA lesions occurs after radiation and it is generally accepted that such lesions are pivotal in determining cell killing by ionizing radiation. Damage to the deoxyribose backbone of the DNA disrupts phosphodiester bonds and produces single-strand breaks (ssb). In absolute, this is the more frequent DNA lesion induced by photon energy, but ssb by themselves are not considered as lethal to mammalian cells, because they are rapidly repaired through highly efficient systems (Ward *et al.*, 1985).

DNA base damage is also frequent after radiation (Ward, 1986). As compared to ssb, this DNA lesion is regarded as a possible source of mutation and is repaired through specific DNA repair pathways, such as base excision repair and nucleotide excision repair.

The most relevant DNA lesions, able to produce cell death by ionizing radiation, are heterologous double-strand breaks (dsb). These lesions can originate either from a single severe event involving both DNA strands at the same time, or as a result of two or more independent ssb closely apposed on the deoxyribose backbone of the DNA. Although also this lesion can be repaired, cell survival in normal and cancer cells depends on the dsb frequency as well as on the efficiency of the DNA repair system. At least two distinct pathways, involving some four discrete complexes, facilitate the repair of dsb (Hendrickson, 1997). The first is called non-homologous recombinational repair and does not require homology of DNA sequence for rejoining of breaks. It acts at the transition G_1/S phase of the cell cycle, involving the DNA-dependent protein kinase (DNA-PK) and the RAD 50 complex. The second is referred to as homologous recombinational repair and requires extensive homology between the regions of DNA with a dsb and the repair template. This pathway acts mainly in S phase and requires both the RAD52 and the breast cancer predisposition genes (BRCA) 1/2 complexes. In addition to the absolute number, qualitative differences in the nature of dsb and their proximity to other DNA lesions can hamper the reparability, thereby facilitating the activation of apoptotic pathways. Interestingly, when cell perceives that the rate of DNA damage is increased, it blocks the cell cycle machine in a manner augmenting the activity of the DNA repair systems described above. This phenomenon is obtained through a cascade that is activated within a few minutes after irradiation and where a key role is played by the transcription factor sensitive to the oxidative stress such as AP-1 and NF-kB. A representative outline of such cascade is shown in Figure 1.

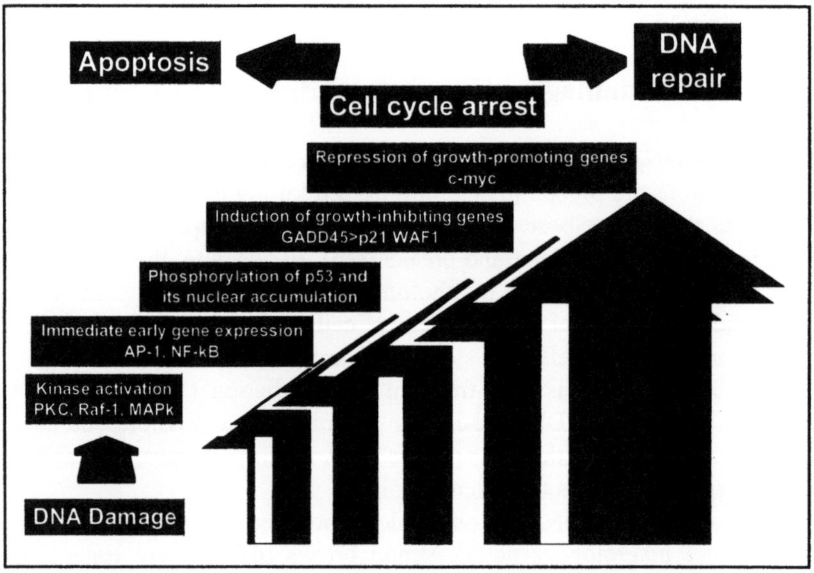

Figure 1. Outline of the biochemical pathway primed by ionizing radiation in human cells.

The effector phase of response to ionizing radiation passes through p53 and its downstream genes such as p21^{waf1}. The relevance of p53, in mediating the biological effect of irradiation, is evidenced by the fact that thymocytes from p53$^{-/-}$ mice are extremely resistant to the cell cycle arrest and induction of apoptosis in response to X-rays (Lowe *et al.*, 1993). p53 acts as a tumor-suppressing gene with a complex mechanism. It binds to promoters of a large number of genes involved in the cell cycle control, DNA repair and apoptosis, resembling a conductor with the role of co-ordinating and regulating the responses to the external signals promoting the opposite programs of proliferation and cell death. If the function of p53 is lacking, an unbalanced ratio between signals leading to cell proliferation and apoptosis occurs, and this step is the "essence" of the oncogenic transformation in the vast majority of solid tumors (Oren, 1992). This explains the high frequency of mutations either at p53 level or at downstream p53-activated genes. Unfortunately, for the same reason, in keeping with the observations from $p53^{-/-}$ mice, solid tumors are reluctant to activate radiation-induced apoptosis. These genetic features, along with additional metabolic considerations reported below, contribute to explain the inherent resistance to radiotherapy of solid tumors.

2.3 Mitochondrial control of radiation-induced cell death

The central dogma of apoptosis is that all the "initiating" pro-apoptotic stimuli converge on the mitochondrial compartment. Thus, although apoptosis can be initiated elsewhere, the execution phase of apoptosis induced by ionizing radiation needs mitochondria. How do DNA lesions trigger mitochondria? Several metabolic pathways connect mitochondria to the nucleus.

Mitochondria are compartmentalised by two membranes. The outer membrane is permeable to small molecules (<6000 Da), whereas the inner membrane is virtually impermeable, with the exception of a small number of metabolites and ions transported through the inner membrane by specific channel systems. The impermeability of the inner membrane allows the formation of a space (matrix) in which the electrochemical gradient is coupled to the ATP production by the mitochondrial respiratory chain. Disruption of the inner membrane dissipates the mitochondrial transmembrane potential ($\Delta\psi_m$) and the pH gradient, with the consequent release of solutes from the matrix. Disruption of the inner membrane represents the transition from initiation to execution phase of apoptosis. This point of no-return from death is due to the escape of constituents from the matrix, as well as release of cytochrome c (Liu *et al.*, 1996) and apoptosis inducing factor (AIF) (Susin *et al.*, 1996). Cytochrome c release activates caspase-9 in concert with the cytosolic factors dATP and Apaf-1, and, subsequently, triggers caspase-3 (Li *et al.*, 1997). Before that mitochondrial inner membrane is disrupted, mitochondria have an alternative way to activate apoptosis, also in the presence of an intact inner membrane. A mega-channel, called permeability transition pore complex (PT), mediates the efflux of solutes through the inner membrane (McEnery *et al.*, 1992). In normal conditions, PT is closed, thereby preventing the leakage of the electrochemical gradient. PT consists of several proteins, including adenine nucleotide translocator (ANT), the voltage-dependent anion channel (VDAC), and the peripheral benzodiazepine receptor (PBR). In pro-apoptotic conditions PT is opened with the consequent activation of the apoptotic pathway. Bcl-2 family proteins regulate the opening/closure of the

PT, representing the actual rheostat through which nucleus modulate sensitivity/resistance to the external stimuli. Some of these proteins, such as Bax and Bak increase the conductivity of PT (Narita *et al.*, 1998), thereby facilitating apoptosis, whereas others (such as Bcl-2 and Bcl-X_L) inhibit cell death by hampering the opening of PT (Shimizu *et al.*, 1998).

Downstream from DNA damage, p53 is accumulated into cell nucleus and activates the transcription of Bax (Findley *et al.*, 1997). Therefore, Bax levels will depend on the length of the DNA repair process. If the amount of Bax overcomes a given threshold, cell will activate apoptosis, thereby removing a cell potentially transformed by incomplete DNA repair.

2.4 Energetic collapse and mitochondrial initiation of apoptosis

Only recently, Taneja *et al.* (2001) has demonstrated that in a cell free system, irradiated mitochondria can initiate the apoptotic cascade without the involvement of the nucleus. This means that mitochondria not only participate to apoptosis in the execution phase, but they can initiate apoptosis by themselves. What are the mechanisms underlying initiation of apoptosis in mitochondria? Mitochondria represent for cells the main energetic source and generation of ATP is coupled to an endogenous ROS production dependent on mitochondrial respiratory chain. When cells are stimulated, ATP consumption is increased and increasing amounts of ADP stimulate a burst of mitochondrial activity, with further increase of endogenous ROS production (Ferlini *et al.*, 1995). Something similar happens during X-ray

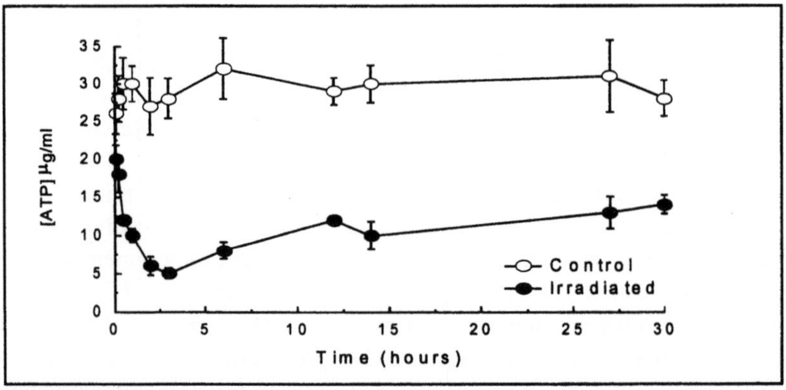

Figure 2. Line chart showing intracellular ATP concentration in freshly isolated peripheral blood mononuclear cells, control (open circle) or irradiated with a dose of 4 Gy X-ray (closed circle). Bars represent SD of triplicates of a single experiment. ATP concentration was normalised for mg of total protein content (Cristiano Ferlini, unpublished observations).

induced apoptosis. As shown in Figure 2, within a few minutes of irradiation with a dose able to half-maximally inhibit cell growth, a slight decrease in the ATP production is noticed.

Due to the lowered efficiency of mitochondria, intracellular ADP concentration increases. The increased amount of ADP stimulates mitochondrial respiration, thereby increasing the endogenous ROS production dependent on the mitochondrial activity. Such a mechanism, within two hours of irradiation, leads to secondary mitochondrial damage and consequent energetic collapse. This secondary oxidative stress explains the delayed depression in ATP production observable after irradiation. Once reached, the fate of the cell depends on: i) the ability in switching to an anaerobic production of ATP via, for example, glycolytic pathways; ii) the capacity to sustain the oxidative stress. The tolerance threshold of ROS is determined by the intracellular thiol levels, which inactivate ROS effects. Both these mechanisms are efficient in solid tumors. The imbalance from blood supply and energetic needs of cancer cells induces within solid tumors a chronic condition of hypoxia. By this way, cancer cells with efficient anaerobic metabolism and high levels of thiols are selected. Along with the genetic properties, these features help to explain the relative radio-resistance of solid tumors (Brown, 1999).

3. CELL DEATH INDUCED BY IONIZING RADIATION

3.1 Apoptotic and non-apoptotic cell death

Radiation-induced apoptosis occurs in normal cells, in hematopoietic tumors where genetic defects of the proteins involved in the apoptotic program, such as p53 or $p21^{waf1}$ are rare. Noteworthy, it takes place also in the stroma of solid tumours, thereby explaining the radiation-induced antiangiogenic effect. On the other hand, cancer cells in non-hematopoietic tumours are often resistant to radiotherapy and exhibit patterns of cell death occurring after multiple rounds of replication. This kind of cell death is referred to either as mitotic (reproductive) or clonogenic cell death (Hendry *et al.*, 1997). The mechanisms underlying this alternative mode of cell death are probably dependent on misrepairing of DNA lesions, with consequent genetic aberrations incompatible with cell survival (Limoli *et al.*, 1997). Thus, mitotic cell death does not follow a co-ordinate program and "per se" can be considered the hallmark of resistance to radiation-induced apoptosis.

The morphology of apoptosis induced by ionizing radiation is not different from that found in other kinds of apoptosis. A representative image of this sequence is shown in Figure 3. The first sign of cellular stress is represented by chromatin condensation into crescents along the nuclear envelope and in perinucleolar positions. This pattern is in common with all the stressing conditions and is promptly reversible if cell can tolerate the stress. The point of no-return from death is determined by the nuclear shrinkage, whereas the typical extreme chromatin condensation and nuclear fragmentation are late events in the apoptotic process.

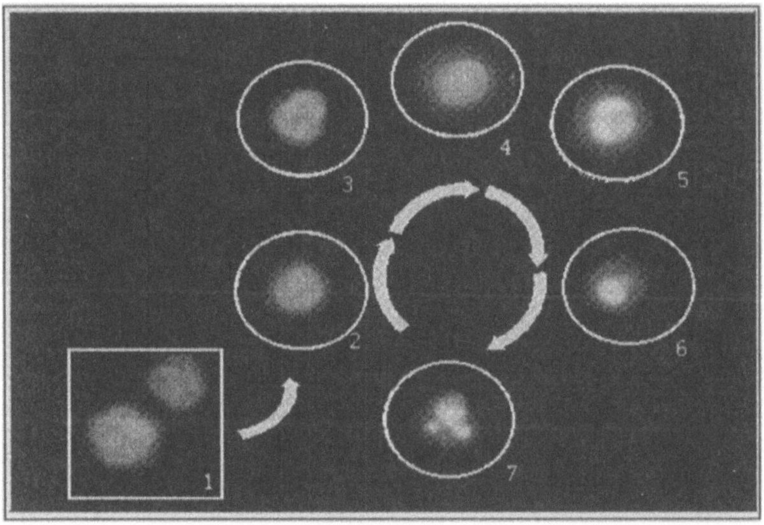

Figure 3. Representative morphology of Jurkat cells undergoing apoptosis after X-radiation (dose 8 Gy) stained with the green florescent DNA dye YOYO-1 and visualised at 800X. From 1 (control cells) to 7 the sequence of morphological features is depicted. The first sign of cellular stress is represented by chromatin condensation into crescents along the nuclear envelope and in perinucleolar positions (step 2). Chromatin condensation further increases (step 3) and nuclear shrinkage occurs (step 4). Condensation still continues and is followed by marginalisation along the nuclear membrane with a further nuclear shrinkage (step 5). Subsequently, chromatin is all condensed in a single nuclear pole (step 6). Condensed chromatin is later fragmented (step 7), yielding the typical apoptotic Figure.

3.2 Chromatin changes induced by ionizing radiation

The ultimate event in the apoptotic process is the activation of endonucleases cutting DNA in fragments of 50 kB and, later, in smaller fragments, responsible for the appearance of the classical DNA laddering on agarose gel electrophoresis (Wyllie, 1980). This event is concomitant with the late phases of apoptosis and requires functionality of the pathways of the execution phase of apoptosis. Therefore, in a number of apoptotic models, we do not observe DNA laddering.

In our laboratory, we focused our attention on the earlier chromatin changes, occurring before that endonucleases can attack DNA (Ferlini *et al.*, 1996). Indeed, as mentioned above, chromatin structure defends the genome by the action of ROS and other potentially mutagenic compounds. In physiologic conditions, DNA is "attackable" either after cell activation and entry into the proliferative phases of the cell cycle, or in M phase when chromatin is fully packaged into chromosomes. The opposite situation occurs in G_0 phase of the cell cycle. Complete absence of the proliferative activity provides chromatin structure greatest protection against DNA injury. Thus, to evaluate changes in chromatin structure during early radiation-induced apoptosis, we used freshly isolated peripheral blood lymphocytes which are synchronised in G_0 phase of the cell cycle. Using this model, we isolated and stained nuclei in a time course study with propidium iodide (PI) as a red DNA dye, and with fluorescein-isothiocyanate (FITC) as a basophilic green stain for nuclear proteins. One hour after irradiation the first change we noticed in treated cells was an increase of FITC fluorescence. Since such increase was not dependent on an actual augmentation of the nuclear protein content, we concluded that changes of chromatin structure increases accessibility of nuclear basic proteins (histones). From six to sixteen hours from irradiation, also DNA stainability increases, generating the transient upward shift of the DNA peak. This "hyperdiploid" peak was observable also with another small DNA dye ethidium bromide (EB), but not with the much bulkier 7-aminoactinomycin D (7-AAD), thereby confirming that changes in chromatin structure expose DNA binding sites for small florescent intercalant dyes. Later, after activation of endonucleases, apoptotic cells lost DNA, thereby generating the classical flow cytometric "hypodiploid" peak, hallmark of the apoptotic process in a vast number of models. Remarkably, this phenomenon, linked to an actual lowering of the DNA content, was detected by the three DNA stains regardless of their molecular mass. Interestingly, when radiation is lowered to a non-lethal dose, we observed similar changes to those occurring during the earlier phases of apoptosis, such as the increase of stainability of nuclear basic proteins by FITC, suggesting that DNA repair and apoptosis follow initially common pathways that only later will diverge.

3.3 Sequence of events during apoptosis by ionizing radation

In the previous paragraph changes occurring at the level of chromatin structure have been described. All these events cannot be considered from those taking place in the cytoplasm. Indeed, it is very likely that, albeit classical definition of apoptosis is based upon morphological criteria involving mainly the nucleus, cytoplasmic events are pivotal in determining the cellular fate after irradiation.

A key feature of apoptosis is that, similar to the cell cycle, the duration of apoptosis is variable and the process is asynchronous in most cell populations. Consequently, there are variable proportions of cells in distinct phases of apoptosis and a sample, collected from a culture in which apoptosis has been induced, will contain cells in all the phases of apoptosis. This could represent a major technical problem in the investigation of apoptosis and requires measurement of metabolic apoptotic changes concomitantly with established markers of apoptosis progression. Using the same cellular model described above, we investigated metabolic alterations induced by X-rays using flow cytometry and a panel of fluorochromes

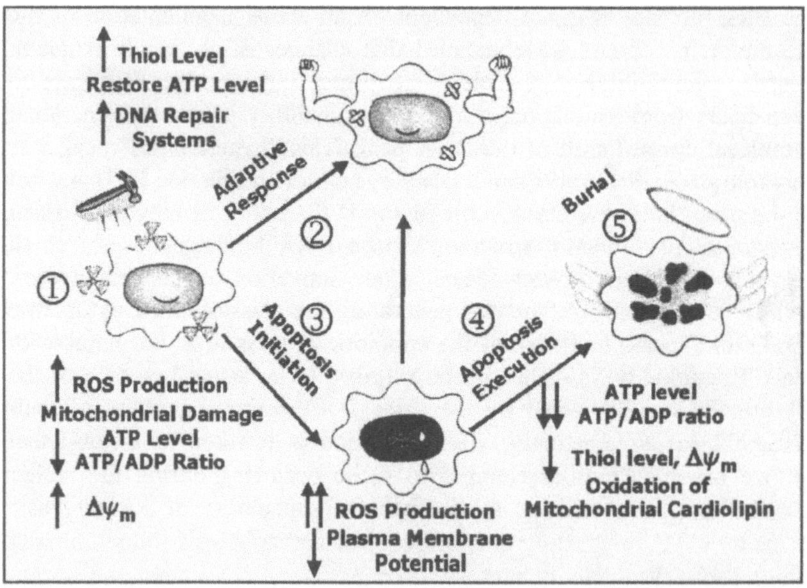

Figure 4. Outline of the cellular respose to ionizing radiation. 1.Early changes induced by radiation. 2. Adaptive response. 3. Initiation of apoptosis. 4. Execution of apoptosis. 5. Late phases of apoptosis. The point of no return from death is the transition from 3 to 4.

designed to evaluate $\Delta\psi_m$, mitochondrial structure, plasma membrane potential and cellular proxidant/antioxidant balance (Ferlini *et al.*, 1999). As a marker of apoptosis progression, we used the staining with EB. In living cells, intercalation within nuclear DNA is prevented by the chromatin structure and by the plasma membrane impermeability. During the apoptotic process, the initiation phase of apoptosis takes place in EB-negative cells, whereas the transition to the execution is accompanied to a faint EB staining, possibly dependent on the first changes in chromatin structure described above. Brighter EB fluorescence, due to the lack of plasma membrane selectivity, is the final burial step. Thus, taking advantage of the multivariate flow cytometric approach, we established the sequence of the metabolic changes associated with X-ray induced apoptosis. This sequence is summarised in Figure 4.

As anticipated, the role played by ROS is pivotal. A dose dependent increase of ROS production was noticed also after sixteen hours from X-ray irradiation. Moreover, a further increase of ROS is the hallmark of the transition from initiation to execution phase of apoptosis, as demonstrated by the fact that EB-positive cells undergo a further ROS increase. The augmentation of proxidants is mirrored by a parallel diminution of the intracellular thiols, reaching its minimum in cells brightly stained by EB. At high X-ray dose, a cell population EB-negative with a consistent increase of thiols emerges. This means that some cells react to the augmented proxidant level and stimulate counteracting mechanisms useful to tolerate a sustained oxidative stress.

Changes in the plasma membrane potential are observed in both EB-negative and EB-positive cells. This alteration reflects the lower ATP availability, because the maintenance of the plasma membrane potential via the electrogenic transporter Na^+/K^+ ATPase is an energy demanding process, consuming 2/3 of the cellular ATP production. It is noteworthy that an increase in the intracellular Na^+ concentration is essential in mediating changes in chromatin structure, stressing the importance of the link between phenomena occurring at cytoplasmic and nuclear levels.

Changes in mitochondrial structure are very relevant during X-ray induced apoptosis. A few hours after irradiation, a hyperpolarisation of $\Delta\psi_m$ is noticed. This likely represents the attempt to restore the depleted ATP levels, stimulating the oxidative burst of surviving mitochondria. If this secondary oxidative stress overcomes the threshold given by mitochondrial thiols, mitochondrial cardiolipin is oxidized and mitochondrial inner membrane allows the leakage of $\Delta\psi_m$ with the consequent initiation of the execution phase.

4. RESISTANCE TO IONIZING RADIATION

4.1 Mechanisms of radioresistance

In principle, the ability to kill cancer cells by photon energy is a question of dose. The problem is that normal cells are often more sensitive than cancer cells to irradiation, so frustrating the administration of an effective radiation dose. Thus, the sensitivity of cancer cells to killing by ionizing radiation is a critical determinant of the probability of cure in patients receiving radiotherapy. Although a single predictive factor of radioresistance has been not yet identified, some relevant mechanisms have been described. Such mechanisms can be divided in two distinct categories: inherent factors depending on the genetic alterations underlying oncogenic transformation of a given cancer; acquired factors determined by the fact that previous treatments act as biological modifiers of the tumour response to ionizing radiation. The knowledge of such mechanisms is important to develop strategies to restore radiosensitivity or to improve the therapeutic index of cancer treatment by ionizing radiation.

4.2 Inherent mechanisms of radioresistance

Several genetic alterations can influence radiosensitivity of cancer cells and radioresponsive tumors are known to allow more easily radiation-induced apoptosis. Therefore, the proper functioning of the apoptotic machinery regulates radiosensitivity. As described above, a crucial role is played by p53, and its downstream genes $p21^{waf1}$ and bax, in activating the apoptotic process. Those cancers with mutations at these levels are expected to be radioresistant.

An additional genetic factor of radioresistance is the presence of activated oncogenes, such as Ras (Gupta *et al.*, 2001). Such oncogene triggers the activation of Raf and, at least three sequential kinase cascades, including MAPK, SAPK/JNK and the p38 pathway. Ras also signals by directly binding and activating the catalytic p110 unit of the phosphoinositide-3-kinase (PI3K), that, phosphorylating phosphatidylinositol phosphates, activates Akt, also known as protein kinase B. This latter mechanism appears as the most relevant mediator of Ras-induced radioresistance. Noteworthy, PI3K can be activated independently of RAS activation. Epidermal growth factor (EGF) can signal through PI3K, and overexpression of the EGF receptor (EGFR) has been correlated with clinical radioresistance in patients with head and neck cancers (Maurizi *et al.*, 1996), and astrocytic gliomas (Zhu *et al.*, 1996).

Finally, growth factors and cytokines induced by radiation treatment can contribute in determining radioresistance. They can stimulate both survival and apoptotic pathways, depending on the integrity of the biochemical signals downstream of the proliferative activity of Myc. In those tumors, where growth factors and cytokines stimulate cell proliferation without activating cell death, we anticipate radioresistance.

Along with these genetic features, a low growth rate and a low blood supply (conditions often present in solid tumors) can affect sensitivity to ionizing radiation with the following mechanisms (Brown, 1999): reduction of the rate of cells in the most sensitive phases of the cell cycle; low efficiency of ROS formation within the irradiated tumor due to hypoxia; clonal selection of cancer cells highly resistant to oxidative stress/apoptosis and with efficient anaerobic metabolism.

4.3 Adaptive response to ionizing radiation

DNA damage induces an adaptive response conserved from prokaryotics to human cells. Since 1977 Samson and Cairns (1977) demonstrated that bacteria exhibit an adaptive response to alkylating agents via the induction of the DNA repair systems. The existence of inducible DNA repair pathways in mammalian cells was later reported by Olivieri *et al.* (1984) also in response to ionizing radiation. In brief, a pretreatment with "sustainable" doses of photon energy results in a reduction in the yield of DNA alterations induced by a subsequent treatment with a higher dose. The same protective effect can be obtained with pretreatment with non toxic doses of H_2O_2 that reduce the lethal effect of gamma rays, stressing the importance of the oxidative stress as the inducer of the adaptive response. Noteworthy, oxidative stress is a common pathway of both radio- and chemo-therapy so that this explains why we often observe cross-resistance to both treatments in relapsing patients. Which are the mediators of the adaptive response? At the nuclear level, DNA damage stimulates *de novo* synthesis of the DNA repair pathways, whereas at the mitochondrial level antioxidant systems are enhanced as well as antiapoptotic proteins such as bcl-2 and bcl-X_L are overexpressed (Skov *et al.*, 1999).

4.4 Chemical modifiers of radiosensitivity

Several compounds have been investigated as modulators of the biological response to ionizing radiation. Two approaches have been pursued: increase of the tolerance to ionizing radiation for normal tissues (radioprotectors), increased biological activity of radiation toward cancer cells (radiosensitizers).

The research of radioprotectors was initiated for military purposes by the US Army at the Walter Reed Army Research Institute and tested approximately 4400 compounds up to 1973. The most interesting drug developed was WR-2721 [S-2(3-aminopropylamino)ethylphosphorothioic acid], thereafter named as amifostine. Amifostine is a pro-drug that generates a free thiol (WR-1065) after dephosphorylation by alkaline phosphatase, thereby increasing the intracellular pool of thiols and the resistance to the oxidative stress. Due to lower levels of alkaline phosphatase in tumor vessels, amifostine underwent clinical trials as a selective protector of normal tissues and not tumors. Although clinical studies confirmed a certain degree of radio and chemoprotection, amifostine has undesirable side effects such as nausea, vomiting and serious hypotension. Moreover, some doubts concerning the selectivity for normal tissues and overall influence on the therapeutic index, cooled enthusiasm for the drug.

The most interesting results in terms of clinical impact have been obtained when radio- and chemotherapy have been combined. In a first approach, both treatments were combined following empirical criteria. More recently, improved knowledge of the mechanism of action of ionizing radiation and the development of drugs targeting specific enzymatic activities have further enhanced the clinical prospective of the integration between physical and medical treatment. Some drugs act by directly increasing ROS production within the most radioresistant hypoxic tumor cells. This is the case of Mitomycin C, Tirapazamine and AQ4N, a di-N-oxide analog of mitoxantrone. These drugs are metabolically activated after the activity of reductase systems, which are most efficient in hypoxic cells. An alternative approach is to combine radiotherapy with antitumour drugs blocking cells at the most radiosensitive cell cycle phases. The prototype of this mechanism is represented by paclitaxel that disrupts normal microtubule turnover and arrests the cell cycle in M phase where segregated chromosomes are highly sensitive to the biological activity of photon energy. Finally, the most interesting application of the recent discoveries about the mechanisms of radioresistance is represented by the use as radiosensitizers of the drugs targeting the signal of the epidermal growth factor receptor.

5. CONCLUSIONS

In this chapter we have summarised the mechanisms through which ionizing radiation kills cancer and normal cells and the strategies useful to increase the curative potential of photon energy. If we consider clinical efficacy and the incidence of side effects visible at the inception of radiotherapy, we are aware of the improvements acquired through the investigation of the cell killing by ionizing radiation. But if we consider the distress resulting from clinical failure of current cancer treatments, we believe that much work still needs to be done. It is likely that in future novel techniques for DNA screening of individual cancers, such as DNA microarray technology, will allow identification of single genetic defects generating apoptosis-resistance in a specific cancer. Physical and medical treatment will be tailored to the specific molecular disorder of the individual case, thereby restoring the sensitivity to apoptosis in cancer cells. Therefore, we are convinced that, never as in this case, the study of cell death will contribute to life and wealth.

ACKNOWLEDGMENTS

We gratefully acknowledge the technical assistance of Mr. Roberto Vitalone, Mr. Angelo Di Pietro, Mrs. Cristiana Gaggini and Mrs. Federica Pignatelli. We also are indebted to Mrs. Simona Ferrarini for the kind drawing of Figure 4 and to Mrs. Ylenia Tiburzi for the secretarial assistance.

REFERENCES

Brown, J.M., 1999, The hypoxic cell: a target for selective cancer therapy--eighteenth Bruce F. Cain Memorial Award lecture, *Cancer Res.* **59**: 5863-5870.

Ferlini, C., Biselli, R., Nisini, R., and Fattorossi, A., 1995, Rhodamine 123: a useful probe for monitoring T cell activation, *Cytometry* **21**: 284-293.

Ferlini, C., Biselli, R., Scambia, G., and Fattorossi, A., 1996, Probing chromatin structure in the early phases of apoptosis, *Cell.Prolif.* **29**: 427-436.

Ferlini, C., De-Angelis, C., Biselli, R., Distefano, M., Scambia, G., and Fattorossi, A., 1999, Sequence of metabolic changes during X-ray-induced apoptosis, *Exp.Cell.Res.* **247**: 160-167.

Findley, H.W., Gu, L., Yeager, A.M., and Zhou, M., 1997, Expression and regulation of Bcl-2, Bcl-xl, and Bax correlate with p53 status and sensitivity to apoptosis in childhood acute lymphoblastic leukemia, *Blood* **89**: 2986-2993.

Gupta, A.K., Bakanauskas, V.J., Cerniglia, G.J., Cheng, Y., Bernhard, E.J., Muschel, R.J., and McKenna, W.G., 2001, The Ras radiation resistance pathway, *Cancer Res.* **61**: 4278-4282.

Hendrickson, E.A., 1997, Cell-cycle regulation of mammalian DNA double-strand-break repair, *Am.J.Hum.Genet.* **61**: 795-800.

Hendry, J.H. and West, C.M., 1997, Apoptosis and mitotic cell death: their relative contributions to normal-tissue and tumour radiation response, *Int.J.Radiat.Biol.* **71**: 709-719.

Li, P., Nijhawan, D., Budihardjo, I., Srinivasula, S.M., Ahmad, M., Alnemri, E.S., and Wang, X,. 1997, Cytochrome c and dATP-dependent formation of Apaf-1/caspase-9 complex initiates an apoptotic protease cascade, *Cell* **91**: 479-489.

Limoli, C.L., Kaplan, M.I., Corcoran, J., Meyers, M., Boothman, D.A., and Morgan, W.F., 1997, Chromosomal instability and its relationship to other end points of genomic instability, *Cancer Res.* **57**: 5557-5563.

Liu, X., Kim, C.N., Yang, J., Jemmerson, R., and Wang, X., 1996, Induction of apoptotic program in cell-free extracts: requirement for dATP and cytochrome c, *Cell* **86**: 147-157.

Lowe, S.W., Schmitt, E.M., Smith, S.W., Osborne, B.A., and Jacks, T., 1993, p53 is required for radiation-induced apoptosis in mouse thymocytes, *Nature* **362**: 847-849.

Maurizi, M., Almadori, G., Ferrandina, G., Distefano, M., Romanini, M.E., Cadoni, G., Benedetti, P.P., Paludetti, G., Scambia, G., and Mancuso, S., 1996, Prognostic significance of epidermal growth factor receptor in laryngeal squamous cell carcinoma, *Br.J.Cancer* **74**: 1253-1257.

McEnery, M.W., Snowman, A.M., Trifiletti, R.R., and Snyder, S.H., 1992, Isolation of the mitochondrial benzodiazepine receptor: association with the voltage-dependent anion channel and the adenine nucleotide carrier, *Proc.Natl.Acad.Sci.U.S.A.* **89**: 3170-3174.

Narita, M., Shimizu, S., Ito, T., Chittenden, T., Lutz, R.J., Matsuda, H., and Tsujimoto, Y., 1998, Bax interacts with the permeability transition pore to induce permeability transition and cytochrome c release in isolated mitochondria, *Proc.Natl.Acad.Sci.U.S.A.* **95**: 14681-14686.

Olivieri, G., Bodycote, J., and Wolff, S., 1984, Adaptive response of human lymphocytes to low concentrations of radioactive thymidine, *Science* **223**: 594-597.

Oren, M., 1992, p53: the ultimate tumor suppressor gene?, *FASEB J.* **6**: 3169-3176.

Roots, R. and Okada, S., 1975, Estimation of life times and diffusion distances of radicals involved in x-ray-induced DNA strand breaks of killing of mammalian cells, *Radiat.Res.* **64**: 306-320.

Samson, L. and Cairns, J., 1977, A new pathway for DNA repair in *Escherichia coli*, *Nature* **267**: 281-283.

Shimizu, S., Eguchi, Y., Kamiike, W., Funahashi, Y., Mignon, A., Lacronique, V., Matsuda, H., and Tsujimoto, Y., 1998, Bcl-2 prevents apoptotic mitochondrial dysfunction by regulating proton flux, *Proc.Natl.Acad.Sci.U.S.A.* **95**: 1455-1459.

Skov, K.A., 1999, Radioresponsiveness at low doses: hyper-radiosensitivity and increased radioresistance in mammalian cells, *Mutat.Res.* **430**: 241-253.

Susin, S.A., Zamzami, N., Castedo, M., Hirsch, T., Marchetti, P., Macho, A., Daugas, E., Geuskens, M., and Kroemer, G., 1996, Bcl-2 inhibits the mitochondrial release of an apoptogenic protease, *J.Exp.Med.* **184**: 1331-1341.

Taneja, N., Tjalkens, R., Philbert, M.A., and Rehemtulla, A., 2001, Irradiation of mitochondria initiates apoptosis in a cell free system, *Oncogene* **19**: 167-177.

Ward, J.F., 1986, Mechanisms of DNA repair and their potential modification for radiotherapy, *Int.J.Radiat.Oncol.Biol.Phys.* **12**: 1027-1032.

Ward, J.F., Blakely, W.F., and Joner, E.I., 1985, Mammalian cells are not killed by DNA single-strand breaks caused by hydroxyl radicals from hydrogen peroxide, *Radiat.Res.* **103**: 383-392.

Wyllie, A.H., 1980, Glucocorticoid-induced thymocyte apoptosis is associated with endogenous endonuclease activation, *Nature* **284**: 555-556.

Zhu, A., Shaeffer, J., Leslie, S., Kolm, P., and El-Mahdi, A.M., 1996, Epidermal growth factor receptor: an independent predictor of survival in astrocytic tumors given definitive irradiation, *Int.J.Radiat.Oncol.Biol.Phys.* **34**: 809-815.

Chapter 10

Ceramidases in the Regulation of Ceramide Levels and Function

Samer El Bawab[*], Cungui Mao[#], Lina M. Obeid[#], and Yasuf A. Hannun[*]
[*]Department of Biochemistry and Molecular Biology; [#] Department of Medicine, Medical University of South Carolina, 114 Doughty Street, Charleston, SC 29425, USA

1. CERAMIDE METABOLISM

Ceramide (Cer) occupies a central role in sphingolipid metabolism (Hannun and Luberto, 2000; Merrill and Wang, 1992). Ceramide metabolism commences in the de novo pathway with the condensation of serine and palmitoyl CoA, resulting in the formation of 3-ketodihydrosphingosine (Merrill and Wang, 1992), which in turn serves as a precursor to dihydrosphingosine (Figure 1). Dihydrosphingosine is then acylated at the 2-amino position to cause the formation of dihydroceramide. It is accepted that the sphingolipid 4-5 double bond is inserted at this point, causing the formation of Cer. In turn, Cer is the precursor of mammalian complex sphingolipids, which are distinguished by substitutions at the 1-hydroxyl position of ceramide. For example, sphingomyelin contains a choline phosphate head group, whereas cerebroside contains a glucosyl or galactosyl subunit. Other complex neutral and acidic glycolipids are based on the cerebrosides through sequential additions of glycose units at the 1-hydroxyl position (Hannun and Luberto, 2000). Reciprocally, the catabolism of complex sphingolipids proceeds in a stepwise fashion through the action of several hydrolytic enzymes that result in the eventual formation of Cer.

Abbreviations: CL, cardiolipin; CDase, ceramidase; Cer, Ceramide; dhCer, dihydroceramide; FB1, fumonisin B1; PA, phosphatidic acid; PS, phosphatidylserine; phytoCer, phytoceramide; SPH, sphingosine; S1P, sphingosine-1-phosphate.

Subcellular Biochemistry, Volume 36: Phospholipid Metabolism in Apoptosis.
Edited by Quinn and Kagan. Kluwer Academic/Plenum Publishers, New York, 2002

Through the action of ceramidases (CDases), Cer is catabolyzed by deacylation to produce sphingosine (SPH) and free fatty acid. Sphingosine can be further phosphorylated to produce sphingosine-1-phosphate (S1P) through the action of sphingosine kinase (Hannun and Luberto, 2000; Merill and Wang, 1992).

Importantly, many of the enzymes of Cer metabolism appear to be physiologically regulated. Thus, regulation of Cer metabolizing enzymes may result in regulation of Cer levels which, in turn, will modulate ceramide-mediated biology. Therefore, many of these enzymes emerge as important candidates in the regulation of ceramide-mediated responses and for pharmacological development of novel chemotherapeutic agents.

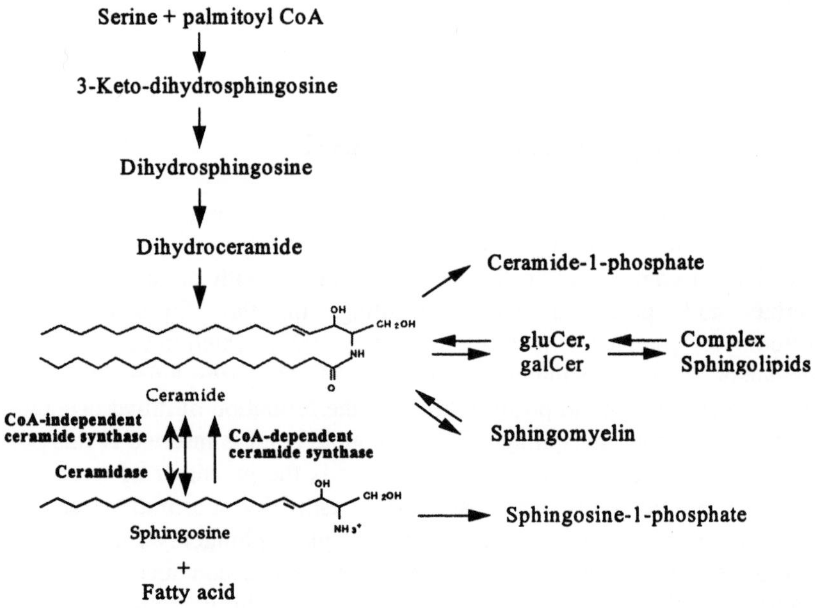

Figure 1. Scheme of sphingolipid metabolism and the central role of ceramide and ceramidases.

2. MECHANISMS OF CERAMIDE ACTION

Ceramide has been shown to activate serine/threonine protein phosphatase termed CAPP (Dobrowsky and Hannun, 1992), and multiple lines of evidence point to an important role for these phosphatases in mediating apoptosis in response to ceramide in mammalian cells (Dobrowsky *et al.*, 1993; Wolff *et al.*, 1994; Reyes *et al.* 1996) and in mediating growth suppression in response to ceramide in yeast cells (Fishbein *et al.,* 1993; Nickels *et al.,* 1996). CAPP has been recently purified and identified as both PP2A and PP1 (Galdari *et al.,* 1998). Some of the noted substrates include c-jun (Reyes *et al.,* 1996), Akt/PKB (Salinas *et al.,* 2000), PKCα (Lee *et al.,* 1996), and bcl-2 (Ruvolo *et al.,* 1999), all important regulators of growth and apoptosis. In addition, Cer has been shown to activate a serine/threonine protein kinase shown later to be the kinase suppressor of ras (Basu *et al.,* 1998), and to activate atypical forms of PKC such as PKCζ (Muller *et al.,* 1995). At the cellular level, ceramide has been shown to inhibit phospholipase D, down regulate c-myc, induce release of PGE2, and regulate a multitude of protein kinases and other signaling pathways, all consistent with roles for ceramide in growth regulation (Hannun and Luberto, 2000; Mathias *et al.,* 1998).

3. CERAMIDE AND APOPTOSIS

Many extracellular agents or inducers of cell injury or stress such as tumor necrosis factor alpha (TNFα), fas ligands, γ-interferon, interleukin-1 (IL-1), nerve growth factor (NGF), radiation, and several chemotherapeutic agents (Hannun, 1996; Mathias *et al.,* 1998) cause the accumulation of endogenous Cer through activation of sphingomyelinases and/or the de novo pathway of sphingolipid synthesis (Perry *et al.,* 2000, Kroesen *et al.,* 2001). The changes in endogenous levels of Cer in response to these agents occur prior to the onset of the first biochemical signs of apoptosis such as activation of caspases (Hannun, 1996; Dbaibo *et al.,* 1997).

In addition, more recent studies are beginning to clearly show that Cer is necessary for several cell responses: 1) increase of intracellular levels of Cer through inhibition of CDases induces cell death (Bielawska *et al.,* 1996); 2) the suppression of Cer formation prevents, at least in part, the cellular response in many cell types. For example inhibitors of the de novo pathway or of neutral sphingomyelinase can inhibit apoptosis in response to several chemotherapeutic agents, angiotensin II, TNFα, and B cell activation (Hannun and Luberto, 2000; Mathias *et al.,* 1998, Kroesen *et al.,* 2001). In some cell types, the formation of Cer in response to extracellular agents (such as retinoic acid, etoposide, angiotensin II, IgM, or daunorubicin) is inhibited by Fumonisin B1 (FB1), an inhibitor of Cer synthase (Bose *et al.,* 1995; Kalen *et al.,* 1992; Plo *et al.,* 1999; Liao *et al.,* 1999; Lehtonen *et al.,* 1999; Xu *et al.,* 1998). This inhibitor also attenuates the apoptotic response of these agents, strongly suggesting an important role for Cer in apoptosis; 3) Ionizing radiation has been shown to cause Cer formation through activation of an acid sphingomyelinase, and animals deficient in this sphingomyelinase become resistant to high doses of irradiation, again demonstrating an important role for Cer in apoptosis (Santana *et al.*, 1996a); and 4) In recent studies (Lavie *et al.*, 1997; Liu *et al.*, 1999), Cabot and co-workers have shown clearly that overexpression of glucosylceramide synthase, which utilizes Cer to generate cerebroside, attenuates the formation of Cer in response to TNFα and to several chemotherapeutic agents. This leads to significant attenuation of apoptosis induced by these agents; again clearly implicating the accumulation of Cer in the apoptotic responses and drug sensitivity of cancer cells.

In particular, in many malignant cell lines (such as leukemia and breast carcinoma cells), Cer rapidly and specifically induces apoptosis, whereas dihydroceramide and other related lipids are inactive (Obeid *et al.*, 1993). Ceramide has been shown to activate proteases of the ICE family (caspases), especially caspase-3, a PARP-cleaving protease (Smyth *et al.*, 1996; Mizushima *et al.*, 1996). Importantly, activation of caspase-3 by Cer and induction of apoptosis are inhibited by overexpression of bcl-2 (Smyth *et al.*, 1996), and bcl-2 does not reduce the levels of Cer produced in response to extracellular agents (Zhang *et al.*, 1996). Taken together, these results suggest that the generation of Cer occurs upstream of the activation of proteases and upstream of the point at which bcl-2 interferes with this activation. Also, activation of PKC attenuates the ability of Cer to induce apoptosis (but does not attenuate other responses to Cer, including the ability of Cer to cause cell cycle arrest). Finally, it has been shown recently that activation of p53 is necessary for Cer formation in response to DNA-damaging agents, suggesting a role for Cer down stream of p53 (Dbaibo *et al.*, 1998).

All together, these results suggest that the increase of Cer levels is part of an orchestrated biochemical process in which Cer accumulation is critical to cell death in particular and to stress responses in general.

4. CERAMIDASES

Ceramidases are enzymes that cleave the N-acyl linkage of Cer into SPH and free fatty acid. They are an emerging class of enzymes composed of multiple isoforms. Historically, these isoforms have been classified as acid, neutral or alkaline, based on the pH optimum of their activities although some isoforms show activity in a broad range. With the recent cloning of several isoforms from yeast, bacteria, and mammals, a genetical distinction and classification of these enzymes can now be employed.

4.1 Acid ceramidase

Acid CDase was first described by Gatt (1963, 1966) in rat brain. The enzyme was found to have a pH optimum of ~5, to be independent of $MgCl_2$, $CaCl_2$ and $MnCl_2$, and the molecular mass was estimated to be ~150 kDa. Three years latter, the same group reported a 200-fold purification of the enzyme from the 27,000 x g membranes of rat brains (Yavin and Gatt, 1969). Subsequently, acid CDase was partially purified from human placenta ~300-fold by Chen *et al.* (1981). The purified enzyme had a pH optimum similar to that obtained from rat brain, but the molecular mass was estimated to be ~47 KDa.

The enzyme has been purified to homogeneity from human urine by the group of Sandhoff (Bernardo *et al.*, 1995) following a purification protocol used for acid sphingomyelinase. The estimated molecular mass of the purified enzyme of ~ 50 kDa on SDS-PAGE (Table I) could be reduced into two subunits of ~ 13 kDa and 40 kDa. The 40 kDa subunit was found to be glycosylated, as treatment with endoglycosidase H or peptido-N-glycanase F reduced the molecular mass to ~30 and ~ 27 kDa, respectively. Using peptide sequences obtained from the purified protein, degenerative oligonucleotides were synthesized and used to screen human fibroblast and pituitary libraries (Koch *et al.*, 1996). A full length cDNA was then assembled from several clones with an open reading frame encoding 395 amino acids. The full length was transiently expressed in Cos-1 cells, and acid CDase activity was found to increase 10-fold. Recently acid CDase was also cloned from mouse tissue and the gene identified (Li *et al.*, 1998). Northern blot analysis of various murine tissues showed highest mRNA levels in kidney and brain, and almost no expression was observed in testis or skeletal muscle. In a very recent study, acid CDase activity has been shown to be activated by Saposins-A, -C, and -D (Linke *et al.*, 2001). In addition, acidic phospholipids such as bis(monoacylglycero)phosphate (BMP) and phosphatidylinositol (PI) were also found to activate this CDase. Acid CDase is located in the lysosomes, and it plays a role in the catabolic pathway of Cer. The inherited deficiency of this lysosomal enzyme causes Farber's disease (Sugita *et al.*, 1975), in which Cer accumulates in lysosomes resulting in this multiorgan disorder.

4.2 Neutral and alkaline ceramidases

A neutral activity has been described in liver plasma membranes (Slife *et al.*, 1989) and in rat intestinal brush border membranes (Nilsson, 1969); little is known about this enzyme. Recently a 70 kDa neutral isoform has been purified and cloned from the culture medium of the bacterium *Pseudomonas aeruginosa* (Okino *et al.*, 1998; Okino *et al.*, 1999). The bacterial enzyme had a pH optimum of 8-9 and appears to require calcium for its activity. The enzyme showed a wide specificity for various Cers (Table I). Using a different purification protocol, the same group also purified this enzyme from mouse liver membranes (Tani *et al.*, 2000a). Contrary to the Pseudomonas enzyme, the activity of the mouse enzyme was not affected by calcium, its molecular mass was estimated to be 94 kDa, and the pH optimum was in the neutral range ~7.5. In another study, the purified protein was sequenced and the cloning of this mouse enzyme from mouse liver and brain was also reported (Tani *et al.*, 2000b). Overexpression of the cloned enzyme increased CDases activity by 900-fold, and Northern blot analysis revealed that the mRNA was expressed widely in mouse tissues, with higher expression in liver and kidney (Tani *et al.*, 2000b).

A membrane bound CDase was purified 22,300-fold to apparent homogeneity from rat brain (El Bawab *et al.*, 1999). The purified enzyme showed an apparent molecular weight of 90 kDa as estimated by SDS-PAGE under reducing conditions and 95 kDa by gel filtration chromatography on Superose 12. The isoelectic point was determined to be ~ 6.5. Using C16-Cer as substrate, the enzyme showed a broad pH optimum in the neutral-alkaline range (Table I). In this study, a Triton X-100/Cer mixed micelle assay was developed, and the enzyme exhibited Michaelis-Menten kinetics. This CDase hydrolyzes Cers preferentially over dihydroceramides. The activity of the purified CDase did not require cations, and it was inhibited by reducing agents. Anionic phospholipids, such as phosphatidic acid (PA), phosphatidylserine (PS), and cardiolipin (CL) stimulated the activity, whereas SPH acted as a competitive inhibitor with a K_i of 1 μM.

Using peptide sequences obtained from the purified rat brain enzyme, the cloning of the human isoform was accomplished (El Bawab *et al.*, 2000). The deduced amino acid sequence of the protein was homologous to three putative proteins from *Arabidopsis thaliana*, *Mycobacterium tuberculosis*, and *Dictyostelium discoideum*, as well as to the purified and cloned enzymes by Ito and co-workers (Okino *et al.*, 1999; Tani *et al.*, 2000b) from *Pseudomonas aeruginosa* and from mouse liver and brain. Figure 2 represents the alignment of all these homologous proteins. Several blocks of amino acids were highly conserved in all of these proteins. Data base search indicated that this CDase gene is absent in the yeast *S. cerevisiae*. Thus, it is very intriguing that lower organisms such as *Mycobacterium tuberculosis* and *Pseudomonas aeruginosa* harbor this CDase specific gene in their genome, while the eukaryotic genome of S. cerevisiae does not. It would be interesting to determine if this is related to the pathogenicity of *Pseudomonas aeruginosa* and *Mycobacterium tuberculosis*.

In addition, El Bawab *et al.* (2000) reported the presence at the N-terminus of the human protein of a signal peptide followed by a putative mitochondrial targeting sequence. A GFP-CDase fusion protein was constructed to investigate the localization of this human enzyme. The results showed that the GFP-CDase fusion protein presented a mitochondrial localization pattern and colocalized with mitochondrial-specifc probes. This putative mitochondrial targeting sequence (amino acids 38-66) is partially absent in the mouse protein and totally absent in all other CDases (Figure 2). Thus, it would be of great interest to know the localization of these other isoforms which lack partially or totally this putative targeting sequence. Northern blot analysis of multiple human tissues showed the presence of a major band corresponding to a size of 3.5 kb, and this major band was ubiquitously expressed with higher levels in kidney, skeletal muscle and heart. The enzyme was then overexpressed in HEK 293 and MCF7 cells and CDase activity (at pH 9.5) increased by 50- and 12-fold, respectively.

4.3 Alkaline ceramidases

Two alkaline CDases were purified from Guinea pig skin epidermis. These two enzymes were membrane bound, and their estimated molecular masses on SDS-PAGE were 60 and 148 kDa, respectively (Yada *et al.*, 1995). No other studies followed on these two proteins.

Very recently, two yeast (*S. cerevisiae*) alkaline ceramidases, phytoceramidase (YPC1p) and dihydroceramidase (YDC1p), were cloned and partially characterized by Mao *et al.* (2000a, b). YPC1p was cloned as a high copy suppressor of the growth inhibition by FB1 as it has fumonisin resistant ceramide synthase activity. The second alkaline ceramidase, YDC1p was identified by sequence homology to YPC1p.

YPC1p and YDC1p have 48% identity over the entire protein sequence, and consist of 316 and 317 amino acids, respectively. YPC1p has a predicted molecular weight of 36.4 kDa and an isoelectric point of 8.5 whereas YDC1p has a molecular weight of 37.2 kDa and an isoelectric point of 6.9. Both YPC1p and YDC1p are integral membrane proteins with several putative transmembrane domains. Sequence analysis shows that YPC1p but not YDC1p contains an ATP binding site shared by diacylglycerol kinase. Both YPC1p and YDC1p have a Golgi-ER retrieval sequence and several phosphorylation sites. Whether the activities of these two alkaline ceramidases are regulated by protein phosphorylation has yet to be investigated. Both enzymes are localized to the ER as revealed by GFP tagging.

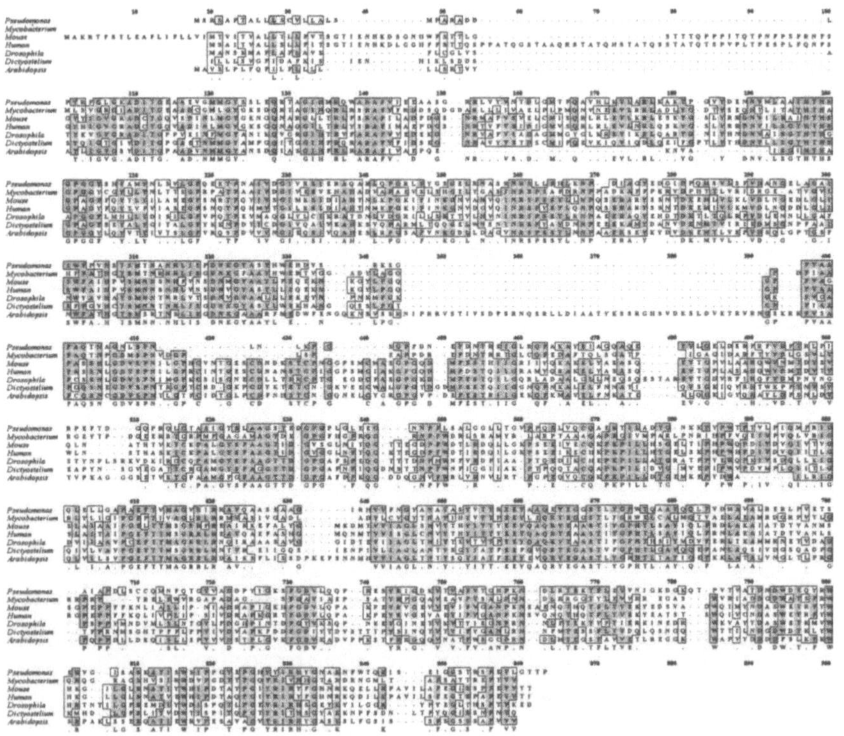

Figure 2. Alignment of the human mitochondrial ceramidase to homologous ceramidases from various organisms. Sequences of ceramidases were aligned using ClustalW algorithm of the Mac Vector program. Amino acids identical are shaded in grey and consensus residues are shown under the alignment. Amino acids of the same group are shaded in white.

YPC1p and YDC1p have different substrate specificity such that YPC1p prefers phytoceramide (phytoCer) over dihydroceramide (dhCer) whereas YDC1p prefers dhCer over phytoCer, however, neither enzyme uses the most common mammalian type ceramide having a 4-5 trans double bond on the sphingoid base as substrate. Both enzymes have a narrow pH optimum of 9.4-10, hence, are classified as alkaline ceramidases. Calcium ions activate but are not absolutely required for the activities of both enzymes. Zn^{2+} and Cu^{2+} inhibit the activities of both enzymes. None of the sphingoid bases inhibit the activities of YPC1p and YDC1p.

YPC1p exhibits reverse activity in cells and *in vitro*. This activity is not inhibited by FB1 and is fatty acyl-CoA independent, suggesting that the reverse activity of YPC1p is distinct from the FB1 inhibitable ceramide synthase activity which uses fatty acyl-CoA as substrate, but not a free fatty acid. YDC1p has only minor *in vitro* reverse activity which catalyzes the formation of dihydroceramide, but not phytoceramide. Its reverse activity was also shown in cells by endowing resistance to FB1 with a lesser potency than YPC1p.

Both YPC1p and YDC1p have an important role in regulating metabolism of ceramide and sphingolipids. Under normal yeast culture conditions, overexpression of YPC1p causes a decrease in phytoCer, dhCer, and yeast complex sphingolipids, with a concomitant increase in phytosphingosine, dihydrosphingosine, and their phosphates. Overexpression of YDC1p has a similar effect, but only on dhCer or dhCer-related lipids. Deletion of YPC1p or YDC1p have opposite effects on sphingolipids metabolism. Thus, deletion of YPC1p and YDC1p cause an increase in the biosynthesis of all complex sphingolipids. These results suggest that under normal conditions, YPC1p and YDC1p, are ceramidases; however they act as CoA-independent ceramide synthases in the presence of FB1, i.e. they are regulated according to substrate availability. Deletion of YPC1p, YDC1p, or both revealed that these genes are not essential for yeast viability. Interestingly, deletion of YDC1p, but not YPC1p, reduced heat tolerance, suggesting that hydrolysis of dhCer may be involved in heat stress response.

A database search reveals that YPC1p and YDC1p are not homologous to any proteins with known functions, but are homologous to putative proteins from *Arabidoposis, C. elegans*, peptides deduced from EST sequences of human, mouse, pig, zebra fish, and human genomic sequences. A human homologue has been identified and its cDNA has been cloned. Preliminary results show that this human homologue is also an alkaline ceramidase that selectively hydrolyzes phytoCer.

Ceramidases have been very poorly studied enzymes, and it is only recently, with the purification of several isoforms, that detailed biochemical characterization has been performed. The presence of several CDase isoforms reveals that the control of Cer levels is very tightly regulated. In addition, the presence of CDases in different cell compartments suggests significant compartmentation of sphingolipid metabolism and the existence of topologically restricted pathways of Cer production and degradation.

5. REVERSE ACTIVITIES OF CERAMIDASES

Recent studies on CDases have revealed the complex nature of these enzymes. In the original report on CDase, Gatt (1966) proposed that a single protein catalyzes the hydrolysis of Cer (CDase activity) and the synthesis of Cer (Cer synthase activity). By comparing the rates of Cer synthesis and the rates of hydrolysis of oleoyl-CoA to oleic acid by the action of thioesterases in their reaction, in a following report the authors concluded that this Cer synthesis reaction proceeds through a CoA-independent mechanism, by using the released oleic acid (Yavin and Gatt, 1969). It should be noted, that the synthetic activity in these studies was performed at pH 7.4 (due to concerns with the solubility of sodium cholate); thus it was not clear whether the observed reverse activity was caused by the contamination with reverse activities from neutral or alkaline CDases. The reverse activity of CDases was recently confirmed by two groups for CDases isolated from yeast and from mouse (Mao *et al.*, 2000a, Tani *et al.*, 2000b).

Detailed characterization of CDase reverse activity was reported recently using rat brain purified enzyme (El Bawab *et al.*, 2001). Using SPH and palmitic acid as substrates, it was shown that the enzyme exhibited Michaelis-Menten kinetics. The enzyme did not use palmitoyl-CoA as substrate and acted through a CoA-independent mechanism as it has been suggested by Gatt (1966) and reported by Mao *et al.* (2000a) and by Tani *et al.* (2000a). Importantly, and as compared to the CoA-dependent Cer synthase, this activity was not inhibited by FB1. In addition, cell labeling studies with [^{3}H]-SPH, showed that this reverse activity contributes to Cer formation in cells. This was particularly clear when these labeling experiments were performed in the presence of FB1 to inhibit the CoA-dependent ceramide synthase activity. Kinetic and inhibition studies performed on the rat purified enzyme indicated that the reverse reaction follows a random sequential mechanism.

The reverse activity showed a narrow pH optimum in the neutral range, and there was very low activity in the alkaline range, in contrast to the CDase activity. This has also been shown for the purified enzyme from *Pseudomonas aeruginosa* (Kita *et al.*, 2000). The reverse activity has been shown to be modulated by phospholipids, in particular, the effect of PA and CL were intriguing. Whereas both lipids stimulated CDase activity, they inhibited, in the same range of concentrations, the reverse reaction in a dose dependent manner. Based on these results, we suggest possible regulation of this enzyme by intracellular pH and/or by interaction with CL and PA.

It is important to note that these reverse activities of CDases are distinct from the major CoA-dependent ceramide synthase activity that is present in the ER and is thought to be responsible for de novo synthesis of Cer. Further, this reverse activity of CDases, may emerge as a new pathway of Cer formation, different from the known de novo or sphingomyelinases pathways.

Table 1. Biochemical characteristics of known ceramidases

Enzyme	Acid ceramidase	Mitochondrial ceramidase	Alkaline ceramidase	Alkaline ceramidase	Alkaline ceramidase	Alkaline ceramidase
Species	*Human*	*Human*	*Mouse*	*Pseudomonas*	*Yeast*	*Yeast*
Accession #	U70063	AF250847	U52513	AB028646	AF191745	AF214455
MW (KDa)	50	90	94	70	37	37
pH optimum	4-5	7-10	7-8	8-9	9-10	9-10
Substrate specificity	Cer > dhCer	Cer > dhCer	Cer > dhCer	Cer = dhCer > phytoCer	phytoCer > dhCer	dhCer > phytoCer
Cofactors	SAP-A, -C, -D BMP, PI	CL, PA, PS	calcium	calcium	not identified	not identified
Reverse activity	not identified	+	+	+	+	+/-
Reference	Bernardo *et al.* (1995); Linke *et al.* (2001)	El Bawab *et al.* (1999, 2000)	Tani *et al.* (1999, 2000)	Okino *et al.* (1998, 1999)	Mao *et al.* (2000a)	Mao *et al.* (2000b)

6. REGULATION OF CERAMIDE FUNCTION THROUGH CERAMIDASES

Ceramidases are emerging as key enzymes in the regulation of Cer levels and consequently the regulation of Cer mediated biology. Prolonged inhibition of CDases leads to accumulation of Cer and to growth suppression on its own or to potentiation of the effects of Cer on apoptosis. For example, studies using the alkaline CDase inhibitor (N-D-erythro-2-(N-myristoylamino)-1-phenyl-1-propanol (D-MAPP)) have shown that treatment of HL-60 cells with this inhibitor increases Cer levels starting 3h after treatment, resulting in a concentration and time-dependent growth suppression accompanied by an arrest in the G0/G1 phase of the cell cycle of these cells (Bielawska *et al.,* 1996). Other inhibitors of CDases have also been used. N-oleoylethanolamine (NOE) has been shown to potentiate the effect of many agonists through accumulation of Cer. As an example, in the B-lymphocyte cell line WEHI 231, treatment with anti-IgM for >24h, increased Cer levels 3-4 fold, and cells underwent apoptosis. Treatment of these cells with NOE increased both Cer formation and apoptosis and accelerated apoptosis induced by anti-IgM (Wiesner *et al.,* 1997).

In addition, CDase may exert important functions in the regulation of its immediate product (SPH) or the down-stream metabolite (S1P). Current understanding indicates that the major pathway for the formation of SPH is via the degradation of Cer and not from the de novo pathway (Michel *et al.,* 1997). This suggests that CDases are also the key enzymes to regulate levels of SPH and/or S1P. Indeed, several reports have shown the involvement of CDases in the regulation of the intracellular levels of Cer, SPH and /or S1P levels.

Coroneos *et al.* (1995) were the first to show the involvement of CDases in agonist mediated cell responses. Thus, using rat mesangial cells as a model, they showed that platelet-derived growth factor (PDGF) increased CDases activities measured in cell homogenate at pH 7.4 and 9.0 but not at pH 4.5. This activation of CDases led to an increase of SPH (and/or S1P) levels and to proliferation of these cells. Interestingly, this was shown to be induced specifically by growth factors and not by inflammatory cytokines such as IL-1 and TNFa (Coroneos *et al.,* 1995); thus suggesting that CDases may provide a pro-growth stimulus. Another exciting observation from this study was the inhibition of the PDGF-induced CDase activity by the tyrosine kinase inhibitor, genistein, suggesting a possible role of tyrosine phosphorylation in the regulation of CDases.

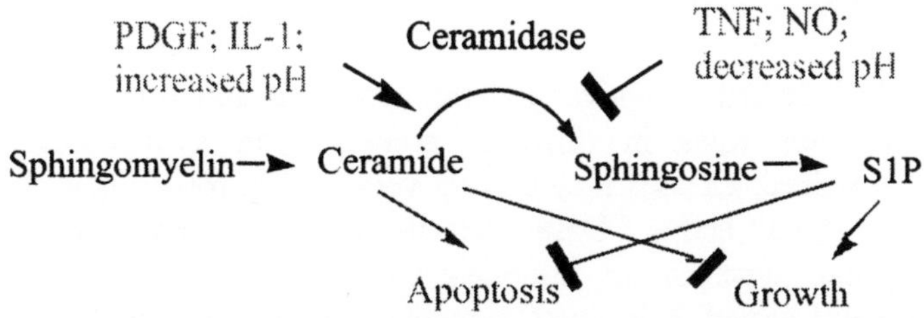

Figure 3. Scheme of ceramidases regulation in agonist mediated cell responses.

It has been shown by multiple groups that TNFα and IL-1 cause activation of sphingomyelinases (Ballou *et al.*, 1992; Mathias *et al*, 1993; Laulederkind *et al.*, 1995; Welsh, 1996; Masamune *et al.*, 1996; Santana *et al.*, 1996b) and result in an increase of intracelluar Cer levels. In certain cell types, recent studies suggested a role for CDases in regulating the net levels of Cer in response to these two cytokines. For example, it has been shown in rat hepatocytes that IL-1β at low concentration activates sphingomyelinases and CDases, resulting in the formation of SPH (or metabolites), while high concentrations of IL-1b, stimulated only sphingomyelinases resulting in the accumulation of Cer (Nikolova-Karakashian *et al.*, 1997). At the physiological level, the high and low treatment with IL-1β resulted in different biological responses. At high concentrations of IL-1β, an induction of a1-acid glycoprotein (AGP) was observed, and this was associated with Cer effects. At low concentrations of IL-1, a down regulation of cytochrome P450 2C11 (CYP2C11) expression was observed, which in this case was associated with SPH (or metabolites) effects. Regulation of CDases by tyrosine phosphorylation was also proposed in this study, as genistein inhibited and vanadate increased IL-1-induced CDase activity.

Similar observations on the role of CDases were also obtained in rat renal mesangial cells stimulated with TNFα or nitric oxide donors. Both of these agonists were shown to stimulate sphingomyelinases, but only nitric oxide donors inhibited CDases and resulted in an increase in Cer levels and DNA fragmentation (Huwiler *et al.*, 1999).

In another line of investigation, Auge *et al.* (1999) showed that treatment of vascular smooth muscle cells with oxidized LDL stimulated sphingomyelinases, CDases, and sphingosine kinase activities, leading to the production of S1P which as these authors suggested, promotes the proliferation of these cells.

7. CONCLUSIONS

The studies described above underscore the potential importance of CDases and their roles in different process such as apoptosis and proliferation. Indeed, CDases may serve a very critical role in how the cell handles a Cer signal: active CDase attenuates a Cer signal and generates a SPH/S1P signal, whereas an inhibited CDase results in a significant accumulation of Cer and in driving Cer-mediated responses. Ample evidence suggests that Cer and S1P exert distinct functions. S1P is known as an important regulator of growth and may serve to promote growth and mitogenesis (Spiegel, 1999), and to specifically counter the apoptotic effect of Cer and Cer-inducing agents (Olivera *et al.,* 1999). Therefore, activation of CDase physiologically may result in promotion of growth. Reciprocally, inhibition of CDase would result in inhibition of growth or apoptosis. Thus, CDases may function as key "switches" in sphingolipid-mediated cell regulation.

The localization of the different CDases isoforms in different cellular compartments is another important observation. For instance, the exclusive presence of a mitochondrial CDase suggests the existence of a specific pool of Cer in mitochondria. Further, the role of mitochondria in apoptosis and the existence of a mitochondrial pathway of cell death have been well established and documented (Budihardjo *et al.,* 1999). According to this emerging understanding, many inducers of apoptosis (including TNFα, chemotherapeutic agents, and Fas ligand) not only activate caspases, but also "activate" mitochondria and cause release of cytochrome c and activation of down stream caspases. Interestingly, most of these agents also cause accumulation of Cer. Together, these observations raise important possibilities on direct interactions between Cer and mitochondria. They also raise possibilities of specific functions of this mitochondrial CDase in cell regulation and in particular in the regulation of apoptosis.

Acid CDase is localized in the lysosomes and it is suggested that it plays an important role in the catabolic pathway of sphingolipids. It has been also suggested recently, to play a role in apoptosis in response to TNFα induced death of L929 cells (Strelow *et al.,* 2000).

The physiological roles of the yeast (and their human homologues) alkaline ceramidases are yet to be established. Since these enzymes regulate a series of bioactive lipids such as ceramide, sphingosine, and sphingosine-1-P, we predict that they have an important role in cell regulatory events.

In conclusion, the CDases field is only at its beginning. The cloning of several isoforms and the availability of molecular and pharmacologic tools will probably impact the advancement of this field in future studies.

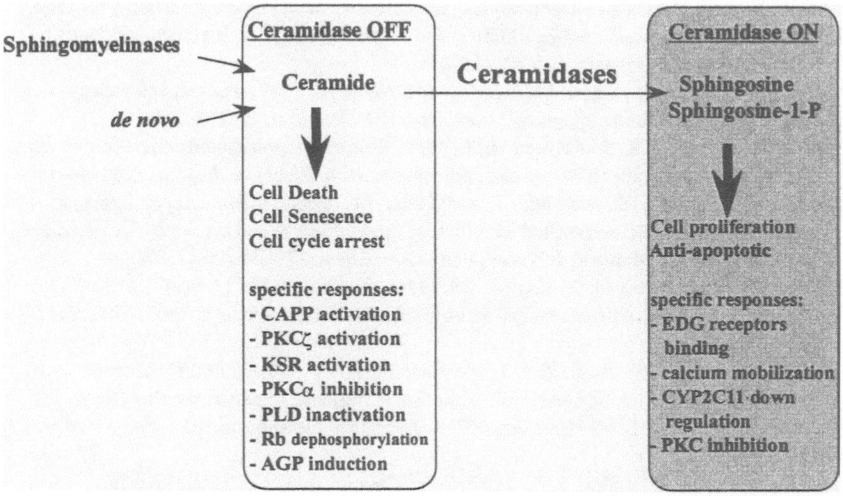

Figure 4. Proposed role for ceramidases as "switches" between cell death and cell proliferation.

ACKNOWLEDGMENTS

This work was supported by NIH grant GM 43825.

REFERENCES

Auge, N., Nikolova-Karakashian, M., Carpentier, S., Parthasarathy, S, Negre-Salvayre, A., Salvayre, R., Merrill, A.H. Jr., and Levade, T., 1999, Role of sphingosine 1-phosphate in the mitogenesis induced by oxidized low density lipoprotein in smooth muscle cells via activation of sphingomyelinase, ceramidase, and sphingosine kinase. *J. Biol. Chem.* **274**: 21533-21538.

Ballou, L.R., Chao, C.P., Holness, M.A., Barker, S.C., and Raghow, R., 1992, Interleukin-1-mediated PGE2 production and sphingomyelin metabolism. Evidence for the regulation of cyclooxygenase gene expression by sphingosine and ceramide. *J.Biol.Chem.* **267**: 20044-20050.

Basu, S., Bayoumy, S., Zhang Y., Lozano, J, Kolesnick, R., 1998, BAD enables ceramide to signal apoptosis via Ras and Raf-1. *J. Biol. Chem.* **273**: 30419-30426.

Bernardo, K., Hurwitz, R., Zenk, T., Desnick, R.J., Ferlinz, K., Schuchman, E.H., and Sandhoff, K., 1995, Purification, characterization, and biosynthesis of human acid ceramidase. *J. Biol. Chem.* **270**: 11098-11102.

Bielawska, A., Greenberg, M.S., Perry, D., Jayadev, S., Shayman, J.A., McKay, C., and Hannun, Y.A., 1996, (1S,2R)-D-erythro-2-(N-myristoylamino)-1-phenyl-1-propanol as an inhibitor of ceramidase. *J. Biol. Chem.* **271**: 12646-12654.

Bose, R., Verheij, M., Haimovitz-Friedman, A., Scotto, K., Fuks, Z. and Kolesnick, R., 1995, Ceramide synthase mediates daunorubicin-induced apoptosis: An alternative mechanism for generating death signals. *Cell* **82**: 405-414.

Budihardjo, I., Oliver, H., Lutter, M., Luo, X., and Wang, X., 1999, Biochemical pathways of caspase activation during apoptosis. *Annu. Rev. Cell Dev. Biol.* **15**: 269-290.

Chen, W.W., Moser, A.B., and Moser, H.W., 1981, Role of lysosomal acid ceramidase in the metabolism of ceramide in human skin fibroblasts. *Arch. Biochem. Biophys.* **208**: 444-455.

Coroneos, E., Martinez, M., McKenna, S. and Kester, M., 1995, Differential regulation of sphingomyelinase and ceramidase activities by growth factors and cytokines. Implications for cellular proliferation and differentiation. *J.Biol.Chem.* **270**: 23305-23309.

Dbaibo, G.S., Pushkareva, M.Y., Rachid, R.A., Alter, N., Smyth, M.J., Obeid, L.M., and Hannun, Y.A., 1998, p53-dependent ceramide response to genotoxic stress. *J.Clin.Invest.* **102**: 329-339.

Dbaibo, G.S., Perry, D.K., Gamard, C.J., and Hannun, Y.A., 1997, Cytokine response modifier A (CrmA) inhibits ceramide formation in response to tumor necrosis factor (TNF)-a: CrmA and Bcl-2 target distinct components in the apoptotic pathway. *J.Exp.Med.* **185**: 481-490.

Dobrowsky, R.T., and Hannun, Y.A., 1992, Ceramide stimulates a cytosolic protein phosphatase. *J.Biol.Chem.* **267**: 5048-5051.

Dobrowsky, R.T., Kamibayashi, C., Mumby, M.C., and Hannun, Y.A., 1993, Ceramide activates heterotrimeric protein phosphatase 2A. *J.Biol.Chem.* **268**: 15523-15530.

El Bawab, S., Bielawska, A., Hannun, Y.A., 1999, Purification and characterization of a membrane-bound non-lysosomal ceramidase from rat brain. *J. Biol.Chem.* **274**: 27948-27955.

El Bawab, S., Roddy, P., Qian, T., Bielawska, A., Lemasters, J.J., and Hannun, Y.A., 2000, Molecular cloning and characterization of a human mitochondrial ceramidase. *J. Biol. Chem.* **275**: 21508-21513.

El Bawab, S., Birbes, H., Roddy, P., Szulc, Z.M., Bielawska, A., and Yusuf A. Hannun., 2001, Biochemical characterization of the reverse activity of rat brain ceramidase: A CoA-independent and fumonisin B1 insensitive ceramide synthase. *J. Biol. Chem.* published February 8, 2001 as 10.1074/jbc.M009331200.

Fishbein, J.D., Dobrowsky, R.T., Bielawska, A., Garrett, S. and Hannun, Y.A., 1993, Ceramide-mediated biology and CAPP are conserved in Saccharomyces cerevisiae. *J. Biol. Chem.* **268**: 9255-9261.

Galadari, S., Kishikawa, K., Kamibayashi, C., Mumby, M.C., and Hannun, Y.A., 1998, Purification and characterization of ceramide-activated protein phosphatases. *Biochemistry* **37**: 11232-11238.

Gatt, S., 1963, Enzymatic hydrolysis of sphingolipids. 1. Hydrolysis and synthesis of ceramides by an enzyme from rat brain. *J. Biol. Chem.* **238**: 3131-3133.

Gatt, S., 1966, Enzymatic hydrolysis of sphingolipids. *J. Biol. Chem.* **241**: 3724-3730.

Hannun, Y.A, and Luberto, C., 2000, Ceramide in the eukaryotic stress response. *Trends Cell. Biol.* **10**: 73-80.

Hannun, Y.A., 1996, Functions of ceramide in coordinating cellular responses to stress. *Science* **274**: 1855-1859.

Huwiler, A., Pfeilschifter, J., and Van den Bosch, H., 1999, Nitric oxide donors induce stress signaling via ceramide formation in rat renal mesangial cells. *J. Biol. Chem.* **274**: 7190-7195.

KalŽn, A., Borchardt, R.A. and Bell, R.M., 1992, Elevated ceramide levels in GH4C1 cells treated with retinoic acid. *Biochim. Biophys. Acta* **1125**: 90-96.

Kita, K., Okino, N., and Ito, M., 2000, Reverse hydrolysis reaction of a recombinant alkaline ceramidase of Pseudomonas aeruginosa. *Biochim. Biophys. Acta.* **1485**: 111-120.

Koch, J., Gartner, S., Li, C.M., Quintern, L.E., Bernardo, K., Levran, O., Schnabel, D., Desnick, R.J., Schuchman, E.H., and Sandhoff, K., 1996, Molecular cloning and characterization of a full-length complementary DNA encoding human acid ceramidase. Identification of the first molecular lesion causing Farber disease. *J. Biol. Chem.* **271**: 33110-33115.

Kroesen, B., Pettus, B., Luberto, C., Busman, M., Sietsma, H., De Leij, L., and Hannun, Y.A., 2001, Induction of apoptosis through B-cell receptor crosslinking occurs via de novo generated C16 ceramide and involves mitochondria. *J. Biol. Chem.* published January 17, 2001 as 10.1074/jbc.M009517200.

Lavie Y., Cao H.t., Volner A., Lucci A., Han T.Y., Geffen V., Giuliano A.E., and Cabot M.C., 1997, Agents that reverse multidrug resistance, tamoxifen, verapamil, and cyclosporin A, block glycosphingolipid metabolism by inhibiting ceramide glycosylation in human cancer cells. *J. Biol. Chem.* **272**: 1682-1687.

Laulederkind, S.J.F., Bielawska, A., Raghow, R., Hannun, Y.A. and Ballou, L.R., 1995, Ceramide induces interleukin 6 gene expression in human fibroblasts. *J. Exp. Med.* **182**: 599-604.

Lee J.Y., Hannun Y.A., Obeid L.M., 1996, Ceramide inactivates cellular protein kinase C alpha. *J. Biol. Chem.* **271**: 13169-13174.

Lehtonen, J.Y.A., Horiuchi, M., Daviet, L., Akishita, M. and Dzau, V.J., 1999, Activation of the de novo biosynthesis of sphingolipids mediates angiotensin II type 2 receptor-induced apoptosis. *J. Biol. Chem.* **274**: 16901-16906.

Li, C.M., Hong, S.B., Kopal, G., He, X., Linke ,T., Hou, W.S., Koch, J., Gatt, S., Sandhoff, K., and Schuchman, E.H., 1998, Cloning and characterization of the full-length cDNA and genomic sequences encoding murine acid ceramidase. *Genomics.* **50**: 267-274.

Liao, W.C., Haimovitz-Friedman, A., Persaud, R.S., McLoughlin, M., Ehleiter, D., Zhang, N., Gatei, M., Lavin, M., Kolesnick, R., and Fuks., Z., 1999, Ataxia telangiectasia-mutated gene product inhibits DNA damage-induced apoptosis via ceramide synthase. *J. Biol. Chem.* **274**: 17908-17917.

Linke, T., Wilkening, G., Sadeghlar, .F, Mozcall, H., Bernardo, K., Schuchman, E., and Sandhoff, K., 2001, Interfacial Regulation of Acid Ceramidase Activity. Stimulation of ceramide degradation by lysosomal lipids and Sphingolipid Activator Proteins. *J. Biol. Chem.* **276**: 5760-5768.

Liu, Y.Y., Han, T.Y., Giuliano, A.E. and Cabot, M.C., 1999, Expression of glucosylceramide synthase, converting ceramide to glucosylceramide, confers adriamycin resistance in human breast cancer cells. *J. Biol. Chem.* **274**: 1140-1146.

Mao, C., Xu, R., Bielawska, A., and Obeid, L.M., 2000a, Cloning of an alkaline ceramidase from Saccharomyces cerevisiae. An enzyme with reverse (CoA-independent) ceramide synthase activity. *J. Biol. Chem.*, **275**: 6876-6884.

Mao, C., Xu, R., Bielawska, A., Szulc, Z.M., and Obeid, L.M., 2000b, Cloning and characterization of a Saccharomyces cerevisiae alkaline ceramidase with specificity for dihydroceramide. *J. Biol. Chem.* **275**: 31369-31378.

Masamune, A., Igarashi, Y. and Hakomori, S.. 1996, Regulatory role of ceramide in interleukin (IL)-1 -induced E-selectin expression in human umbilical vein endothelial cells. Ceramide enhances IL-1 action, but is not sufficient for E-selectin expression. *J. Biol. Chem.* **271**: 9368-9375.

Mathias, S., Pena, L.A., and Kolesnick, R.N., 1998, Signal transduction of stress via ceramide. *Biochem. J.* **335**: 465-80.

Mathias, S., Younes, A., Kan, C.C., Orlow, I., Joseph, C. and Kolesnick, R.N., 1993, Activation of the sphingomyelin signaling pathway in intact EL4 cells and in a cell-free system by IL-1. *Science* **259**: 519-522.

Merrill, A.H.Jr., and Wang, E., 1992, Enzymes of ceramide biosynthesis. *Methods Enzymol.* **209**: 427-437.

Michel, C., Van Echten-Deckert, G., Rother, J., Sandhoff, K., Wang, E, and Merrill, A.H Jr., 1997, Characterization of Ceramide Synthesis. A Dihydroceramide desaturase intoduces the 4,5-trans-double bond of sphingosine at the level of dihydroceramide. *J. Biol. Chem.* **272**: 22432-22437.

Mizushima, N., Koike, R., Kohsaka, H., Kushi, Y., Handa S., Yagita, H., Miyasaka, N., 1996, Ceramide induces apoptosis via CPP32 activation. *FEBS Lett.* **395**: 267-271.

Muller, G., Ayoub, M., Storz, P., Rennecke, J., Fabbro, D., Pfizenmaier, K., 1995, PKC zeta is a molecular switch in signal transduction of TNF-alpha, bifunctionally regulated by ceramide and arachidonic acid. *EMBO J.* **14**: 1961-1969.

Nickels, J.T. and Broach, J.R., 1996, A ceramide-activated protein phosphatase mediates ceramide-induced G1 arrest of Saccharomyces cerevisiae. *Genes & Dev.* **10**: 382-394.

Nikolòva-Karakashian, M., Morgan, E.T., Alexander, C., Liotta, D.C., Merrill, A.H.Jr., 1997, Bimodal regulation of ceramidase by interleukin-1beta. Implications for the regulation of cytochrome P450 2C11. *J. Biol. Chem.* **272**: 18718-18724.

Nilsson, A., 1969, The presence of spingomyelin- and ceramide-cleaving enzymes in the small intestinal tract. *Biochim. Biophys. Acta.***176**: 339-347.

Obeid, L.M., Linardic, C.M., Karolak, L.A. and Hannun, Y.A., 1993, Programmed cell death induced by ceramide. *Science* **259**: 1769-1771.

Okino, N., Tani, M., Imayama, S., and Ito, M., 1998, Purification and characterization of a novel ceramidase from Pseudomonas aeruginosa. *J. Biol. Chem.* **273**: 14368-14373.

Okino, N., Ichinose, S., Omori, A., Imayama, S., Nakamura, T., Ito, M., 1999, Molecular cloning, sequencing, and expression of the gene encoding alkaline ceramidase from Pseudomonas aeruginosa. Cloning of a ceramidase homologue from Mycobacterium tuberculosis. *J. Biol. Chem.* **274**: 36616-36622.

Olivera, A., Kohama, T., Edsall, L., Nava, V., Cuvillier, O., Poulton, S., and Spiegel, S., 1999, Sphingosine kinase expression increases intracellular sphingosine-1-phosphate and promotes cell growth and survival. *J. Cell. Biol.* **147**: 545-558.

Perry, D.K., Carton, J., Shah, A.K., Meredith, F., Uhlinger, D.J., and Hannun, Y.A., 2000, Serine palmitoyltransferase regulates de novo ceramide generation during etoposide-induced apoptosis. *J. Biol. Chem.* **275**: 9078-9084.

Plo, I., Ghandour, S., Feutz, A. C., Clanet, M., Laurent, G., and Bettaieb, A., 1999, Involvement of de novo ceramide biosynthesis in lymphotoxin-induced oligodendrocyte death. *Neuroreport* **10**: 2373-2376.

Reyes, J.G, Robayna, I.G, Delgado, P.S, Gonzalez, I.H, Aguiar, J.Q, Rosas, F.E, Fanjul, L.F, and Galarreta, C.M., 1996, c-Jun is a downstream target for ceramide-activated protein phosphatase in A431 cells. *J. Biol. Chem.* **271**: 21375-21380.

Ruvolo, P.P, Deng, X, Ito, T, Carr, B.K, and May, W.S., 1999, Ceramide induces Bcl2 dephosphorylation via a mechanism involving mitochondrial PP2A. *J. Biol. Chem.* **274**: 20296-20300.

Salinas, M., Lopez-Valdaliso, R., Martin, D., Alvarez, A., and Cuadrado, A., 2000, Inhibition of PKB/Akt1 by C2-ceramide involves activation of ceramide-activated protein phosphatase in PC12 cells. *Mol. Cell. Neurosci.* **15**: 156-169.

Santana, P., Pena, L.A., Haimovitz-Friedman, A., Martin, S., Green, D., McLoughlin, M., Cordon-Cardo, C., Schuchman, E.H., Fuks, Z., Kolesnick R., 1996a, Acid sphingomyelinase-deficient human lymphoblasts and mice are defective in radiation-induced apoptosis. *Cell* **86**: 189-199.

Santana, P., Llanes, L., Hernandez, I., Gonzalez-Robayna, I., Tabraue, C., Gonzalez-Reyes, J., Quintana, J., Estevez, F., Ruiz de Galarreta, C.M., and Fanjul, L.F., 1996b, Interleukin-1 stimulates sphingomyelin hydrolysis in cultured granulosacells: Evidence for a regulatory role of ceramide on progesterone and prostaglandin biosynthesis. *Endocrinology* **137**: 2480-2489.

Slife, C.W., Wang, E., Hunter, R., Wang, S., Burgess, C., Liotta, D.C., and Merrill, A.H.Jr., 1989, Free sphingosine formation from endogenous substrates by a liver plasma membrane system with a divalent cation dependence and a neutral pH optimum. *J. Biol. Chem.* **264**: 10371-10377.

Smyth, M.J., Perry, D.K., Zhang, J., Poirier, G.G., Hannun, Y.A. and Obeid, L.M., 1996, prICE: a downstream target for ceramide-induced apoptosis and for the inhibitory action of bcl-2. *Biochem. J.* **316**: 25-28.

Spiegel, S., 1999, Sphingosine 1-phosphate: a prototype of a new class of second messengers. *J. Leukoc. Biol.* **65**: 341-4.

Strelow, A., Bernardo, K., Adam-Klages, S., Linke ,T., Sandhoff, K., Kronke ,M., and Adam, D., 2000, Overexpression of acid ceramidase protects from tumor necrosis factor-induced cell death. *J. Exp. Med.* **192**: 601-612.

Sugita, M., Dulaney, J.T., Moser, H.W., 1972, Ceramidase deficiency in Farber's disease (lipogranulomatosis). *Science* **178**: 1100-1102.

Tani, M., Okino, N., Mitsutake, S., Tanigawa, T., Izu, H., and Ito, M., 2000a, Purification and characterization of a neutral ceramidase from mouse liver. A single protein catalyzes the reversible reaction in which ceramide is both hydrolyzed and synthesized. *J. Biol. Chem.* **275**: 3462-3468.

Tani, M., Okino, N., Mori, K., Tanigawa, T., Izu, H., and Ito, M., 2000b, Molecular cloning of the full-length cDNA encoding mouse neutral ceramidase. A novel but highly conserved gene family of neutral/alkaline ceramidases. *J. Biol. Chem.* **275**: 11229-11234.

Welsh, N., 1996, Interleukin-1-induced ceramide and diacylglycerol generation may lead to activation of the c-Jun NH2-terminal kinase and the transcription factor ATF2 in the insulin-producing cell line RINm5F. *J. Biol. Chem.* **271**: 8307-8312.

Wiesner, D.A., Kilkus, J.P., Gottschalk, A.R., Quint‡ns, J. and Dawson, G., 1997, Anti-immunoglobulin-induced apoptosis in WEHI 231 cells involves the slow formation of ceramide from sphingomyelin and is blocked by bcl-XL. *J. Biol. Chem.* **272**: 9868-9876.

Wolff, R.A., Dobrowsky, R.T., Bielawska, A., Obeid, L.M. and Hannun, Y.A., 1994, Role of ceramide-activated protein phosphatase in ceramide-mediated signal transduction. *J. Biol. Chem.* **269**: 19605-19609.

Xu, J., Yeh, C.H., Chen, S.W., Luming He, Sensi, S.L., Canzoniero, L.M.T., Choi, D.W., and Hsu., C.Y., 1998, Involvement of de novo ceramide biosynthesis in tumor necrosis factor-a cycloheximide-induced cerebral endothelial cell death. *J. Biol. Chem.* **273**: 16521-16526.

Yada, Y., Higuchi, K., and Imokawa, G., 1995, Purification and biochemical characterization of membrane-bound epidermal ceramidases from guinea pig skin. *J. Biol. Chem.* **270**: 12677-12684.

Yavin, E., and Gatt, S., 1969, Enzymatic hydrolysis of sphingolipids. 8. Further purification and properties of rat brain ceramidase. *Biochemistry* **8**: 1692-1698.

Zhang, J., Alter, N., Reed, J.C., Borner, C., Obeid, L.M., and Hannun, Y.A., 1996, Bcl-2 interrupts the ceramide-mediated pathway of cell death. *Proc. Natl. Acad. Sci. USA* **93**: 5325-5328.

Chapter 11

Effect of Ceramides on Phospholipid Biosynthesis and Its Implication for Apoptosis

Arie B. Vaandrager and Martin Houweling
Department of Biochemistry and Cell Biology, Faculty of Veterinary Medicine, University of Utrecht, P.O. Box 80176, 3508 TD Utrecht, The Netherlands

1. INTRODUCTION

Ceramide and other sphingolipids are now recognized as novel intracellular signal mediators. One of the important and regulated steps in the metabolism of sphingolipids is the hydrolysis of sphingomyelin into ceramides by sphingomyelinases. Multiple experimental approaches suggest an important role for ceramides in regulating a diversity of responses such as induction of cell differentiation, inhibition of cell proliferation, and induction of programmed cell death (apoptosis). Although it has been shown that ceramides participate in signal transduction by activating specific serine/threonine kinases, or by stimulating protein phosphatases, the mechanism by which ceramides exert their effects on these cellular processes is not completely elucidated yet. The aim of this review is to discuss the cross-talk between ceramides and phospholipid biosynthesis in relation to apoptosis, with special emphasis on the role of phosphatidylcholine depletion and a disturbance in the balance between glycerolipid- and sphingolipid-derived second messengers in triggering programmed cell death.

2. EFFECTS OF CERAMIDES ON PHOSPHOLIPID BIOSYNTHESIS

2.1 Biosynthesis of phosphatidylcholine and phosphatidylethanolamine

Mammalian cell membranes consists of a variety of lipids. The phospholipids, phosphatidylcholine (PC) and phosphatidylethanolamine (PE) are quantitatively the most important lipids in cellular membranes. PC and PE are synthesized via different pathways in mammalian cells.

2.1.1 Phosphatidylcholine synthesis

The majority of PC is synthesized in mammalian cells by the CDP-choline or Kennedy pathway in the endoplasmic reticulum (Figure 1). In this pathway, choline taken up from the external medium or released in the cytosol by breakdown of choline containing compounds, is first converted to phosphocholine by the enzyme choline kinase (CK) (Ishidate, 1997). There are two isoforms of CK cloned which both can convert also ethanolamine to phosphoethanolamine, albeit with a lesser affinity (Aoyama *et al.*, 2000). Alternatively phosphocholine can be generated by enzymes that preferentially phosphorylate ethanolamine and are therefore designated ethanolamine kinases (EK). As y*et al*so two different EKs are known (EKI 1 and 2; Lykidis *et al.*, 2001).

Subsequently, the enzyme CTP: phosphocholine cytidylyltransferase (CT) attaches a CMP-group to phoshocholine to form CDP-choline in a reaction which is thought to be the rate limiting step under physiological conditions (Kent, 1997). CT is also the most strictly regulated enzyme of the CDP-choline pathway. Its activity is modulated by phospholipids, membrane binding, and by phosphorylation (Clement and Kent, 1999; Cornell and Northwood, 2000). Two different isoforms of CT (CTα and CTβ) have been cloned which have a partly different subcellular and tissue distribution (Lykides *et al.*, 1998; 1999).

Finally PC is made by releasing the CMP group in the process of fusing phosphocholine to diacylglycerol (DAG) by the enzyme CDP-choline: 1,2-diacylglycerol cholinephosphotransferase (CPT; McMaster and Bell, 1997). One cloned isoform of CPT seems specific for CDP-choline, whereas another CPT also can synthesize PE from CDP-ethanolamine and DAG (Henneberry and McMaster, 1999; Henneberry *et al.*, 2000).

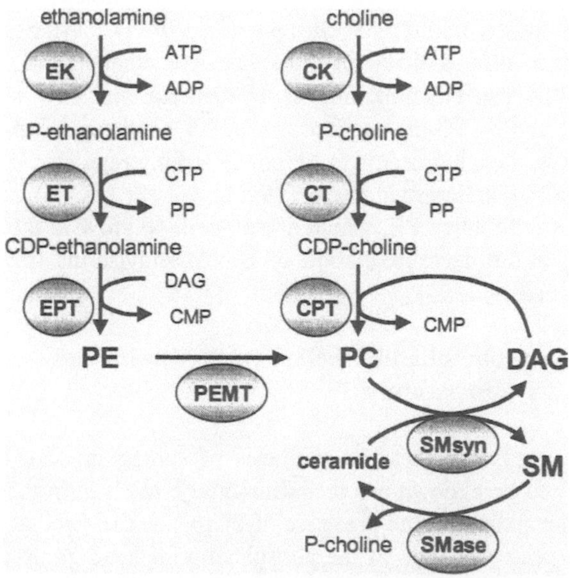

Figure 1. Pathways for the synthesis of phosphatidylcholine, phosphatidylethanolamine and sphingomyelin. Abbreviations: CK, choline kinase; CPT, cholinephosphotransferase; CT, CTP: phosphocholine cytidylyltransferase; DAG, diacylglycerol; PC, phosphatidylcholine; PE, phosphatidylethanolamine; PEMT, phosphatidylethanolamine-N-methyltransferase; SM, sphingomyelin; SMase, sphingomyelinase; SMsyn, sphingomyelin synthase.

A minor pathway to synthesize PC, which is mainly active in liver cells, utilizes the enzyme phosphatidylethanolamine-*N*-methyltransferase (PEMT), which converts phosphatidylethanolamine (PE) to PC by the subsequent transfer of three methyl groups from S-adenosylmethionine (Vance *et al.*, 1997). The PEMT pathway, which links PE synthesis to PC, was found to be critical for PC homeostasis in the liver during choline deficiency (Walkey *et al.*, 1997).

2.1.2 Phosphatidylethanolamine synthesis

PE can be generated by several mechanisms. The CDP-ethanolamine route may be important under physiological conditions *in vivo* when an abundant supply of ethanolamine is present. The CDP-ethanolamine pathway is very similar to the CDP-choline pathway described above (Tijburg *et al.*1989; Bladergroen and Van Golde, 1997). The pathway comprises (i) an ethanolamine kinase (EK) which converts ethanolamine to phosphoethanolamine, (ii) an CTP: phosphoethanolamine cytidyltransferase

(ET), which catalyzes the formation of CDP-ethanolamine, and (iii) an ethanolaminephosphotransferase (EPT), which finally synthesizes PE from DAG and CDP-ethanolamine. As discussed under PC synthesis, four enzymes have been cloned that can phosphorylate ethanolamine, two of which preferentially use ethanolamine as a substrate, and two which are more specific for choline. Only one isoform of ET has been cloned, which contains two active sites, but seems to be not as strictly regulated compared to its counterpart CT (Bladergroen *et al.*, 1999a).

Another pathway to make PE, which allows cells to grow in the absence of ethanolamine, is the decarboxylation of phosphatidylserine (PS) by PS decarboxylase (Voelker, 1997).

2.1.3 Relation of phosphatidylcholine synthesis with the sphingomyelin pathway

It should be noted here that the synthesis of sphingomyelin, the lipid whose synthesis and breakdown are most intricately involved in the process of apoptosis, is dependent on the presence of PC, since the final step in SM synthesis involves the exchange of the phosphocholine head group from PC to ceramide (see Figure 1). Theoretically, therefore, inhibition of PC synthesis may lead to an inhibition of SM generation and an accumulation of pro-apoptotic ceramides. On the other hand, an increased breakdown of SM by SMases as frequently observed in apoptotic cells, was shown to increase the conversion of PC to anti-apoptotic DAG (Sillence and Allan, 1998, for anti-apoptotic action of DAG see below).

2.2 Phosphatidylcholine and phosphatidylethanolamine in cell signaling

PC and PE are the major structural components of mammalian membranes. In addition to their role as structural component of biological membranes and lipoproteins PC and to a lesser extend PE, are important sources for several signal transduction molecules (Exton 1994; Billah and Anthes, 1990; Kiss, 1990; Kester *et al.*, 1989). Agonist-stimulated hydrolysis of PC and PE is particularly important in producing a sustained elevation of diacylglycerol (DAG) levels. After cell stimulation the first wave in DAG, which occurs very rapidly and lasts for about 1 min, comes from the breakdown of phosphatidylinositol (4,5)bisphosphate. This is accompanied by the release of inositol trisphosphate, which triggers Ca^{2+} release from intracellular stores. The simultaneous generation of DAG and Ca^{2+} leads to the activation of the protein kinase C isoforms α, β and γ, which have been implicated in the regulation of cell division (Merrill and

Jones, 1990; Nishizuka 1992 and 1995). The second, more sustained increase in DAG comes from the hydrolysis of PC by PC-specific PLC (direct-route), or indirectly via the sequential activation of phospholipase D (PLD) and phosphatidic acid phosphohydrolase (PAP) to generate PA and DAG, respectively. The second phase of DAG production proceeds without an elevation in the concentration of intracellular Ca^{2+} and might be related to the activation of Ca^{2+}-independent protein kinase C isoforms (Wakelam, 1998).

Recently it was shown that DAG species generated by PLC are primarily polyunsaturated, whereas the ones generated via the combined action of PLD and PAP are saturated/monounsaturated. The latter DAG species were unable to activate protein kinase C in porcine aortic endothelial cells (Pettitt *et al.* 1997). Whether this is a general phenomenon remains to be unravelled.

In contrast to the large amount of studies establishing the role of phospholipids in cell proliferation and differentiation, information on a role of a disturbance in phospholipid (bio)synthesis in apoptosis is scarce. An exception to this is the large amount of research studying the role of phosphatidylserine (PS) in apoptosis. Active maintenance of membrane phospholipid asymmetry is universal in normal cell membranes and its disruption with subsequent externalization of PS is a hallmark of apoptosis (Verhoven *et al.*, 1995; Martin *et al.*, 1995). PS exposure on the outer leaflet of the plasma membrane is a surface change common to many apoptotic cells and has several potential biological consequences, one of which is recognition and removal of the apoptotic cell by phagocytes (Fadok *et al.*, 1992). However, beside, the well established role of PS in apoptosis, the interest in the role of PC depletion in relation to apoptosis has been growing during the last five years (see Section 3).

2.3 Sphingolipids (ceramides) in cell signaling

In analogy with the central role that DAG plays in glycerolipid metabolism, ceramide is an equally critical component in sphingolipid metabolism and signaling (Hannun, 1994; Ghosh *et al.*, 1997). Ceramide serves as a precursor of sphingomyelin (SM), cerebrosides, sulfatides and gangliosides and can be degraded to the PKC inhibitor sphingosine (Nishizuka, 1995; Merrill and Jones, 1990). Although less abundant than glycerolipids, sphingolipids are found ubiquitously in eukaryotic cell membranes and lipoproteins. The key role of SM turnover in signal transduction pathways was first recognised in human promyelocytic leukemia HL-60 cells treated with 1α,25-dihydroxyvitamin D3 (vit. D3) (Okazaki *et al.*, 1989). Since then, also other agonists were found to exert effects through activation of the so-called SM cycle, which is characterized by reversible degradation of SM and concomitant increase in ceramide levels. Ceramide turned out to be the central regulator molecule mediating the effects of these agonists, i.e. inhibition of cell growth, induction of differentiation and apoptosis (Hannun, 1994; Kim *et al.,* 1991; Mathias *et al.*, 1993; Skowronski *et al.*, 1996).

Short-chain ceramides have been found to be valuable tools in studying the action of naturally-occurring ceramides and the mechanisms involved in transducing signals from the plasma membrane to the cell interior (Hannun 1994; Skowronski *et al.*, 1996 and references therein). Exogenously added short-chain ceramides mimic the effects of various agonists like vit. D3 (Okazaki *et al.*, 1990), tumor necrosis factor (TNF-α) (Kim *et al.*, 1991; Obeid *et al.*, 1993) and γ-interferon (Kim *et al.*, 1991) by impairing cell proliferation and inducing differentiation and apoptosis and inhibiting agonist-stimulated PC-PLD (Gomez Munoz *et al.*, 1994; Venable *et al.*, 1994) and PE-PLD (Kiss and Deli, 1995) activity. The mechanism by which short-chain ceramides exert their effects on cells is unclear. It has been proposed that they act as lipid second messengers by altering the phosphorylation state and hence the activity of certain crucial enzymes (Mathias *et al.*, 1998; Hannun 1996 and Chapter 10, this volume). Alternatively, short-chain ceramides may exert part of their effect on cells by interfering with glycerolipid pathways, which is discussed in detail in the next section.

2.4 Cross-talk between ceramide and phospholipid biosynthesis

Recently, it has been shown that cell-permeable ceramides dramatically inhibited the synthesis of the two major membrane phospholipids, PC and PE (Bladergroen *et al.*, 1999b; Allan, 2000). The inhibition of phospholipid synthesis was rapid, within 2 h, and resulted in massive apoptosis after 16-24 h. The mechanism by which short-chain ceramides exert their effect on phospholipid synthesis is possibly cell type dependent. In baby-hamster kidney (BHK) fibroblasts PC synthesis was reduced at the level of CT, the putative rate-determining enzyme in the CDP-choline pathway (Allan, 2000). This conclusion was based solely on radio-label studies in combination with an earlier published observation (Wieder *et al.*, 1995) showing that C_2-SM (the SM generated from C_2-ceramide by SM synthase, which was actively synthesized in the BHK-cells) inhibited CT activity *in vitro*. On the other hand, data obtained from studies with rat-2 fibroblasts clearly showed that short-chain ceramides regulate the synthesis of PC and PE mainly at the final step of the CDP-pathways. This conclusion was based on the following observations: (a) incorporation of [^{3}H]-choline into PC and

[^{3}H]-ethanolamine into PE was inhibited by C_6-ceramide in a dose and time-dependent manner; (b) delayed disappearance of label from CDP-choline and CDP-ethanolamine in pulse-chase experiments indicated impaired conversion of these CDP-metabolites to PC and PE by CPT and EPT, respectively; (c) the activities of CPT and EPT are decreased upon C_6-ceramide treatment (see Figure 2; adapted from Bladergroen *et al.* 1999b). In contrast to BHK cells the activity of CT was not affected significantly in rat-2 fibroblasts by short-chain ceramides.

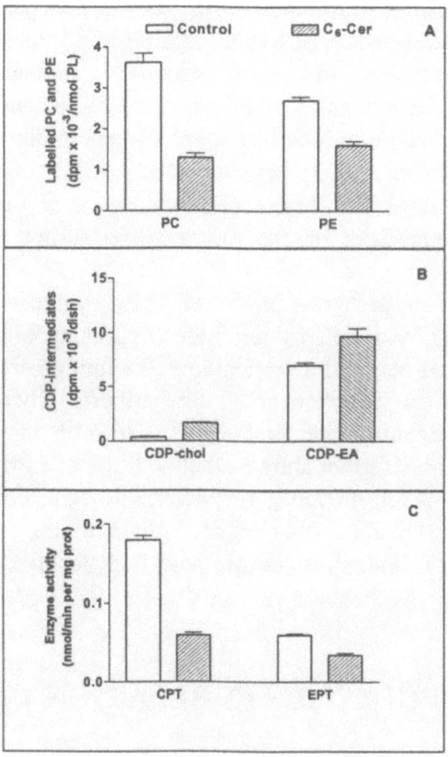

Figure 2. Effect of C_6-ceramide on phosphatidylcholine and phosphatidylethanolamine synthesis in rat-2 fibroblasts. Cells were treated for 2 h in the absence (open bars) or presence (hatched bars) of 25μM C6-ceramide and the following parameters were determined: i) the incorporation of [^{3}H]choline and [^{3}H]ethanolamine into phosphatidylcholine (PC) and phosphatidylethanolamine (PE), respectively (panel A) and in CDP-choline and CDP-ethanolamine, respectively (panel B) and ii) the *in vitro* activity of choline- and ethanolaminephosphotransferase (CPT and EPT) (panel C).

This is in agreement with data showing that natural ceramides do not have any effect on CT activity *in vitro* (Sohal and Cornell, 1990). The mechanism by which cell-permeable ceramides exert their effect on CPT and EPT activity and whether natural or intracellular generated ceramides have a similar effect on phospholpid synthesis is at present unknown.

Decreased PC and PE synthesis by ceramides in combination with the observations that inhibition of PC synthesis induces apoptosis, suggests that in some cell types the effect of ceramides on apoptosis may be exerted at least in part via a disturbance in PC synthesis (see also Section 3).

Treatment of Rat-2 fibroblasts with various cell-permeable ceramides resulted, as discussed above, in a rapid inhibition of phospholipid synthesis. An intriguing observation was that C_2-ceramides inhibited PC biosynthesis without affecting the synthesis of PE via the CDP-ethanolamine route. C_2-ceramide are well known inducers of apoptosis and inhibit DNA- and protein synthesis (Obeid *et al.*, 1993; Jayadev *et al.*, 1995). C_2-ceramide induced apoptosis in the rat-2 fibroblasts (Houweling *et al.*, unpublished data), suggesting that inhibition of PE biosynthesis is not a prerequisite for apoptosis.

A recent report suggests that inhibition of PC synthesis constitutes one of the primary events by which C_2-ceramide triggers apoptosis (Ramos *et al.*, 2000). Treatment of cerebral granule neurons with C_2-ceramide resulted in a rapid (within 1 hour) reduction in PC biosynthesis, whereas only 6 h after exposure to the agonists the first significant drop in cell viability was observed. The authors further showed that addition of exogenous PC resulted in a dose-dependent full prevention of neuronal death. This was specific for PC, because addition of other glycerolipids like, PE, PS, phosphatidylinositol and phosphatidic acid had no effect on C_2-ceramide-induced apoptosis.

3. PHOSPHOLIPIDS IN RELATION TO APOPTOSIS

3.1 Diminished phosphatidylcholine synthesis may lead to apoptosis

The most convincing example that inhibition of PC synthesis can induce apoptosis comes from a cell line with a temperature-sensitive defect in one of the enzymes in the CDP-choline pathway (Cui *et al.*, 1996). The MT58 cell line is a Chinese hamster ovary (CHO) derived cell line with a mutation in CTP: phosphocholine cytidylyltransferase (CT), which renders the enzyme unstable at elevated temperatures. MT58 cells can grow and sustain normal PC levels at the permissive temperature of 33°C. However, shifting these cells to 40°C leads to a 50 % drop in PC levels and a subsequent growth arrest and apoptosis (Esko *et al.*, 1981; Cui *et al.*, 1996). Addition of exogenous PC or lyso PC could rescue the MT58 cells at the non-permissive temperature (Esko *et al.*, 1982), suggesting that indeed a lack of PC or a PC derived compound caused the observed apoptosis. As expected transfection with recombinant wild type CTα or CTβ also allowed MT58 cells to grow normally at the non-permissive temperature (Houweling *et al.*, 1995, Lykidis *et al.*, 1999).

However, surprisingly, transfection with the other PC synthesising enzyme, PEMT, which converts PE into PC, could not sustain growth of the MT58 cells at 40°C (Houweling *et al.*, 1995). It was speculated that PC synthesized by the PEMT pathway was somehow not equivalent to PC made in the CDP-choline pathway. PC derived from PE might have a different subcellular localization or fatty acid composition compared to that synthesized from CDP-choline and DAG (Houweling *et al.*, 1995, DeLong *et al.*, 1999). Alternatively it was suggested that the PEMT pathway in the transfected MT58 cells could not synthesize enough PC to meet the increased demand for this phospholipid at the elevated temperature of 40°C (Waite and Vance, 2000). The mechanism by which an inhibition of PC synthesis in MT58 cells leads to apoptosis is as yet unknown.

Another straightforward approach to lower the amount of PC in a cell is by growing them in a choline-deficient medium. In line with a causative relationship between inhibition of PC synthesis and apoptosis, choline deficiency was shown to induces apoptosis in cell lines as well as *in vivo* (Holmes-McNary *et al.*, 1997). In PC12 cells, choline deficiency-induced apoptosis was associated with a 50 % reduction in PC levels, but also with a considerable decrease in the level of the other choline-containing lipid, SM (Yen *et al.*, 1999). Concomitant with the decreased levels of PC and SM there was an increase in the amounts of their precursors or breakdown products DAG and ceramide, respectively. The increase in ceramide might contribute to the observed choline deficiency-provoked apoptosis, as it was well observed before the onset of the cell death and exogenous cell-permeable ceramides could also induce apoptosis in PC 12 cells. Alternatively, generation of ceramides might be an early process in choline deficiency-induced apoptosis as its formation and the decrease in SM, but not PC, levels could be prevented by stopping the apoptotic process through addition of caspase inhibitors. The formation of ceramides by activation of SMase as a consequence, rather than as a mediator of apoptosis was also observed in other cells (Tepper *et al.*, 1999).

Furthermore, an increasing body of circumstantial evidence is emerging for a role of diminished PC synthesis in apoptosis from observations, that a number of apoptosis-inducing compounds can inhibit PC synthesis. However, it should be kept in mind that most of these compounds might not be completely specific and may also affect other pro-apoptotic pathways, besides PC synthesis. In the following section we will discuss some of the apoptotic compound that affect PC synthesis in more detail.

3.2 Enzymes in the CDP-choline pathway are targets for other pro-apoptotic compounds besides ceramides

3.2.1 Choline kinase inhibitors

It was noticed that a number of tumour cells, especially those transformed with the ras oncogene, have relatively high levels of phosphocholine, the first intermediate in the CDP-choline pathway. Prevention of the generation of phosphocholine by addition of specific inhibitors of choline kinase stopped the growth of the tumor cells and forced them into apoptosis (Hernandez-Alcoceba *et al.*, 1997). The most potent inhibitors studied so far are derivatives from hemicholium-3 (HC-3), which are active in *ex vivo* and in *in vitro* assays in the micromolar range (Figure 3). Recently these potent choline kinase inhibitors were also shown to be active *in vivo* against human tumours transplanted into mice (Hernandez-Alcoceba *et al.*, 1999). The mechanism of the induction of growth arrest and apoptosis by the choline kinase inhibitors is not clear. They might interfere with the generation of the major membrane phospholipid PC, as a disturbance in the synthesis of PC by inhibition of CT or choline deficiency was shown to lead to apoptosis as discussed above. However, no depletion of PC was observed in cells treated with the choline kinase inhibitors (Hernandez-Alcoceba *et al.*, 1999). Therefore, a low level of phosphocholine, which itself was shown to act as a mitogen, was suggested to prevent cell growth in these cells (Hernandez-Alcoceba *et al.*, 1999). The absence of an effect of the choline kinase inhibitors on PC levels might be due to the fact that CT rather than choline kinase is the rate limiting step in PC synthesis. The presence of multiple enzymes that can phosphorylate choline with different sensitivities towards the choline kinase inhibitors may also explain their inability to lower PC levels. As discussed above, there are at least two isoforms of CK, which can both also convert ethanolamine to phosphoethanolamine, and there are also at least two different ethanolamine kinases, which can also form phosphocholine (Aoyama *et al.*, 2000; Lykidis *et al.*, 2001).

3.2.2 CTP: Phosphocholine cytidylyltransferase inhibitors

A number of lysophospholipid analogs with anti-neoplastic activity were shown to inhibit CT, the putative rate-limiting enzyme in the CDP-choline pathway, and to induce apoptosis in tumor cells (Figure 3). These anticancer lipids include, 1-O-octadecyl-2-O-methyl-rac-glycero-3-phosphocholine (ET-18-OCH_3), and hexadecylphosphocholine (HePC), which were both tested with encouraging results in clinical settings as a purging agent in autologous bone marrow transplantation and in the treatment of skin metastases of breast cancer, respectively (Ruiter *et al.*, 2001). The choline head group was shown to be essential for the inhibition of CT and the antiproliferative effect of these alkyl-lysophospholipids as hexadecyl-phosphoserine and hexadecylphosphoethanolamine were not active as inhibitors of PC synthesis and do not induce apoptosis (Geilen *et al.*, 1994). The inhibition of PC synthesis by the alkyl-lysophospholipids was suggested to cause, or at least contribute to, the induction of apoptosis, as addition of lysoPC prevented the cytotoxic effect of both HePC and ET-18-OCH_3 presumably by providing a source for PC synthesis independent of CT activity (Boggs *et al.*, 1998). In case of HePC treatment, lysoPC could even restore normal cell proliferation.

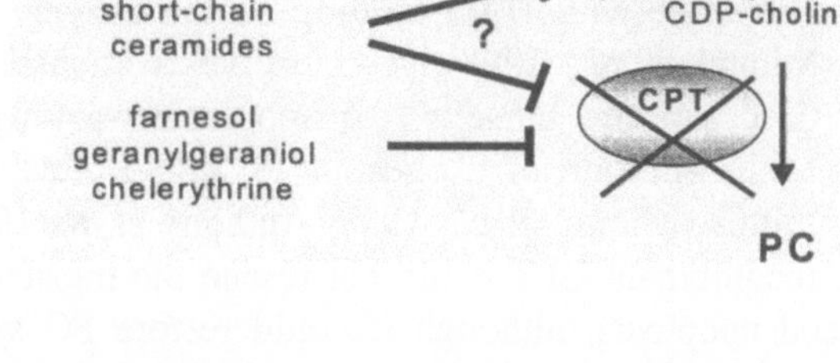

Figure 3. Inhibition of phosphatidylcholine biosynthesis by apoptosis-inducing compounds. The target enzyme for the inhibition of PC biosynthesis is shown for several compounds that have further in common that they all induce apoptosis. Abbreviations are as follows: CK, choline kinase; CPT, cholinephosphotransferase; CT, CTP: phosphocholine cytidylyl-transferase; PC, phosphatidylcholine.

In contrast, LysoPC could not reverse the cytostatic effect of ET-18-OCH_3 (Boggs *et al.*, 1998). Likewise overexpression of CT could prevent the apoptosis induced by ET-18-OCH_3, but could not restore proliferation of the HeLa cells in the presence of this ether lipid (Baburina and Jackowski, 1998).

Besides inhibition of CT, these alkyl-lysophospholipids were shown to affect a number of other processes in biological membranes, which may contribute to their apoptotic action (Ruiter *et al.*, 2001). Other targets of the ether lipids include: stimulation of pro-apoptotic ceramide formation; (Wieder *et al.*, 1996); stimulation of the cellular stress-related, apoptosis inducing, SAP/JNK pathway (Gajate *et al.*, 1998, Ruiter *et al.*, 1999); stimulation of FAS clustering (Gajate *et al.*, 2000); inhibition of the Akt/PK-B survival pathway, inhibition of the MAP/ERK (Zhou *et al.*, 1996); inhibition of phospholipase C (Powis *et al.*, 1992) and protein kinase C activation (Uberall *et al.*, 1991); and stimulation and subsequent downregulation of phospholipase D (Wieder *et al.*, 1996; Lucas *et al.*, 2001). MAP/ERK, phospholipase C and D and protein kinase C are all enzymes implicated in mitogenic signaling. From most of above mentioned effects of the ether phospholipids on membrane processes it is not known whether they are related to their effects on PC synthesis.

3.2.3 Cholinephosphotransferase inhibitors

The cytotoxic drugs farnesol, geranylgeraniol, and chelerythrine were shown to induce apoptosis and to inhibit PC synthesis at the level of CPT (Figure 3) (Voziyan *et al.*, 1993; Miquel *et al.*, 1998; Anthony *et al.*, 1999). It was suggested that these drugs directly decreased the activity of CPT (Miquel *et al.*, 1998; Anthony *et al.*, 1999). Farnesol and geranylgeraniol were found to compete with DAG binding on the enzyme in microsomal preparations (Miquel *et al.*, 1998). A causative relationship between the inhibition of CPT and the induction of apoptosis was suggested by the observation that the addition of exogenous PC could prevent apoptosis by these CPT inhibitors (Miquel *et al.*, 1998; Anthony *et al.*, 1999). However, expression of recombinant CPT could not rescue the transfected cells from farnesol-induced apoptosis, although it could restore PC synthesis in the presence of farnesol in these cells (Wright *et al.*, 2001). This suggests that in case of farnesol, the inhibition of CPT is not absolutely required for the induction of apoptosis, but it does not exclude a contribution of the inhibition of PC synthesis to the induction of apoptosis. Furthermore, Wright *et al.* (2001) could not establish a direct inhibition of recombinant CPT by farnesol in a mixed micell assay, suggesting that the inhibition of CPT in intact cells or isolated microsomes is indirect.

3.3 Diminished phosphatidylethanolamine synthesis is not known to lead to apoptosis

To our knowledge there are no studies showing a direct link between a disturbance in PE synthesis via either the CDP-ethanolamine route or PS decarboxylation and the induction of apoptosis. On the other hand, PE has, although rarely, been implicated as a functional component of the apoptotic process. An example is a publication by Emoto *et al.* (1997). Using Ro09-0198, a peptide that specifically recognizes PE, the authors showed PE exposure in CTLL-2 cells undergoing apoptosis. The exposure of PE correlated well with PS exposure on the outer leaflet. A similar finding was observed in cardiomyocytes, where PE and PS redistribution from the inner- to the outer leaflet precedes reperfusion-induced apoptosis (Maulik *et al.*, 1998). Whereas, the exposure of PS has been linked to phagocytosis of the apoptotic cells by macrophages (Fadok *et al.*, 1992), no physiological role for the exposure of PE in apoptosis was implicated.

4. CROSS-TALK BETWEEN DIACYLGLYCEROL AND CERAMIDE IN THE ONSET OF APOPTOSIS

4.1 The intracellular diacylglycerol-ceramide balance and its relation to apoptosis

An intriguing hypothesis is that the balance between mitogenic and anti-mitogenic lipid second messengers determines the fate of cells towards proliferation, cell arrest and apoptosis (Figure 4). A major player in the mitogenic response is the glycerolipid-breakdown product DAG, whereas ceramides formed upon stimulation of sphingomyelin (SM) hydrolysis has an anti-mitogenic effect. The role of DAG and ceramides in cell signaling has been well established as was discussed in paragraph 2.2 and 2.3, respectively. Ceramide elevation induces cell death in many cell types, and this action can be over come by the presence of DAG. This is of particular interest, as several anti-tumour drugs that prolong their effect via the SM-cycle have been shown to increase DAG production at the same time (Bettaïeb *et al.*, 1999).

The hypothesis that the intracellular DAG to ceramide ratio is a critical determinant in the fate of cells towards cell division, cell cycle arrest or cell death (Figure 4) was recently tested in a T lymphocyte model (Flores *et al.*,

2000). Kit 225 cells, an interleukin 2 (IL-2) dependent human T lymphocyte cell line, was cultured under different conditions so that the cells were in three distinct states; apoptosis, cell arrest and proliferation. Upon withdrawal of IL-2 from the culture medium the cells rapidly entered into apoptosis. This results in a shift of the DAG to ceramide ratio from 1 to 0.1, 72 h after IL-2 withdrawal. On the other hand, addition of IL-2 results in T-lymphocyte proliferation and a shift in the DAG: ceramide ratio from 1 to 10. The ratio of PC to SM, the possible precursors of DAG and ceramide, respectively, changed accordingly (Flores *et al.*, 2000). This suggests that the tightly regulated balance between mitogenic lipids (DAG and PC) and anti-mitogenic lipids (ceramide and SM) would be the indicator of the cellular state. Is there also evidence indicating that this balance not only acts as an indicator of the cellular state, but also as an active component in the decision of cellular fate?

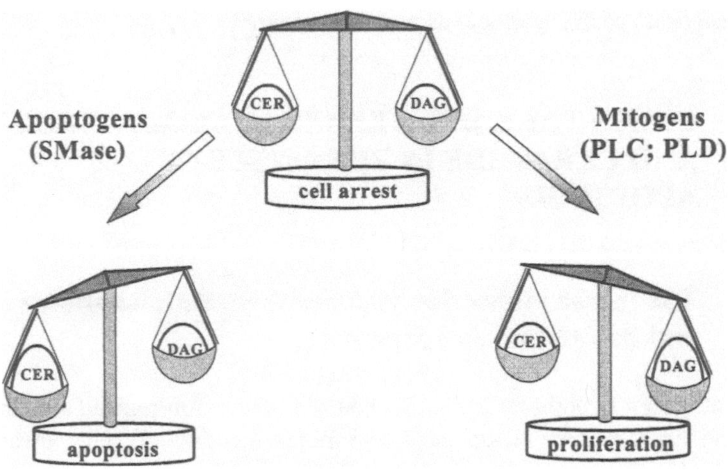

Figure 4. Cartoon showing the role of the diacylglycerol to ceramide ratio in determining the fate of cells. Abbreviations: CER, ceramide; DAG, diacylglycerol; PLC, phospholipase C; PLD, phospholipase D; SMase, sphingomyelinase.

One of the first reports showing that DAG attenuates ceramide induced apoptosis used two human myeloid leukemia cell lines, namely HL-60 and U937. Exposure of HL-60 and U937 cells to SMase or 10μM cell-permeable ceramides for 6 h resulted in DNA-laddering, a typical hallmark of apoptosis. This could be abolished by di-octanoylglycerol, a cell permeable

DAG analog, and pharmacological protein kinase C activators (Jarvis *et al.*, 1994a). These data suggest that the cytoprotective effect of DAG derive directly from activation of protein kinase C. This is strengthened by the fact that specific protein kinase C inhibitors induce apoptosis ((Jarvis *et al.*, 1994b). The mechanism was further explored in 1-β-D-arabino-furanosylcytosine (ara-C)-treated HL-60 cells. Ara-C is an anti-cancer drug that stimulates the formation of both ceramides and DAG in these cells (Whitman *et al.*, 1997). DAG formed in response to ara-C treatment activates a signaling pathway via protein kinase C that antagonizes apoptosis triggered by the increase in intracellular ceramides upon ara-C treatment. Other studies have implicated protein kinase C as a negative key regulator of ceramide-induced apoptosis (Mansat *et al.*, 1997). Protein kinase C-dependent inhibition of apoptosis is possibly exerted through activation of Bcl-2, a proto-oncogene that functions as an apoptosis suppressor in many systems (Yang and Korsmeyer, 1996; Vaux *et al.*, 1988). Mutation of putative protein kinase C phosphorylation sites in Bcl-2 showed that the S70A Bcl-2 mutant, a serine which is phosphorylated by protein kinase C *in vitro*, was not phosphorylated upon stimulation with different agents, resulting in the loss to prolonge cell survival in comparison with the wild-type Bcl-2 (Ito *et al.*, 1997).

4.2 Disadvantage of the antagonizing effect of diacylglycerol on ceramide-induced apoptosis in oncology

For some of the most active anti-tumour compounds used in clinical oncology (eg., daunorubicin (DNR) and mitoxanthrone (MTX)) as well as for ionizing radiation (IR) it was generally thought that cytotoxicity mediated by these methods results from DNA damage. However, IR and the anti-cancer drugs also induce apoptotic signaling at a clinical relevant setting or concentration, respectively (Haimovitz-Friedmann *et al.*, 1994; Bettïaeb *et al.*, 1999). Present, knowledge does not allow discrimination whether apoptosis is a consequence of drug-induced DNA lesions or represents an independent cytotoxic mechanism triggered by a specific signaling pathway. DNR, MTX, IR and a variety of anti-tumour drugs including other anthracyclines and alkylating agents do not only trigger ceramide generation, but also increase DAG levels and protein kinase C activity (Rubin *et al.*, 1992; Uckun *et al.*, 1993).

From the discussion above, it is clear that DAG plays a critical role in the negative control of chemotherapeutic-induced apoptosis mediated by ceramide by potentially intervening upstream of its generation (by blocking

SMase activity (Mansat *et al.*, 1997) and downstream (by blocking the apoptotic effect of ceramide). However, the cross-talk between ceramides and DAG opens new intriguing avenues of research and may lead to better pharmacological manipulation of key steps in the apoptosis/survival signaling cascade resulting in an increased chemosensitivity of neoplastic cells.

5. CONCLUSION

Regulation of PC and SM synthesis is of major importance to cells, as these phospholipids serve as precursors for important bio-active lipids, such as DAG and ceramides. Recent literature suggests that the overall signal that a cell receives leading to cell proliferation, differentiation or death, depends upon the balance in the production of bio-active lipids. Therefore, a disturbance in the biosynthesis of PC by anti-cancer drugs, which either directly inhibit enzymes in the CDP-choline route or which exert their effect on PC synthesis via intracellular generated molecules as ceramides, can perturb the balance between mitogenic and anti-mitogenic second messenger molecules and thus influence cellular fate. A better understanding of the interaction between the different metabolites that can be generated from PLC, PLD and SMase pathways will be essential to elucidate the mechanism that regulates cell proliferation and death, which may lead to better pharmacological treatment of neoplastic cells.

REFERENCES

Allan, D., 2000, Lipid metabolic changes caused by short-chain ceramides and the connection with apoptosis. *Biochem. J.* **345**: 603-610

Anthony, M.L., Zhao, M., and Brindle, K.M., 1999, Inhibition of phosphatidylcholine biosynthesis following induction of apoptosis in HL-60 cells. *J. Biol. Chem.* **274**: 19686-19692

Aoyama, C., Yamazaki, N., Terada, H., and Ishidate, K., 2000, Structure and characterization of the genes for murine choline/ethanolamine kinase isozymes alpha and beta. *J. Lipid. Res* **41**: 452-464

Baburina, I., and Jackowski, S., 1998, Apoptosis triggered by 1-O-octadecyl-2-O-methyl-rac-glycero-3-phosphocholine is prevented by increased expression of CTP: phosphocholine cytidylyltransferase. *J. Biol. Chem.* **273**: 2169-2173

Bettaïeb, A., Plo, I., Mansat-de Mas, V., Quillet-Mary, A., Levade, T., Laurent, G., and Jaffrézou, J-P., 1999, Daunorubicin- and Mitoxantrone-triggered phosphatidylcholine hydrolysis: implication in drug-induced ceramide generation and apoptosis. *Mol. Pharm.* **55**: 118-125

Billah, M.M., and Anthes, J.C., 1990, The regulation and cellular functions of phosphatidylcholine hydrolysis. *Biochem. J.* **269**: 281-291

Bladergroen, B.A., and van Golde, L.M., 1997, CTP: phosphoethanolamine cytidylyltransferase. *Biochim. Biophys. Acta.* **1348**: 91-99

Bladergroen, B.A., Houweling, M., Geelen, M.J., and van Golde, L.M., 1999a, Cloning and expression of CTP: phosphoethanolamine cytidylyltransferase cDNA from rat liver. *Biochem. J.* **343**: 107-114

Bladergroen, B.A., Bussière, M., Klein, W., Geelen, M.J.H., van Golde, L.M.G., and Houweling, M., 1999b, Inhibition of phosphatidylcholine and phosphatidylethanolamine biosynthesis in rat-2 fibroblasts by cell-permeable ceramides. *Eur. J. Biochem.* **264**: 152-160

Boggs, K., Rock, C.O., and Jackowski, S., 1998, The antiproliferative effect of hexadecylphosphocholine toward HL60 cells is prevented by exogenous lysophosphatidylcholine. *Biochim. Biophys. Acta.* **1389**: 1-12

Clement, J.M., and Kent, C., 1999, CTP: phosphocholine cytidylyltransferase: insights into regulatory mechanisms and novel functions. *Biochem. Biophys. Res. Commun.* **257**: 643-650

Cornell, R.B., and Northwood, I.C., 2000, Regulation of CTP: phosphocholine cytidylyltransferase by amphitropism and relocalization. *Trends Biochem. Sci.* **25**: 441-447

Cui, Z., Houweling, M., Chen, M.H., Record, M., Chap, H., Vance, D.E., and Terce, F., 1996, A genetic defect in phosphatidylcholine biosynthesis triggers apoptosis in Chinese hamster ovary cells. *J. Biol. Chem.* **271**: 14668-14671

Delong, C.J., Shen, Y-J., Thomas, M.J., and Cui, Z., 1999, Molecular distinction of phosphatidyl-choline synthesis between the CDP-choline pathway and phosphatidylethanolamine methylation pathway. *J. Biol. Chem.* **274**: 29683-29688

Emoto, K., Toyama-Sorimachi, N., Karasuyama, H., Inoue, K., and Umeda, M., 1997, Exposure of phosphatidylethanolamine on the surface of apoptotic cells. *Exp. Cell Res.* **232**: 430-434

Esko, J.D., Wermuth, M.M., and Raetz, C.R., 1981, Thermolabile CDP-choline synthetase in an animal cell mutant defective in lecithin formation. *J. Biol. Chem.* **256**: 7388-7393

Esko, J.D., Nishijima, M., and Raetz, C.R., 1982, Animal cells dependent on exogenous phosphatidylcholine for membrane biogenesis. *Proc. Natl. Acad. Sci. USA.* **79**: 1698-1702

Exton, J. H., 1994, Phosphatidylcholine breakdown and signal transduction, *Biochim. Biophys. Acta* **1212**: 26-42.

Fadok, V.A., Voelker, D.R., Campbell, P.A., Cohen, J.J., Bratton, D.L., and Henson, P.M., 1992 Exposure of phosphatidylserine on the surface of apoptotic lymphocytes triggers specific recognition and removal by macrophages. *J. Immunol.* **148**: 2207-2216

Flores, I., Jones, D.R., and Mérida, I., 2000, Changes in the balance between mitogenic and anti-mitogenic lipid second messengers during proliferation, cell arrest, and apoptosis in T lymphocytes. *Faseb J.* **14**: 1873-1875

Gajate, C., Santos-Beneit, A., Modolell, M., and Mollinedo, F., 1998, Involvement of c-Jun NH2-terminal kinase activation and c-Jun in the induction of apoptosis by the ether phospholipid 1-O-octadecyl-2-O-methyl-rac-glycero-3-phosphocholine. *Mol Pharmacol.* **53**: 602-612

Gajate, C., Fonteriz, R.I., Cabaner, C., Alvarez-Noves, G., Alvarez-Rodriguez, Y., Modolell, M., and Mollinedo, F., 2000, Intracellular triggering of Fas, independently of FasL, as a new mechanism of antitumor ether lipid-induced apoptosis. *Int J Cancer.* **85**: 674-682

Geilen, C.C., Haase, A., Wieder, T., Arndt, D., Zeisig R. and Reutter W., 1994, Phospholipid analogs: side chain- and polar head group-dependent effects on phosphatidylcholine biosynthesis. *J. Lipid Res.* **35**: 625-632

Ghosh, S., Strum, J. C., and Bell, R. M., 1997, Lipid biochemistry: functions of glycerolipids and sphingolipids in cellular signaling. *FASEB J.* **11**: 45-50

Gomez Munoz, A., Martin, A., O'Brien, L., and Brindley, D. N., 1994, Cell-permeable ceramides inhibit the stimulation of DNA synthesis and phospholipase D activity by phosphatidate and lysophosphatidate in rat fibroblasts. *J. Biol. Chem.* **269**: 8937-8943

Haimovitz_Friedman, A., Kan, C.C., Ehleiter, D., Persaud, R.S., McLoughlin, M., Fuks, Z., and Kolesnick R.N., 1994, Ionizing radiation acts on cellular membranes to generate ceramide and initiate apoptosis. *J. Exp. Med.* **180**: 525-535

Hannun, Y.A., 1994, The sphingomyelin cycle and the second messenger function of ceramide. *J. Biol. Chem.* **269**: 3125-3128

Hannun, Y.A., 1996, Functions of ceramide in coordinating cellular responses to stress. *Science* **274**: 1855-1859

Henneberry, A.L., and McMaster, C.R., 1999, Cloning and expression of a human choline/ethanol-aminephosphotransferase: synthesis of phosphatidylcholine and phosphatidylethanolamine. *Biochem J.* **339**: 291-298

Henneberry, A.L., Wistow, G., and McMaster, C.R., 2000, Cloning, genomic organization, and characterization of a human cholinephosphotransferase. *J. Biol. Chem.* **275**: 29808-29815

Hernandez-Alcoceba, R., Saniger, L., Campos, J., Nunez, M.C., Khaless, F., Gallo, M.A., Espinosa, A., and Lacal, J.C., 1997, Choline kinase inhibitors as a novel approach for antiproliferative drug design. *Oncogene* **15**: 2289-2301

Hernandez-Alcoceba, R., Fernandez, F., and Lacal, J.C., 1999, *In vivo* antitumor activity of choline kinase inhibitors: a novel target for anticancer drug discovery. *Cancer Res.* **59**: 3112-3118

Holmes-McNary, M.Q., Loy, R., Mar, M.H., Albright, C.D., and Zeisel, S.H., 1997, Apoptosis is induced by choline deficiency in fetal brain and in PC12 cells. *Brain Res. Dev. Brain Res.* **18**: 9-16

Houweling, M., Cui, Z., and Vance, D.E., 1995, Expression of phosphatidylethanolamine N-methyltransferase-2 cannot compensate for an impaired CDP-choline pathway in mutant Chinese hamster ovary cells. *J. Biol. Chem.* **270**: 16277-16282

Ishidate, K.. 1997, Choline/ethanolamine kinase from mammalian tissues. *Biochim. Biophys. Acta* **1348**: 70-78

Ito, T., Deng, X., Carr, B., and Stratford May, W., 1997, Bcl-2 phosphorylation required for anti-apoptosis function. *J. Biol. Chem.* **272**: 11671-11673

Jarvis, W.D., Fornari Jr., F.A., Browning, J.L., Gewirtz, D.A., Kolesnick, R.N., and Grant, S., 1994a, Attenuation of ceramide-induced apoptosis by diglyceride in human myeloid leukemia cells. *J. Biol. Chem.* **269**: 31685-31692

Jarvis, W.D., Turner, A.J., Povirk, LF, Traylor, R.S., and Grant, S., 1994b, Induction of apoptotic DNA fragmentation and cell death in HL-60 human promyelocytic leukemia cells by pharmacological inhibitors of protein kinase C. *Cancer Res.* **54**: 1707-1714

Jayadev, S., Liu, B., Bielawska, A.E., Lee, J.Y., Nazaire, F., Pushkareva, M.Yu., Obeid, L.M., and Hannun, Y.A., 1995, Role for ceramide in cell cycle arrest. *J. Biol. Chem.* **270**: 2047-2052

Kent C., 1997, CTP: phosphocholine cytidylyltransferase. *Biochim. Biophys. Acta* **1348**: 79-90

Kester, M., Simonson, M. S., Mene, P., and Sedor, J. R., 1989, Interleukin-1 generates transmembrane signals from phospholipids through novel pathways in cultured rat mesangial cells. *J. Clin. Invest.* **83**: 718-723

Kim, M. Y., Linardic, C., Obeid, L., and Hannun, Y., 1991, Identification of sphingomyelin turnover as an effector mechanism for the action of tumor necrosis factor alpha and gamma- interferon. Specific role in cell differentiation. *J. Biol. Chem.* **266**: 484-489

Kiss, Z., 1990, Effects of phorbol ester on phospholipid metabolism. *Prog. Lipid Res.* **29**: 141-166

Kiss, Z., and Deli, E., 1995, Preferential inhibition of phorbol ester-induced hydrolysis of phosphatidylethanolamine by N-acetylsphingosine in NIH 3T3 fibroblasts. *FEBS Lett.* **365**: 146-148

Lucas L., Hernandez-Alcoceba, R., Penalva, V., and Lacal, J.C., 2001, Modulation of phospholipase D by hexadecylphosphorylcholine: a putative novel mechanism for its antitumoral activity. *Oncogene* **20**: 1110-1117

Lykidis, A., Murti, K.G., and Jackowski, S., 1998, Cloning and characterization of a second human CTP: phosphocholine cytidylyltransferase. *J. Biol. Chem.* **273**: 14022-14229

Lykidis, A., Baburina, I., and Jackowski, S., 1999, Distribution of CTP: phosphocholine cytidylyltransferase (CCT) isoforms. Identification of a new CCTbeta splice variant. *J. Biol. Chem.* **274**: 26992-27001

Lykidis, A., Wang, J., Karim, M.A., and Jackowski, S., 2001, Overexpression of a mammalian ethanolamine-specific kinase accelerates the CDP-ethanolamine pathway. *J. Biol. Chem.* **276**: 2174-2179

Mansat, V., Laurent, G., Levada, T., Bettaïeb, A., and Jaffrézou, J-P., 1997, The protein kinase C activators phorbol esters and phosphatidylserine inhibit neutral sphingomyelinase activation, ceramide generation and apoptosis triggered by daunorubicin. *Cancer Res.* **57**: 5300-5304

Mathias, S., Younes, A., Kan, C. C., Orlow, I., Joseph, C., and Kolesnick, R. N., 1993, Activation of the sphingomyelin signaling pathway in intact EL4 cells and in a cell-free system by IL-1 beta. *Science* **259**: 519-522

Mathias, S., Pena, L.A., and Kolesnick, R.N., 1998, Signal transduction of stress via ceramide *Biochem. J.* **335**: 465-480

Martin, S.J., Reutelingsperger, C.P.M., McGahon, A.J., Rader, A.J., van Schie, R.C.A.A., LaFace, D.M., and Green, D.R., 1995, Early redistribution of plasma membrane phosphatidyl-serine is a general feature of apoptosis regardless of the initiating stimulus: inhibition by overexpression of Bcl_2 and Abl. *J. Exp. Med.* **182** 1545-1556

Maulik, N., Kagan, M.V., Tyurin, V., and Das, D.K., 1998 Redistribution of phosphatidyl-ethanolamine and phosphatidyserine precedes reperfusion-induced apoptosis. *Am. J. Physiol.* **274**: H242-H248

McMaster, C.R., and Bell, R.M., 1997, CDP-choline: 1,2-diacylglycerol cholinephospho-transferase. *Biochim. Biophys. Acta* **1348**: 100-110

Merrill, A.H., and Jones, D.D., 1990, An update of the enzymology and regulation of sphingomyelin metabolism. *Biochim. Biophys. Acta* **1044**: 1-12

Miquel, K., Pradines, A., Terce, F., Selmi, S., and Favre, G., 1998, Competitive inhibition of choline phosphotransferase by geranylgeraniol and farnesol inhibits phosphatidylcholine synthesis and induces apoptosis in human lung adenocarcinoma A549 cells. *J Biol Chem.* **273**: 26179-26186

Nishizuka, Y., 1992, Intracellular signaling by hydrolysis of phospholipids and activation of protein kinase C. *Science* **258**: 607-614

Nishizuka, Y., 1995, Protein kinase C and lipid signaling for sustained cellular responses. *FASEB J.* **9**: 484-496

Obeid, L.M., Linardic, C.M., Karolak, L.A., and Hannun, Y.A., 1993, Programmed cell death induced by ceramide. *Science* **259**: 1769-1771

Okazaki, T., Bell, R.M., and Hannun, Y.A., 1989, Sphingomyelin turnover induced by vitamin D3 in HL-60 cells. Role in cell differentiation. *J. Biol. Chem.* **264**: 19076-19080

Okazaki, T., Bielawska, A., Bell, R.M., and Hannun, Y. A., 1990, Role of ceramide as a lipid mediator of 1a, 25-dihydroxyvitamin D3-induced HL-60 cell differentiation. *J. Biol. Chem.* **265**: 15823-15831

Pettitt, T.R., Martin, A., Horton, T., Liossis, C., Lord, J.M., and Wakelam M.J.O., 1997, Diacyl-glycerol and phosphatidate generated by phospholipases C and D, respectively, have distinct fatty acid compositions and functions. *J. Biol. Chem.* **272**: 17354_17359

Powis G., Seewald, M.J., Gratas, C., Melder, D., Riebow, J., and Modest, E.J., 1992, Selective inhibition of phosphatidylinositol phospholipase C by cytotoxic ether lipid analogs. *Cancer Res.* **52**: 2835-2840

Ramos, B., Salido, G.M., Campo, M.L., and Claro, E., 2000, Inhibition of phosphatidylcholine synthesis precedes apoptosis induced by C_2-ceramide: protection by exogenous phosphatidylcholine. *Neuroreport* **11**: 3103-3108

Rubin, E., Kharbanda, S., Gunji, H., Weichselbaum, R., and Kufe, D., 1992, cis-Diammine-dichloroplatinum(II) induces c-jun expression in human myeloid leukemia cells: potential involvement of a protein kinase C-dependent signaling pathway. *Cancer Res.* **52**: 878-882

Ruiter, G.A., Zerp, S.F., Bartelink, H., van Blitterswijk, W.J., and Verheij M., 1999, Alkyl-lysophospholipids activate the SAPK/JNK pathway and enhance radiation-induced apoptosis. *Cancer Res.* **59**: 2457-2463

Ruiter, G.A., Verheij, M., Zerp, S.F. and van Blitterswijk, W.J., 2001, Alkyl-lysophospholipids as anticancer agents and enhancers of radiation-induced apoptosis. *Int. J. Radiat. Oncol. Biol. Phys.* **49**: 415-419

Sillence, D.J., and Allan, D., 1998, Utilization of phosphatidylcholine and production of diradylglycerol as a consequence of sphingomyelin synthesis. *Biochem J.* **331**: 251-256

Skowronski, E.W., Kolesnick, R.N., and Green, D.R., 1996, Fas-mediated apoptosis and sphingo-myelinase signal transduction: the role of ceramide as a second messenger for apoptosis. *Cell Death and Differentiation* **3**: 171-176

Sohal, P.S., and Cornell, R.B., 1990, Sphingosine inhibits the activity of rat liver CTP: phospho-choline cytidylyltransferase. *J. Biol. Chem.* **265**: 11746-11750

Tepper, A.D., de Vries, E., van Blitterswijk, W.J., and Borst, J., 1999, Ordering of ceramide formation, caspase activation, and mitochondrial changes during CD95- and DNA damage-induced apoptosis. *J Clin Invest.* **103**: 971-978

Tijburg, L.B.M., Geelen, M.J.H., and Van Golde, L.M.G., 1989, Regulation of the biosynthesis of triacylglycerol, phosphatidylcholine and phosphatidylethanolamine in the liver. *Biochim. Biophys. Acta* **1004**: 1-19

Uberall, F., Oberhuber, H., Maly, K., Zaknun, J., Demuth, L., and Grunicke, H.H., 1991, Hexadecyl-phosphocholine inhibits inositol phosphate formation and protein kinase C activity. *Cancer Res.* **51**: 807-812

Uckun, F.M., Schieven, G.L., Tuel-Ahlgren, L.M., Dibirdik, I., Myers, D.E., Ledbetter, J.A., and Song, C.W., 1993, Tyrosine phosphorylation is a mandatory proximal step in radiation_induced activation of the protein kinase C signaling pathway in human B-lymphocyte precursors. *Proc. Natl. Acad. Sci. U.S.A.* **90**: 252-256

Vance, D.E., Walkey, C.J., and Cui Z., 1997, Phosphatidylethanolamine N-methyltransferase from liver. *Biochim. Biophys. Acta* **1348**: 142-150

Vaux, D.L., Cory, S., and Adams, J.M., 1988, Bcl-2 gene promotes hemeopoietic cell survival and cooperates with c_myc to immortalize pre-B cells. *Nature* **325**: 440-442

Verhoven, B., Schlegel, R.A., and Williamson, P., 1995, Mechanisms of phosphatidylserine exposure, a phagocyte recognition signal, on apoptotic T lymphocytes. *J. Exp. Med.* **182**: 1597-1601

Venable, M.E., Blobe, G.C., and Obeid, L.M., 1994, Identification of a defect in the phospholipase D/diacylglycerol pathway in cellular senescence. *J. Biol. Chem.* **269**: 26040-26044

Voelker, D.R., 1997, Phosphatidylserine decarboxylase. *Biochim. Biophys. Acta* **1348**: 236-244

Voziyan, P.A., Goldner, C.M., and Melnykovych, G., 1993, Farnesol inhibits phosphatidylcholine biosynthesis in cultured cells by decreasing cholinephosphotransferase activity. *Biochem. J.* **295**: 757-762

Waite, K.A., and Vance, D.E., 2000, Why expression of phosphatidylethanolamine N-methyltransferase does not rescue Chinese hamster ovary cells that have an impaired CDP-choline pathway. *J. Biol. Chem.* **275**: 21197-21202

Wakelam, M.J.O., 1998, Diacylglycerol-when is it an intracellular messenger? *Biochim. Biophys. Acta* **1436**: 117-126

Walkey, C.J., Donohue, L.R., Bronson, R., Agellon, L.B., and Vance, D.E., 1997, Disruption of the murine gene encoding phosphatidylethanolamine N-methyltransferase. *Proc. Natl. Acad. Sci. U.S.A.* **94**: 12880-12885

Whitman, S.P., Civoli, F., and Daniel, L.W., 1997, Protein kinase CβII activation by 1-β-D-arabino-furanosylcytosine is antagonistic to stimulation of apoptosis and BCL-2α down-regulation. *J. Biol. Chem.* **272**: 23481-23484

Wieder, T., Perlitz, C., Wieprecht, M., Huang, R.T., Geilenm C.C. and Orfanosm, C,E., 1995, Two new sphingomyelin analogs inhibit phosphatidylcholine biosynthesis by decreasing membrane-bound CTP: phosphocholine cytidylyltransferase levels in HaCaT cells. *Biochem. J.* **311**: 873-879

Wieder, T., Zhang, Z., Geilen, C.C., Orfanos, C.E., Giuliano, A.E., and Cabot, M.C., 1996, The antitumor phospholipid analog, hexadecylphosphocholine, activates cellular phospholipase D. *Cancer Lett.* **100**: 71-79

Wright, M.M., Henneberry, A.L., Lagace, T.A., Ridgway, N.D., and McMaster, C.R., 2001, Uncoupling Farnesol Induced Apoptosis from its Inhibition of Phosphatidylcholine Synthesis. *J. Biol. Chem.* **276**: 25254-25261.

Yang, E., and Korsmeyer, S.J., 1996, Molecular thanatopsis: a discourse on the BCL2 family and cell death. *Blood* **88**: 386-401

Yen, C.L., Mar, M.H., and Zeisel, S.H., 1999, Choline deficiency-induced apoptosis in PC12 cells is associated with diminished membrane phosphatidylcholine and sphingomyelin, accumulation of ceramide and diacylglycerol, and activation of a caspase. *FASEB. J.* **13**: 135-142

Zhou, X., Lu, X., Richard, C., Xiong, W., Litchfield, D.W., Bittman, R., and Arthur, G., 1996, 1-O-octadecyl-2-O-methyl-glycerophosphocholine inhibits the transduction of growth signals via the MAPK cascade in cultured MCF-7 cells. *J. Clin. Invest.* **98**: 937-944

Chapter 12

Acid Sphingomyelinase-derived Ceramide Signaling in Apoptosis

Erich Gulbins[*] and Richard Kolesnick[§]
[*]Department of Immunology, St. Jude Children's Research Hospital, Memphis, TN 38105; *[§]Laboratory of Signal Transduction, Memorial Sloan Kettering Cancer Center, New York, NY 10021*

1. INTRODUCTION

Biological membranes of eukaryotic cells primarily consist of glycerophospholipids, sphingolipids and cholesterol. Sphingomyelin, which is comprised of a highly hydrophobic ceramide moiety and a hydrophilic phosphorylcholine headgroup, is the most prevalent cellular sphingolipid (Figure 1). Ceramide is the amide ester of the sphingoid base D-erythro-sphingosine and a fatty acid usually of C_{16} through C_{26} chain length (Figure 1). Sphingomyelin is synthesized predominantly on the luminal side of the Golgi apparatus, but also in the plasma membrane (Jeckel *et al.*, 1990; Futerman *et al.*, 1990; Andrieu-Abadie *et al.*, 1998). The two pools are connected by vesicular transport. Sphingomyelin is almost exclusively located in the anti-cytosolic leaflet of biological membranes resulting in lipid bilayer asymmetry (Emmelot and Van Hoeven, 1975). Hydrogen bonds and hydrophobic van der Waal interactions mediate the tight interaction between the cholesterol sterol ring system and the ceramide moiety of sphingomyelin. The lateral association of sphingolipids in the cell membrane is further promoted by hydrophilic interactions between the sphingolipid headgroups.

Subcellular Biochemistry, Volume 36: Phospholipid Metabolism in Apoptosis.
Edited by Quinn and Kagan. Kluwer Academic/Plenum Publishers, New York, 2002

Sphingomyelin

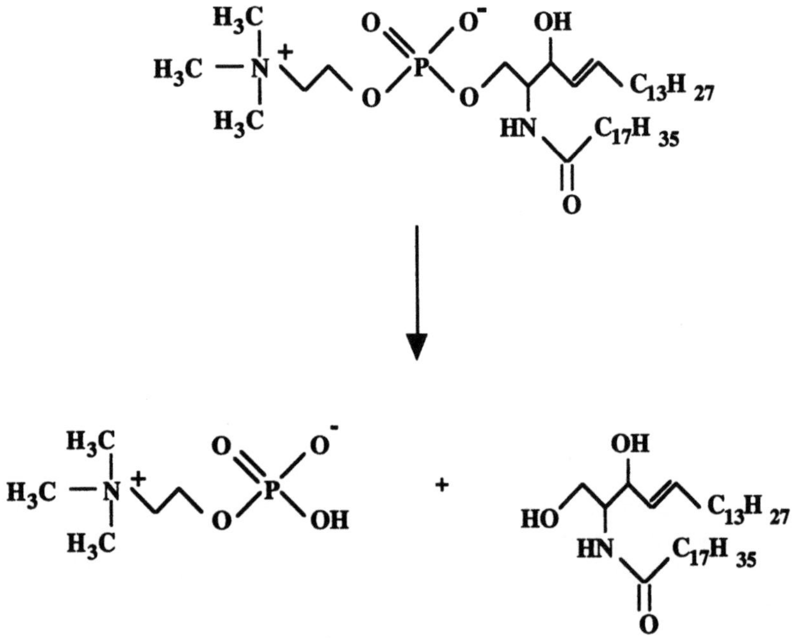

Phosphorylcholine **Ceramide**

Figure 1. Displayed are the structures of sphingomyelin and ceramide. Sphingomyelinases function as phospholipases C to hydrolyze sphingomyelin.

The association of these lipids and the high local concentration result in their separation from phospholipids within the bilayer and the transformation of these sphingolipid- and cholesterol-rich membrane domains into a distinct liquid ordered phase (Harder and Simons, 1997; Brown and London, 1998; Andersen, 1998). The tightly packed sphingomyelin/cholesterol domains have been compared to rafts floating within the phospholipid portion of the cell membrane. Cholesterol seems to function in rafts as a hydrophobic spacer between the bulky sphingolipids required for the generation of the liquid ordered phase (Harder and Simons, 1997; Brown and London, 1998). In accordance with this concept, pharmacological extraction of cholesterol destroys these membrane rafts. This highly ordered biophysical phase results in a relative resistance of these sphingolipid-rich rafts to detergents and, thus, these glycosphingolipid-enriched membrane (GEM) domains have also been named detergent-insoluble glycolipid enriched membranes (DIGs) (Harder and Simons, 1997; Brown and London, 1998).

Metabolites derived from sphingolipids constitute a novel class of cellular signaling molecules involved in apoptosis, cell proliferation and differentiation. Members of this group of signaling molecules include ceramide (Figure 1), sphingosine, sphingosine-1-phosphate, and perhaps ceramide 1-phosphate. In the present overview, we will focus on the function of ceramide in apoptosis. Death of cells by apoptosis is central in the development of the organism to delete excess cells, but also in the physiological turnover of normal tissues to remove aged or damaged cells. Apoptosis, reviewed extensively recently (e.g. Hengartner, 2000), is characterized by the ordered disintegration of intracellular organelles and their packaging into apoptotic bodies, while integrity of the plasma membrane is maintained until late in the process. These events are the result of activation of a family of cysteine aspartate proteases known as caspases.

Ceramide, released by the acid sphingomyelinase or synthesized *de novo* via the ceramide synthase pathway, is critically involved in many forms of apoptosis. This includes apoptosis triggered by receptor molecules, e.g. CD95 or the tumor necrosis factor receptor, during development, or upon induction by environmental stress, e.g. irradiation, heat shock or UV light. These stimuli initiate rapid activation of acid sphingomyelinase and concomitant release of ceramide. Cellular effector mechanisms for ceramide involve direct or indirect regulation of the activity of several proteins, e.g. kinase suppressor of Ras, cathepsin D or phospholipase A_2. This chapter will consider the recently advanced concept suggesting that ceramide modifies distinct domains in the membrane to create rafts, enabling activated receptors to concentrate in clusters. Clustering of receptor and signaling molecules in these rafts seems required for initiation of receptor- and cell type-specific signaling.

2. CERAMIDE AND ACID SPHINGOMYELINASE IN APOPTOSIS

Diverse stimuli have been shown to trigger apoptosis. In general, at least three types of pro-apoptotic stimuli are identified. First, many receptors have been shown to initiate apoptosis. The most important of these belong to the tumor necrosis factor (TNF)/nerve growth factor (NGF) receptor superfamily and include CD95, the Trail-receptor DR4, and the TNF receptor (TNF-R) (Walczak and Krammer, 2000; Holtzman *et al.*, 2000). These receptors are constitutively present or up-regulated on the cell surface upon cellular maturation. Interaction of the cognate receptor with its ligand initiates apoptosis. For example, expression of CD95 on stimulated lymphocytes seems critically involved in regulation of the immune response, since genetic deficiency of CD95 or its ligand in *lpr* or *gld* mice, respectively, results in accumulation of atypical lymphocytes and in the development of an autoimmune-like disorder (Nagata and Suda, 1995). Second, apoptosis can be induced by various pathophysiologic stress stimuli including γ-irradiation, ultraviolet (UV) light, heat shock, cytotoxic drugs, H_2O_2, toxins, bacteria and viruses (for review see Mathias *et al.*, 1998). Third, survival of many cells critically relies on the supply of growth factors and their withdrawal results in apoptosis. For example, interleukin-2 is necessary to prevent apoptosis in lymphocytes, while NGFs are necessary in neurons. These cells require continuous external survival signals to balance pre-existing intracellular death-signaling pathways. Depletion of growth factors very likely also plays a fundamental role in apoptosis during development. Cells provided with the correct mix of survival factors will be maintained, while cells lacking the required growth factors undergo apoptosis and are deleted in the developmental process. Ceramide appears to be involved in all three forms of apoptosis, however, most data exist on the role of ceramide in receptor and stress-mediated apoptosis.

Cellular ceramide is generated either by hydrolysis of sphingomyelin or by *de novo* synthesis. Sphingomyelin hydrolysis is catalyzed by at least three different sphingomyelinases that are discriminated by their optimal pH: acid, neutral, and alkaline sphingomyelinases. In the present manuscript we will focus on acid sphingomyelinase (ASM)-released ceramide.

ASM is the best characterized sphingomyelinase and many studies implicate this enzyme in apoptosis. Upon synthesis, ASM is post-translationally modified by glycosylation, a requirement for functionality (Newrzella and Stoffel, 1996). A lysosomal ASM and a secretory ASM have been distinguished (Schissel *et al.*, 1996). They are derived from the same gene, but differentially processed at the NH2-terminus and display a different glycosylation pattern (Schissel *et al.*, 1998). The glycosylation pattern very likely determines the targeting of the ASM to acidic compartments (lysosomes and endosomes) to regulate sphingomyelin turnover, or into secretory vesicles for secretion into the extracellular space. Recently, we identified a third form of ASM, which binds to the cell surface upon application of the appropriate stimulus (Grassmé *et al.*, 2001). At present, it is unknown whether this surface ASM represents a third isoform or a specialized form of the secretory or lysosomal ASM. Even if the pH at the cell surface or the extracellular space is only slightly acidic, it is very likely that both the secretory and the surface ASM are active: The increase of the pH from 4.5 to neutral values does not alter the activity [V_{max}] of the enzyme, but increases the K_m value, decreasing substrate affinity (Callahan *et al.*, 1983). Since the activity of the ASM is modified by environment (Schissel *et al.*, 1998), the reduction of substrate affinity might be compensated by the presence of extracellular factors such as LDL or by membrane co-factors, restoring ASM activity even at neutral pH values.

ASM is activated by several pro-apoptotic receptors, in particular the TNF-R (Schutze *et al.*, 1992), IL-1 receptor (Mathias *et al.*, 1993) and CD95 (Cifone *et al.*, 1994; Gulbins *et al.*, 1995; Brenner *et al.*, 1998), but the enzyme is also stimulated by non-apoptotic receptors, e.g. CD40 (Grassmé *et al.*, 2001b) or CD28 (Boucher *et al.*, 1995) or by internalization of some bacteria into mammalian cells (Grassmé *et al.*, 1997), an observation discussed in the last part of this review. Most of these stimuli trigger only a 2-4 fold activation of the enzyme. However, since only a small portion of total cellular ASM appears stimulated upon ligation of a cognate receptor, whereas the bulk of the lysosomal ASM might not be regulated, it is likely that the effective local increase in ASM activity is much greater than the measurement of total cellular ASM activity suggests.

CD95 and the TNF-R activate ASM to release ceramide within seconds to minutes upon ligation suggesting that ASM functions in the early signaling of apoptosis. Several genetic studies have provided clear evidence that ASM plays an important role in TNF-R1- and CD95-triggered death, at least in some cell types (Lozano *et al.*, 2001; De Maria *et al.*, 1998; Paris *et al.*, 2001a; Kirschnek *et al.*, 2000). These studies employed B lymphocytes from Niemann-Pick disease Type A (NPDA) patients, which suffer from an inborne defect of ASM, or cells from ASM knock-out mice. Thus, human B-lymphocytes from NPDA patients are resistant to CD95-triggered ceramide generation and apoptosis (Grassmé *et al.*, 2001a; De Maria *et al.*, 1998). Furthermore, recent studies showed a 10-fold reduction of apoptosis in hepatocytes isolated from ASM-deficient mice upon CD95 stimulation compared to normal hepatocytes (Paris *et al.*, 2001a; Kirschnek *et al.*, 2000). Addition of C_{16}-ceramide restored CD95-induced apoptosis providing evidence that ceramide is central for CD95-triggered apoptosis in this population. This notion is supported by *in vivo* studies employing a CD95-dependent autoimmune hepatitis and CD95-dependent death of lymphocytes after anti-CD4 antibody injection (Kirschnek *et al.*, 2000). Both types of CD95-triggered death were deficient in mice lacking ASM. Additional studies show, however, that the ASM is dispensible for apoptosis in some cell types as CD95-triggered apoptosis of thymocytes is normal in ASM-deficient animals consistent with the low ASM content of these cells (Lin *et al.*, 2000).

The function of ASM-released ceramide is not restricted to receptor-initiated signaling, but has been also demonstrated to play a pivotal role in developmental apoptosis in at least one specific cell type. It has been shown that ASM is required for the normal development of the ovaries (Morita *et al.*, 2000). The normal pattern of development of the ovary involves deletion of as many as 80% of all oocytes before birth of the mouse. A similar process occurs in the human female. Genetic deficiency of ASM prevented developmental apoptosis of oocytes resulting in oocytes hyperplasia at birth. Further, oocytes from *asmase*$^{-/-}$ mice, when superovulated from adult females, were resistant to therapeutic levels of daunorubicin, in contrast to wild type littermates. These investigations provide clear genetic evidence for a central role for ASM in developmental and stress-induced deletion of oocytes.

In addition to the function of the ASM in receptor-triggered and developmental apoptosis, the enzyme has a central function in the cellular response to environmental stress stimuli. This review will focus on the examples of ischemia, ultraviolet (UV) light exposure and, in particular, γ-irradiation.

Ischemia is a major cause of some important clinical problems including myocardial infarction and stroke. A recent study on stroke showed that ischemia results in activation of ASM and ceramide release (Yu *et al.*, 2000). The significance of this finding is indicated by the resistance of *asmase*$^{-/-}$ mice to focal cerebral ischemia. Those mice were protected against the middle cerebral artery stroke syndrome showing smaller infarct sizes and less behavioral changes than their wild type littermates. Consistent with these data, hippocampal neurons isolated from *asmase*$^{-/-}$ mice were resistant to hypoxic and excitotoxic stress compared to their wild type siblings. These genetic data suggest that activation of neuronal ASM is central to stress-induced apoptotic death of hippocampal neurons.

In addition, several studies identified ASM involvement in the stress response leading to apoptosis induction upon UVA exposure (Huang *et al.*, 1997). In contrast to wild type B blasts, MS1418 B blasts derived from a patient with Niemann-Pick disease were defective in the ASM activation, ceramide generation and c-Jun N-terminal kinase (JNK) activation that occurred within minutes of UVA exposure, and were defective in apoptosis induction. UVB and UVC responses were normal in MS1418 cells indicating specific utilization of the ASM-mediated signaling mechanism for UVA-induced apoptosis. That JNK was required for UVA-induced apoptosis was derived from studies using *jnk1-/-* and *jnk2-/-* cells which still activated ASM but were found resistant to UVA-induced apoptosis compared to wild type counterparts (Zhang, 2001).

Finally, ASM may be central to the response to γ-irradiation *in vitro* and *in vivo,* in some cells. A requirement for ASM for release of ceramide upon irradiation has been observed in T- and B-lymphocytes (Santana *et al.*,

1996), murine embryonic fibroblasts (Lozano *et al.*, 2001) and endothelial cells (Pena *et al.*, 2000; Paris *et al.*, 2001a). ASM deficiency prevented ceramide release and resulted in resistance of these cells to the ionizing radiation while re-constitution of ASM restored irradiation-triggered ceramide generation and cell death. The significance of ASM for radiation-induced cell death is clearly defined in experiments on ASM knock-out mice (Santana *et al.*, 1996; Pena *et al.*, 2000; Paris *et al.*, 2001b). Mice exposed to γ-irradiation failed to generate ceramide in endothelial cells in the lung (Santana *et al.*, 1996), small intestine (Paris *et al.*, 2001b) or brain (Santana *et al.*, 1996) and were resistant to radiation-induced damage, events readily observed in normal mice. In fact, *asmase*$^{-/-}$ mice were resistant to the GI syndrome, a major limiting toxicity for use of chemotherapy and of radiation to the abdomen (Paris *et al.*, 2001b). Likewise, oocytes within the ovaries of *asmase*$^{-/-}$ mice resisted irradiation if pre-treated with sphingosine 1-phosphate, a ceramide metabolite and antagonist, whereas vehicle-treated oocytes died (Morita *et al.*, 2000). However, in the murine thymus, irradiation-triggered cell death does not seem to involve ASM, since the thymus of *asmase*$^{-/-}$ mice were as sensitive to irradiation as *asmase*$^{+/+}$ littermates (Santana *et al.*, 1996). This indicates different cells employ different mechanisms to trigger apoptosis in response to a single stress.

3. ASM ACTIVATION METHODS

Molecular mechanisms mediating activation of ASM by CD95 or the TNF-R are still only poorly understood. Both receptors contain a short intracellular domain, the death domain (Itoh and Nagata, 1993), which is required for apoptosis (Itoh and Nagata, 1993), and crucial for ASM activation (Jekle *et al.*, 2001). This is consistent with the finding that caspases, which interact with the death domain of CD95 or the TNF-R via the adapter proteins FADD and TRADD, respectively, are involved in ceramide release (Brenner *et al.*, 1998; Schwandner *et al.*, 1998; Jekle *et al.*, 2001; Grullich *et al.*, 2000). Thus, overexpression of FADD or caspase 8, or transfection of a constitutively active caspase 8 mutant, resulted in increased ceramide release suggesting that FADD and caspase 8 regulate ceramide release during some forms of apoptosis (Schwandner *et al.*, 1998; Grullich *et al.*, 2000). A role for caspase 8 in ASM activation and ceramide release upon CD95 triggering is also suggested by the finding that transfection of Crm A, a viral protein blocking some caspases, or treatment of cells with Ac-YVAD-cmk, a pharmacological inhibitor of caspases, prevent ASM stimulation and ceramide release by CD95 (Brenner *et al.*, 1998). However, caspase 8 seems not to be involved in TNF-R-triggered ASM stimulation, which appears mediated by a yet unknown initiator caspase (Schwandner *et al.*, 1998). At present, the intermediates between caspases and ASM are unknown.

Recent studies showed that ASM specifically associates with phosphatidylinositol-3-kinase (PI-3-K) upon NGF stimulation of PC12 cells via TrkA (Bilderback *et al.*, 2001). Association between the two molecules was mediated by the regulatory p85 subunit of PI-3-K and was restricted to membrane rafts. Activation of PI-3-K by NGF triggering resulted in an approximately 50% reduction of ASM activity pointing to a negative regulation of ASM by this mechanism.

4. NEUTRAL AND ALKALINE SPHINGOMYELINASES

In addition to ASM, neutral sphingomyelinase (NSM) is activated by the TNF-R and CD95 (Cifone *et al.*, 1995; Chatterjee, 1994). However, at least for CD95, the NSM functions at a later phase of the apoptotic process, and ceramide released by NSM seems to have a completely different role than ceramide released by ASM. It is likely that NSM, which resides in the cytosol, gains access to sphingomyelin only in the late phase of apoptosis, and may be involved in an amplification loop for the apoptotic process. In addition to its activation by the TNF-R and CD95, NSM is also stimulated by neurotrophic factors, CD40 ligand, L-selectin, daunorubicin, dexamethasone, D-cytosine arabinoside, cell cycle arrest, serum deprivation and cell senescence (for review see Ferlinz *et al.*, 2000). Thus, ceramide released by NSM may serve diverse functions depending on cell type or the context in which the lipid is acting. The tight regulation of the NSM by glutathione suggests the enzyme to be involved in sensing oxidative stress (Liu and Hannun, 1997). A recently identified human alkaline sphingomyelinase (Nyberg *et al.*, 1996) from bile with an enzymatic optimum activity at pH 9.0 does not seem to be involved in cellular signaling.

5. CELLULAR TARGETS OF CERAMIDE IN APOPTOSIS

Ceramide has been shown to regulate directly or indirectly kinase suppressor of Ras (KSR; identical to ceramide-activated protein kinase) (Basu *et al.*, 1998), a ceramide-activated protein phosphatase (Dobrowsky

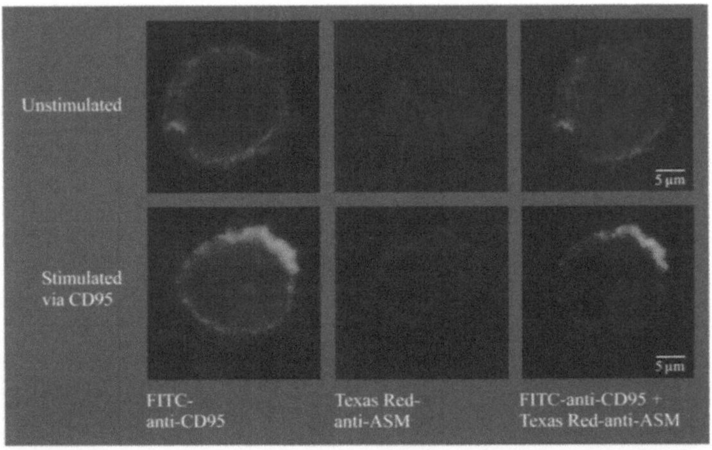

Figure 2. The confocal microscopy results show the clustering of CD95 upon stimulation. Clustered CD95 co-localizes with ASM in the cell membrane. ASM is transported to the cell membrane most likely with intracellular storage vesicles, which fuse with the cell membrane upon CD95 stimulation. Unstimulated cells show a homogenous distribution of CD95 in the cell membrane. Cells were stimulated via CD95 for 2 min or left unstimulated, fixed, permeabilized, stained with a FITC-labeled anti-CD95 and a Texas Red anti-ASM antibody and analyzed by confocal microscopy. The right pictures show the overlay.

and Hannun, 1993), protein kinase C (Muller *et al.*, 1995), c-Raf-1 (Yao *et al.*, 1995), the small G-proteins Ras and Rac (Gulbins *et al.*, 1995; Brenner *et al.*, 1997), Src-like tyrosine kinases (Gulbins *et al.*, 1997), phospholipase A_2 (Huwiler *et al.*, 2001), cathepsin D (Heinrich *et al.*, 1999), Jun-N-terminal kinases (Westwick *et al.*, 1995) and the ion channels Kv1.3 and CRAC (Szabo *et al.*, 1996; Lepple-Wienhues *et al.*, 1999) (Table 1).

Table 2. Shown is a summary of direct or indirect targets of cellular ceramide indicating the multiple functions of this lipid mediator.

Targets of ceramide
Kinase suppressor of Ras (KSR = ceramide-activated protein kinase)*
Ceramide-activated protein phosphatase
Protein kinase C isoforms alpha, delta and zeta*
c-Raf-1*
The small G-proteins Ras and Rac
Src-like tyrosine kinases
Phospholipase A_2*
Cathepsin D*
Jun-N-terminal kinases
Ion channels (Kv1.3 and CRAC)

* indicates putative direct targets of ceramide

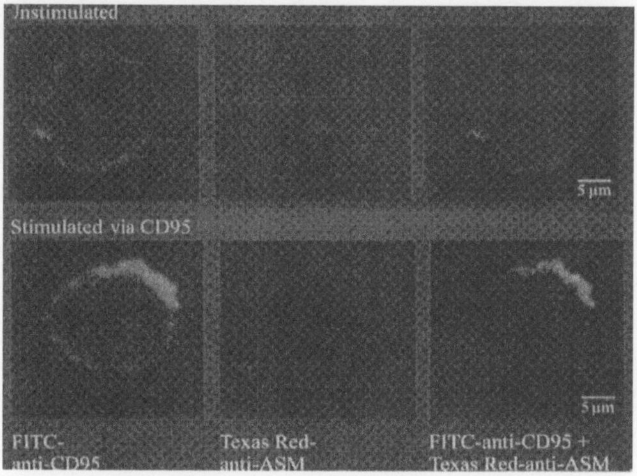

Figure 2. The confocal microscopy results show the clustering of CD95 upon stimulation. Clustered CD95 co-localizes with ASM in the cell membrane. ASM is transported to the cell membrane most likely with intracellular storage vesicles, which fuse with the cell membrane upon CD95 stimulation. Unstimulated cells show a homogenous distribution of CD95 in the cell membrane. Cells were stimulated via CD95 for 2 min or left unstimulated, fixed, permeabilized, stained with a FITC-labeled anti-CD95 and a Texas Red anti-ASM antibody and analyzed by confocal microscopy. The right pictures show the overlay.

The membrane-associated kinase KSR plays an important role in ceramide-mediated regulation of BAD, a pro-apoptotic protein belonging to the Bcl-2-like protein family (Basu *et al.*, 1998). Ceramide indirectly activates BAD via a pathway involving KSR, Ras, c-Raf-1, and MEK-1. Stimulation of this pathway results in Akt inactivation. Since the kinase activity of Akt maintains Bad in the inactive form, inhibition of Akt in turn releases BAD and, finally, permits BAD-triggered cell death.

While the exact mechanism of ceramide-mediated regulation of most of the above proteins is still unknown, ASM-released ceramide binds directly to PLA_2 and cathepsin D (Huwiler *et al.*, 2001; Heinrich *et al.*, 1999). Binding of ceramide to cathepsin D in endosomes triggers autocatalytic cleavage of cathepsin to its active form (Heinrich *et al.*, 1999). However, the physiological role of ceramide binding to cathepsin D and PLA_2 for apoptosis still remains to be determined.

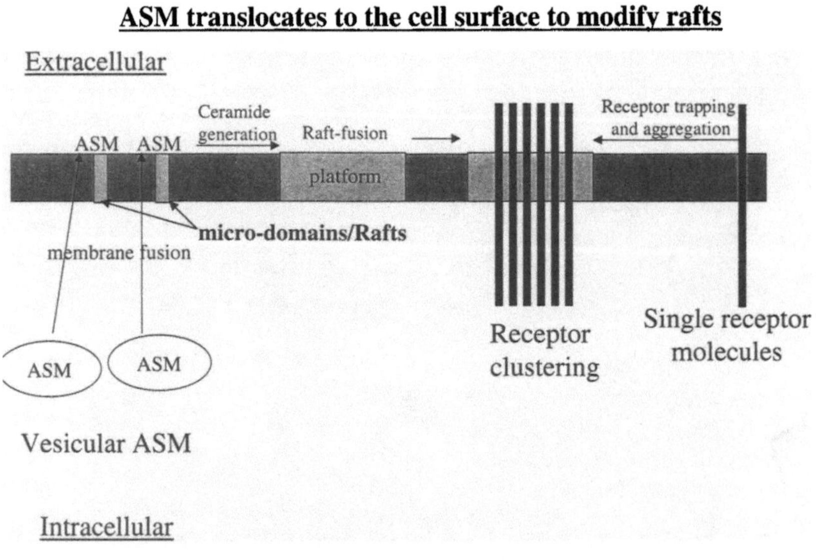

Figure 3. Primary stimulation of CD95 is insufficient to trigger apoptosis but sufficient to initiate ASM transport to the outer leaflet of the cell membrane. Surface ASM releases ceramide from sphingomyelin in the outer membrane leaflet. Ceramide alters pre-existing micro-rafts to fuse to larger platforms and to trap activated CD95. This clustering event of CD95 permits efficient signaling via the receptor possibly by recruiting intracellular signaling molecules to CD95, bringing intracellular signaling molecules into close proximity to each other, excluding inhibitory pathways, altering the affinity/avidity of the receptor for its ligand and/or by increasing the stability of CD95/CD95 ligand interactions.

We have recently suggested a novel concept for the function of ceramide (Figures 2 and 3). Stimulation via CD95 or CD40 with the cognate ligand results in translocation of ASM onto the extracellular leaflet of the cell membrane (Grassmé *et al.*, 2001a; Cremesti *et al.*, 2001). Surface ASM, now in the proximity of sphingomyelin which predominantly resides in the outer plasma membrane of mammalian cells, releases extracellularly-oriented ceramide within sphingolipid rich rafts. The notion of a function of ASM in lipid rafts is consistent with the findings of other groups that demonstrated an ASM-mediated consumption of sphingomyelin and release of ceramide in rafts upon stimulation of several receptors including the IL-1 and p75 NGF receptor (Liu and Anderson, 1995; Bilderback *et al.*, 1997). Accumulation of ceramide in rafts may alter the rafts in several respects based on data derived from biophysical studies using model membrane systems: First, ceramide has the tendency to spontaneously self-aggregate into microdomains (Huang *et al.*, 1999; Veiga *et al.*, 1999; ten Grotenhuis *et al.*, 1996) and, second, ceramide-rich microdomains are capable of fusing into larger platforms (Holopainen *et al.*, 1998). Thus, generation of ceramide within rafts may result in their fusion into larger platforms with altered biophysical properties. Those ceramide rich raft-platforms may then serve to cluster liganded receptor molecules, e.g., CD95 or CD40. Clustering results in high receptor density and may facilitate trans-activation of intracellular signaling molecules associating with the receptor. In addition, trapping of activated receptors in rafts may stabilize the interaction with ligand. Finally, ceramide-enriched rafts may recruit intracellular signaling molecules to the activated, clustered receptor, exclude inhibitory pathways and/or directly alter the affinity/avidity of the receptor for its ligand. A modification of intracellular signaling molecules by ceramide generated in rafts has been recently demonstrated for PI-3-K (Zundel *et al.*, 2000). Those studies identified an ASM- and ceramide-dependent recruitment of caveolin 1 to PI-3-K-receptor complexes within rafts, which blocked the activity of PI-3-K. The inhibition of anti-apoptotic PI-3-K by ceramide through caveolin 1, then sensitized cells to apoptotic stimuli.

The concept of ceramide activity in and modification of rafts also provides an explanation for the activation and function of ASM and ceramide in non-apoptotic signaling pathways that might require clustering of specific receptors. Thus, preliminary studies suggest that ASM signaling in rafts might mediate clustering of CD40, CD28, CD48 or the CFTR molecule.

Further, rafts altered by stimulated ASM activation may be involved in non-receptor-mediated signaling. For instance, UVC triggers aggregation of cell surface CD95 (Rehemtulla *et al.*, 1997), which might be sufficient to initiate signaling of CD95 at least under stress conditions. Likewise, it has been shown that stress stimuli are able to trigger aggregation of the TNF-R to initiate cell death (Boldin *et al.*, 1995). Activation of both receptors by stress stimuli seems to occur in the absence of the CD95-ligand or TNF, respectively. Thus, it is possible that some stress stimuli alter the conformation of CD95 or change the composition of sphingolipid-rich rafts permitting trapping of unliganded CD95 molecules. This may result in low level receptor activation, which even if inefficient, may be sufficient to signal apoptosis, if persistent. This notion is supported by the finding that dominant negative FADD blocks, at least in part, UV-induced apoptosis (Chatterjee and Wu, 2001). The relationship, if any, of CD95 clustering to the aforementioned JNK activation requirement for UVA-induced apoptosis is unknown.

6. CONCLUSIONS

A variety of studies demonstrate a central role for ASM and ceramide in several forms of apoptosis. Ceramide seems to regulate the activity of certain proteins and, thus, may function, in some circumstances, as a second messenger. In addition, the concept of raft modification by ceramide provides a comprehensive model for cellular effects of ceramide, and perhaps a biophysical explanation for the diverse functions of this lipid.

ACKNOWLEDGMENTS

The authors are thankful to Janice Mann for excellent editorial help.

REFERENCES

Andersen, R. G., 1998, The caveolae membrane system. *Annu. Rev. Biochem.* **67**: 199-225

Andrieu-Abadie, N., Carpentier, S., Salvayre, R. and Levade, T., 1998, The tumour necrosis factor-sensitive pool of sphingomyelin is resynthesized in a distinct compartment of the plasma membrane. *Biochem. J.* **333**: 91-97

Basu, S., Bayoumy, S., Zhang, Y., Lozano, J. and Kolesnick, R., 1998, BAD enables ceramide to signal apoptosis via Ras and Raf-1. *J. Biol. Chem.* **273**: 30419-30426

Bilderback, T.R., Gazula, V.R. and Dobrowsky, R.T., 2001, Phosphoinositide 3- kinase crosstalk between Trk A tyrosine kinase and p75 (NTR)-dependent sphingolipid signaling pathways. *Neurochem.* **76**: 1540-1551

Bilderback, T.R., Grigsby, R.J. and Dobrowsky, R.T., 1997, Association of p75 (NTR) with caveolin and localization of neutrophin-induced sphingomyelin hydrolysis to caveolae. *J. Biol. Chem.* **272**: 10922-10927

Boldin, M.P., Mett, I.L., Varfolomeev, E.E., Chumakov, I., Shemer-Avni, Y., Camonis, J.H. and Wallach, D., 1995, Self-association of the "death domain" of the p55 tumor necrosis factor (TNF) receptor and Fas/APO1 prompts signaling for TNF and Fas/APO1 effects. *J. Biol. Chem.* **270**: 387-391

Boucher, L.M., Wiegmann, K., Futterer, A., Pfeffer, K., Machleidt, T., Schutze, S., Mak, T.W. and Kronke, M., 1995, CD28 signals through acidic sphingomyelinase. *J. Exp. Med.* **181**: 2059-2068

Brenner, B., Ferlinz, K., Grassme, H., Weller, M., Koppenhoefer, U., Dichgans, J., Sandhoff, K. Lang, F. and Gulbins, E., 1998, Fas/CD95/Apo-I activates the acidic sphingomyelinase via caspases. *Cell Death Differ.* **5**: 29-37

Brenner, B., Koppenhoefer, U., Weinstock, C., Linderkamp, O., Lang, F. and Gulbins, E., 1997, Fas- or ceramide-induced apoptosis is mediated by a Rac1-regulated activation of Jun N-terminal kinase/p38 kinases and GADD153. *J. Biol. Chem.* **272**: 22173-22181

Brown, D. A. and London, E., 1998, Functions of lipid rafts in biological membranes. *Annu. Rev. Cell. Dev. Biol.* **14**: 111-367

Callahan, J.W., Jones, C.S., Davidson, D.J. and Shankaran, P., 1983, The active site of lysosomal sphingomyelinase: evidence for the involvement of hydrophobic and ionic groups. *J. Neurosci. Res.* **10**: 151-161.

Chatterjee, M. and Wu, S., 2001, Involvement of Fas receptor and not tumor necrosis factor-alpha receptor in ultraviolet-induced activation of acid sphingomyelinase. *Mol. Carcinog.* **30**: 47-55.

Chatterjee, S., 1994, Neutral sphingomyelinase action stimulates signal transduction of tumor necrosis factor-alpha in the synthesis of cholesteryl esters in human fibroblasts. *J. Biol. Chem.* **269**: 879-882

Cifone, M.G., De Maria, R., Roncaioli, P., Rippo, M.R., Azuma, M., Lanier, L.L., Santoni, A. and Testi, R., 1994, Apoptotic signaling through CD95 (Fas/Apo-1) activates an acidic sphingomyelinase. *J. Exp. Med.* **180**: 1547-1552

Cifone, M.G., Roncaioli, P., De Maria, R., Camarda, G., Santoni, A., Ruberti, G. and Testi, R., 1995, Multiple pathways originate at the Fas/APO-1 (CD95) receptor: sequential involvement of phosphatidylcholine-specific phospholipase C and acidic sphingomyelinase in the propagation of the apoptotic signal. *EMBO J.* **14**: 5859-5868

Cremesti, A., Paris, F, Grassmé, H., Holler, N., Tschopp, J., Fuks, Z., Gulbins, E. and Kolesnick, R., 2001, Ceramide enables Fas to cap and kill. *J. Biol. Chem.* **276**: 23954-23961

De Maria, R., Rippo, M.R., Schuchman, E. H. and Testi, R., 1998, Acidic sphingomyelinase (ASM) is necessary for Fas-induced CD3 ganglioside accumulation and efficient apoptosis of lymphoid cells. *J. Exp. Med.* **187**: 897-902

Dobrowsky, R.T. and Hannun, Y.A., 1993, Ceramide-activated protein phosphatase: partial purification and relationship to protein phosphatase 2A. *Adv. Lipid Res.* **25**: 91-104

Emmelot, P. and Van Hoeven, R.P., 1975 Phospholipid unsaturation and plasma membrane organization. *Chem. Phys. Lipids.* **14**: 236-246.

Ferlinz, K., Grassmé, H., Lang, F. and Gulbins, E., 2000, Sphingomyelin breakdown and ceramide release in signal transduction. *Rec. Res. Devel. Immunology* **2**: 189-200.

Futerman, A.H., Stieger, B., Hubbard, A.L. and Pagano, R. E., 1990, Sphingomyelin synthesis in rat liver occurs predominantly at the cis and medial cisternae of the Golgi apparatus. *J. Biol. Chem.* **265**: 8650-8657

Grassmé, H., Gulbins, E., Brenner, B., Ferlinz, K., Sandhoff, K., Harzer, K., Lang, F. and Meyer, T.F., 1997, Acidic sphingomyelinase mediates entry of N. gonorrhoeae into nonphagocytic cells. *Cell* **91**: 605-615

Grassmé, H., Jekle, A., Riehle, A., Schwarz, H., Berger, J., Sandhoff, K., Kolesnick, R. and Gulbins, E., 2001a, CD95 signaling via ceramide-rich membrane rafts. *J. Biol. Chem.* **276**: 20589-20596

Grassmé, H., Jendrossek, V., Bock, J., Riehle, A. and Gulbins E., 2001b, Ceramide-rich Membrane Rafts Mediate CD40 Clustering. *J. Immunol.*, in press

Grullich, C., Sullards, M.C., Fuks, Z., Merrill, A.H. Jr. and Kolesnick, R., 2000, CD95 (Fas/APO-1) signals ceramide generation independent of the effector stage of apoptosis. *J. Biol. Chem.* **275**: 8650-8656

Gulbins, E., Bissonnette, R., Mahboubi, A., Martin, S., Nishioka, W., Brunner, T., Baier, G., Baier-Bitterlich, G., Byrd, C., Lang, F., Kolesnick, R, Altman, A., Green, D., 1995, FAS-induced apoptosis is mediated via a ceramide-initiated RAS signaling pathway. *Immunity* **2**: 341-351

Gulbins, E., Szabo, I., Baltzer, K. and Lang, F., 1997, Ceramide-induced inhibition of T lymphocyte voltage-gated potassium channel is mediated by tyrosine kinases. *Proc. Natl. Acad. Sci. U S A* **94**: 7661-7666

Harder, T. and Simons, K., 1997, Caveolae, DIGs, and the dynamics of sphingolipid-cholesterol microdomains. *Curr. Opin. Cell. Biol.* **9**: 534-542.

Heinrich, M., Wickel, M., Schneider-Brachert, W., Sandberg, C., Gahr, J., Schwander, R., Weber, T., Saftig, P., Peters, C., Brunner, J., Kronke, M. and Schutze, S., 1999, Cathepsin D targeted by acid sphingomyelinase-derived ceramide. *EMBO J.* **18**: 5252-5263

Hengartner, M. O., 2000, The biochemistry of apoptosis. *Nature* **407**: 770-776.

Holopainen, J.M., Subramanian, M. and Kinnunen, P.K., 1998, Sphingomyelinase induces lipid microdomain formation in a fluid phosphatidycholine/sphingomyelin membrane. *Biochemistry* **37**: 17562-17570

Holtzman, M.J., Green, J.M., Jayaraman, S. and Arch, R.H., 2000, Regulation of T cell apoptosis. *Apoptosis* **5**: 459-471

Huang, C., Wy, M., Ding, M., Bowden, G.T. and Dong, Z., 1997, Direct evidence for an important role of sphingomyelinase in ultraviolet-induced ctivation of c-Jun N-terminal kinase. *J. Biol. Chem.* **272**: 27753-27757

Huang, H.W., Goldberg, E.M. and Zidovetzki, R., 1999, Ceramides modulate protein kinase C activity and perturb the structure of Phosphatidylcholine/Phosphatidylserine bilayers. *Biophys. J.* **77**: 1489-1497

Huwiler, A., Johansen, B., Skarstad, A. and Pfeilschifter, J., 2001, Ceramide binds to the CaLB domain of cytosolic phospholipase A_2 and facilitates its membrane docking and arachidonic acid release. *FASEB J.* **15**: 7-9

Itoh, N. and Nagata, S., 1993, A novel protein domain required for apoptosis. Mutational analysis of human Fas antigen. *J. Biol. Chem.* **268**: 10932-10937

Jeckel, D., Karrenbauer, A., Birk, R., Schmidt, R. R. and Wieland, F., 1990, Sphingomyelin is synthesized in the cis Golgi. *FEBS Lett.* **261**: 155-157

Jekle, A., Bock, J., Mueller, C., Kun, J. and Gulbins E. 2001, The CD95 death domain is required for acid sphingomyelinase mediated receptor clustering. In prep.

Kirschnek, S., Paris, F., Weller, M., Grassmé, H., Ferlinz, K., Riehle, A., Fuks, Z., Kolesnick, R. and Gulbins, E., 2000, CD95-mediated apoptosis *in vivo* involves acid sphingomyelinase. *J. Biol. Chem.* **275**: 27316-27323

Lepple-Wienhues, A., Belka, C., Laun, T., Jekle, A., Wieland, U., Welz, M., Heil, L., Kun, J., Busch, G., Weller, M., Bamberg, M., Gulbins, E. and Lang, F., 1999, Stimulation of CD95 (Fas) blocks T lymphocyte calcium channels through sphingomyelinase and sphingolipids. *Proc. Natl. Acad. Sci. U S A* **96**: 13795-13800

Lin, T., Genestier, L., Pinkoski, M.J., Castro, A., Nicholas, S., Mogil, R., Paris, F., Fuks, Z., Schuchman, E.H., Kolesnick, R.N. and Green D.R., 2000, Role of acidic sphingomyelinase in Fas/CD95-mediated cell death. *J. Biol. Chem.* **275**: 8657-8663

Liu, P. and Anderson, R.G., 1995, Compartmentalized production of ceramide at the cell surface. *J. Biol. Chem.* **270**: 27179-27185

Liu, P. and Hannun, Y.A., 1997, Inhibition of the neutral magnesium-dependent sphigomyelinase by glutathione. *J. Biol.* **272**: 16281-16287

Lozano, J., Menendez, S., Morales, A., Ehleiter, D., Liao, W.C., Wagman, R., Haimovitz-Friedman, A., Fuks, Z. and Kolesnick, R., 2001, *Cell* autonomous apoptosis defects in acid sphingomyelinase knockout fibroblasts. *J. Biol. Chem.* **276**: 442-448

Mathias, S., Pena, L.A. and Kolesnick, R.N., 1998, Signal transduction of stress via ceramide. *Biochem. J.* **335**: 465-480

Mathias, S., Younes, A., Kan, C.C., Orlow, I., Joseph, C. and Kolesnick, R.N., 1993, Activation of the sphingomyelin signaling pathway in intact EL4 cells and in a cell-free system by IL-1 beta. *Science* **259**: 519-522

Morita, Y., Perez, G.I., Paris, F., Miranda, S.R., Ehleiter, D., Haimovitz-Friedman, A., Fuks, Z., Xie, Z., Reed, J.C., Schuchman, E.H., Kolesnick, R.N. and Tilly, J.L., 2000, Oocyte apoptosis is suppressed by disruption of the acid sphingomyelinase gene or by sphingosine-1- phosphate therapy. *Nat. Med.* **6**: 1109-1114

Muller, G., Ayoub, M., Storz, P. Rennecke, J., Fabbro, D. and Pfizenmaier, K., 1995, PKC zeta is a molecular switch in signal transduction of TNF-alpha, bifunctionally regulated by ceramide and arachidonic acid. *EMBO J.* **14**: 1961-1969

Nagata, S. and Suda, T., 1995, Fas and Fas ligand: lpr and gld mutations. *Immunol. Today* **16**: 39-43

Newrzella, D. and Stoffel, W., 1996, Functional analysis of the glycosylation of murine acid sphingomyelinase. *J. Biol. Chem.* **271**: 32089-32095

Nyberg, L., Duan, R.D., Axelson, J. and Nilsson, A., 1996, Identification of an alkaline sphingomyelinase activity in human bile. *Biochim. Biophys. Acta.* **1300**: 42-48

Paris, F., Grassmé, H., Cremesti, A., Zager, J., Fong, Y., Haimovitz-Friedman, A., Fuks, Z., Gulbins, E. and Kolesnick, R., 2001a, Natural ceramide reverses Fas resistance of acid sphingomyelinase (-/-) hepatocytes. *J. Biol. Chem.* **276**: 8297-8305

Paris, F., Fuks, Z., Kang, A., Capodieci, P., Juan, G., Ehleiter, D., Haimovitz-Friedman, A., Cordon-Cardo, C. and Kolesnick, R., 2001b, Endothelial apoptosis as the primary lesion initiating intestinal radiation damage in mice. *Science*, *293*: 293-297

Pena, L.A., Fuks, Z. and Kolesnick, R.N., 2000, Radiation-induced apoptosis of endothelial cells in the murine central nervous system: protection by fibroblast growth factor and sphingomyelinase deficiency. *Cancer Res.* **60**: 321-327

Rehemtulla, A., Hamilton, C.A., Chinnaiyan, A.M. and Dixit, V.M., 1997, Ultraviolet radiation-induced apoptosis is mediated by activation of CD-95 (Fas/APO-1). *J. Biol. Chem.* **272**: 25783-25786

Santana, P., Pena, L.A., Haimovitz-Friedman, A., Martin, S., Green, D., McLoughlin, M., Cordon-Cardo, C., Schuchman, E.H., Fuks, Z. and Kolesnick, R., 1996, Acid sphingomyelinase-deficient human lymphoblasts and mice are defective in radiation-induced apoptosis. *Cell* **86**: 189-199

Schissel, S.L., Jiang, X., Tweedie-Hardman, J., Jeong, T., Camejo, E.H., Najib, J., Rapp, J.H., Williams, K.J. and Tabas, I., 1998, Secretory sphingomyelinase, a product of the acid sphingomyelinase gene, can hydrolyze atherogenic lipoproteins at neutral pH. Implications for atherosclerotic lesion development. *J. Biol. Chem.* **273**: 2738-2746

Schissel, S.L., Keesler, G.A., Schuchman, E.H., Williams, K.J. and Tabas, I., 1998, The cellular trafficking and zinc dependence of secretory and lysosomal sphingomyelinase, two products of the acid sphingomyelinase gene. *J. Biol. Chem.* **273**: 18250-18259

Schissel, S.L., Schuchman, E.H., Williams, K.J. and Tabas, I., 1996, Zn^{2+}-stimulated sphingomyelinase is secreted by many cell types and is a product of the acid sphingomyelinase gene. *J. Biol. Chem.* **31**: 18431-18436

Schutze, S., Potthoff, K., Machleidt, T., Berkovic, D., Wiegmann, K. and Kronke, M., 1992, TNF activates NF-kappa B by phosphatidylcholine-specific phospholipase C-induced "acidic" sphingomyelin breakdown. *Cell* **71**: 765-776

Schwandner, R., Wiegmann, K., Bernardo, K., Kreder, D. and Kronke, M., 1998, TNF receptor death domain-associated proteins TRADD and FADD signal activation of acid sphingomyelinase. *J. Biol. Chem.* **273**: 5916-5922

Szabo, I., Gulbins, E., Apfel, H., Zhang, X., Barth, P., Busch, A.E., Schlottmann, K., Pongs, O. and Lang, F., 1996, Tyrosine phosphorylation-dependent suppression of a voltage-gate K^+ channel in T lymphocytes upon Fas stimulation. *J. Biol. Chem.* **271**: 20465-20469

ten Grotenhuis, E., Demel, R.A., Ponec, M., Boer, D.R., van Miltenburg, J.C. and Bouwstra, J.A., 1996, Phase behavior of stratum corneum lipids in mixed Langmuir-Blodgett monolayers. *Biophys. J.* **71**: 1389-1399

Veiga, M.P., Arrondon, J.L. Goni, F.M. and Alonso, A., 1999, Ceramides in phospholipid membranes: effects on bilayer stability and transition to nonlamellar phases. *Biophys. J.* **76**: 342-350

Walczak, H. and Krammer, P.H., 2000, Minireview: The CD95 (APO-1/Fas) and the TRAIL (APO-2L) Apoptosis Systems. *Exp. Cell Res.* **256**: 58-66

Westwick, J.K., Bielawska, A.E., Dbaibo, G., Hannun, Y.A. and Brenner, D.A., 1995, Ceramide activates the stress-activated protein kinases. *J. Biol. Chem.* **270**: 22689-22692

Yao, B., Zhang, Y., Delikat, S., Mathias, S., Basu, S. and Kolesnick, R., 1995, Phosphorylation of Raf by ceramide-activated protein kinase. *Nature* **378**: 307-310

Yu, Z.F., Nikolova-Karakashian, M., Zhou, D., Cheng, G., Schuchman, E.H. and Mattson, M.P., 2000, Pivotal role for acidic sphingomyelinase in cerebral ischemia-induced ceramide and cytokine production, and neuronal apoptosis. *J. Mol. Neurosci.* **15**: 85-97

Zhang, Y., Mattjus, P., Schmid, P.C., Dong, Z., Zhong, S., Ma, W.Y., Brown, R.E., Bode, A.M., Schmid, H.H. and Dong, Z., 2001, Involvement of the acid sphingomyelinase pathway in UVA-induced apoptosis. *J. Biol. Chem.* **276**: 11775-11782

Zundel, W., Swiersz, L.M. and Giaccia, A., 2000, Caveolin 1-mediated regulation of receptor tyrosine kinase-associated phosphatidylinositol 3- kinase activity by ceramide. *Mol. Cell. Biol.* **20**: 1507-1514

Chapter 13

Cellular Signaling by Sphingosine and Sphingosine 1-Phosphate

Their opposing roles in apoptosis

Susan Pyne
Department of Physiology & Pharmacology, Strathclyde Institute for Biomedical Sciences, University of Strathclyde, Glasgow, G4 0NR, Scotland, UK

1. INTRODUCTION

Sphingosine and sphingosine 1-phosphate (S1P[1]) contribute to determining cellular fate, such as survival, proliferation or apoptosis, by regulating specific signaling pathways. These include the caspase cascade (a family of cysteine proteases), which is essential for apoptosis. Sphingosine and S1P also influence the Bcl-2 protein family, which has both pro- and anti-apoptotic members (Strasser *et al.*, 2000). Moreover, these sphingolipids also affect the relative activities of the p42/p44 mitogen-activated protein kinase (p42/p44 MAPK) cascade (which is critical for the initiation of immediate early gene expression and cell cycle progression leading to mitogenesis (Marshall, 1995)) and the stress-activated protein

[1] **Abbreviations** DHS, dihydrosphingosine (sphinganine); DMS, *N,N*-dimethylsphingosine; EDG, endothelial differentiation gene; G(i)PCR, $G_{(i)}$protein coupled receptor; hsp, heat shock protein; HUVEC, human umbilical vein endothelial cells; JNK, c-Jun N-terminal-directed kinase; LDL, low density lipoprotein; LPP, lipid phosphate phosphatase; LPA, lysophosphatidic acid; MAPK, mitogen-activated protein kinase; NO, nitric oxide; NOS, nitric oxide synthase; $p125^{FAK}$, focal adhesion kinase; PDGF, platelet-derived growth factor; PGE_2, prostaglandin E2; PGF_{2a}, prostaglandin F2alpha; PI3K, phosphoinositide 3-kinase; PKC, protein kinase C; PMA, phorbol 12-myristate 13-acetate; SAPK, stress-activated protein kinase; S1P, sphingosine 1-phosphate; SPHK, sphingosine kinase; TNFα, tumor necrosis factor alpha.

Subcellular Biochemistry, Volume 36: Phospholipid Metabolism in Apoptosis.
Edited by Quinn and Kagan. Kluwer Academic/Plenum Publishers, New York, 2002

kinase cascade (SAPK/JNK) (which is involved in programming cells for apoptosis (Verheij *et al.*, 1996)), both of which regulate transcription factor activation. However, other coincident signals also play a role and contribute to the pleiotropic actions of sphingosine and S1P, which can differ, depending upon the cell type.

The balance between opposing cellular signal pathways activated by ceramide/sphingosine and S1P is believed to be critical (Cuvillier *et al.*, 1996). In general, sphingosine is a pro-apoptotic signal and a number of intracellular targets for its action have been suggested (reviewed by Alessenko, 2000). In contrast, S1P usually, although not invariably, opposes the effects of sphingosine by promoting proliferation, cell survival and other biological processes that enhance survival, such as adhesion. S1P is unique in that it may act as *both* an extracellular mediator (agonist) and an intracellular second messenger (reviewed by Pyne and Pyne, 2000). Extracellular effects of S1P in mammalian cells are mediated via a recently identified family of plasma membrane G-protein coupled receptors (GPCRs), whereas specific intracellular sites of action remain to be defined. The ability of S1P to inhibit apoptosis may be important in, for example, wound healing and repair whereas it also plays a role in several pathophysiologic disease states involving proliferation.

This chapter describes the metabolism of sphingosine and S1P, evidence to support their pro- and anti-apoptotic effects and their proposed sites of action on signaling processes that contribute to apoptosis and proliferation.

2. METABOLISM OF SPHINGOSINE AND S1P

The intracellular concentrations of sphingosine and S1P are governed by the activities of enzymes that catalyse their synthesis and removal. These include ceramidase, sphingosine kinase (SPHK), S1P phosphatase and S1P lyase (Figure 1). Several of these enzymes have only recently been cloned and knowledge of their respective roles and regulation is incomplete.

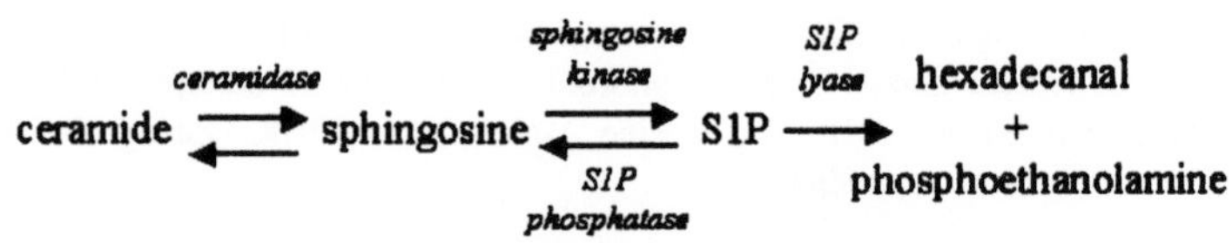

Figure 1 Metabolic pathways of sphingosine and S1P

2.1 Ceramidase

Ceramidase hydrolyses the amide linkage of ceramide to produce a non-esterified fatty acid and sphingosine (Figure 1). To date, three types of mammalian ceramidase have been cloned: acid (Koch *et al.*, 1996), neutral (El Bawab *et al.*, 2000; Tani *et al.*, 2000) and alkaline (Mao *et al.*, 2001), according to the pH optimum for enzyme activity. The enzymes differ in their sub-cellular localization and substrate specificity. For example, acid ceramidase is lysosomal, preferentially uses unsaturated ceramide and is believed to be responsible for ceramide catabolism since dysfunction of this enzyme results in the sphingolipidosis, Farber disease. Neutral ceramidase also preferentially uses unsaturated ceramide but is mitochondrial (El Bawab *et al.*, 2000) whereas alkaline ceramidase preferentially hydrolyses phytoceramide and is found in the Golgi apparatus and endoplasmic reticulum (Mao *et al.*, 2001). Notably, neutral ceramidase also catalyses the reverse reaction, i.e. acylation of sphingosine to ceramide (El Bawab *et al.*, 2001). The precise function of the various ceramidase isoforms is still to be established. However, both neutral and alkaline ceramidases in mammalian cells are regulated by agonists such as platelet-derived growth factor (PDGF), interleukin 1β and tumor necrosis factor α (TNFα) (Coroneos *et al.*, 1995; Nikolova-Karakashian *et al.*, 1997).

2.2 Sphingosine kinase

Sphingosine kinase (SPHK) phosphorylates sphingosine to S1P and, thereby, serves two roles, i.e. to make S1P and to remove sphingosine (Figure 1). Two distinct isoforms of human SPHK, termed hSPHK1 and hSPHK2, have been cloned (Nava *et al*, 2000; Liu *et al.*, 2000). These genes encode proteins of 384 (42.4kDa) and 618 (65.6kDa) amino acids, respectively. Thus, hSPHK2 is much larger than hSPHK1 with differences being a 140 amino acid region at the N-terminal and a 120 amino acid mid-region. The common regions of the isoforms have approximately 78% similarity, including five highly conserved regions (C1-C5) that are also found in the mouse, *Saccharomyces cerevisiae, Schizosaccharomyces pombe* and *Caenorhabditis elegans* enzymes. K_m values for phosphorylation of D-*erythro*-sphingosine, which is the best substrate for SPHK1, are similar (5 and 3.4μM for SPHK1 and SPHK2, respectively). Both isoforms also phosphorylate D-*erythro*-dihydrosphingosine (DHS, sphinganine) but this is a better substrate than sphingosine for SPHK2. *N,N*-dimethylsphingosine (DMS) inhibits both isoforms whereas *threo*-DHS inhibits SPHK1 but is a weak substrate for SPHK2.

The relative importance of the two SPHK isoforms is unclear at present. They differ in their tissue distribution, with SPHK1 mRNA expression being highest in lung, spleen and liver whereas liver and heart contain the highest levels of SPHK2. Additionally, their temporal expression differs in development with high levels of mouse SPHK1 mRNA up to embryonic day 7 followed by a reduction in contrast to SPHK2, which is low to day 7 and increases thereafter to day 17 (Liu *et al.*, 2000).

Regulation of the SPHK isoforms is yet to be established. A wide variety of stimuli increase SPHK activity and elevate S1P levels including growth factors, GPCR agonists, cytokines, phorbol esters, vitamin D_3 and antigen (Pyne and Pyne, 2000). For example, *acute* activation of SPHK occurs in response to PDGF (Olivera and Spiegel, 1993; Pyne *et al.*, 1996), epidermal growth factor (Meyer zu Heringdorf *et al.*, 1999), carbachol (Meyer zu Heringdorf *et al.*, 1998), purinergic agonists (Alemany *et al.*, 2000), TNFα (Xia *et al.*, 1999) and antigen (Choi *et al.*, 1996; Melendez *et al.*, 1998) whereas *chronic* effects have been observed with phorbol 12-myristate 13-acetate (PMA) (Buehrer *et al.*, 1996), Vit D_3 (Kleuser *et al.*, 1998), oxidized LDL (Auge *et al.*, 1999) and forskolin (Machawate *et al.*, 1998). Consensus sequences for calcium/calmodulin binding and potential phosphorylation sites for protein kinases A and C and casein kinase II have been identified. Evidence to support regulation of SPHK by calcium includes the use of ionophores (Alemany *et al.*, 2000) and a PDGFβ receptor mutant, which lacked the docking site (tyrosine 1021) for recruitment of phospholipase Cγ and which failed to activate SPHK in response to PDGF (Olivera *et al.*, 1999). Regulation probably depends upon cell type, which may reflect differing signaling pathways to SPHK or the expression of different SPHK isoforms.

Whether the two SPHK isoforms play similar or distinct roles is yet to be determined. Transfection of HEK 293 cells with a dominant negative form of SPHK1 abolished subsequent TNFα-stimulated S1P formation although basal levels were unaffected (Pitson *et al.*, 2000). Therefore, distinct isoforms of SPHK may regulate separate pools of S1P.

2.3 S1P phosphatase

A murine S1P phosphatase gene, which encodes a 430 amino acid (47.7kDa) protein, has recently been cloned (Mandala *et al.*, 2000). This enzyme bears similarity to a family of lipid phosphate phosphatases (LPPs) (Brindley and Waggoner, 1998) in that it has invariant residues within the three domains of the common predicted catalytic site ((K-X_6-RP)-$X_{16\text{-}37}$-(S-X-H)-$X_{40\text{-}46}$-(R-X_5-H-X_3-D)) although the spacing between the domains differs from the LPPs. In contrast to LPPs, S1P phosphatase is highly substrate specific, inhibited by detergents and has 3-4 additional predicted hydrophobic regions. Its subcellular distribution, membrane topology and orientation of the catalytic site remain unknown. Thus, it remains to be determined whether S1P phosphatase can limit the signaling potential of extracellular S1P at GPCRs (see below) by dephosphorylation. Such a mechanism has been proposed for LPPs with regard to extracellular lysophosphatidic acid (LPA) (Xu *et al*, 2000). However, this remains contentious since over-expression of LPPs limits G_iPCR-mediated signaling *per se*, i.e. in response to G_iPCR agonists that are not substrates for LPPs (Alderton *et al*., 2001).

Multiple isoforms of S1P phosphatase may exist (De Cuester *et al*., 1995) and is supported by the lack of correlation of the highest levels of specific activity (detected in brain, spleen and intestinal mucosa) with mRNA levels (highest in liver and kidney). Alternatively, other regulatory factors may be required (Mandala, 2001). S1P phosphatase may be constitutively active since exogenous [^{3}H]dihydroS1P (1μM) was, in part, dephosphorylated to [^{3}H]sphingosine and subsequently metabolised to [^{3}H]ceramide and [^{3}H]sphingomyelin in human fibroblasts (Van Veldhoven and Mannaerts, 1994). The rate of dephosphorylation of S1P by this phosphatase was calculated to be at least the same as that for its degradation by the S1P lyase.

2.4 S1P lyase

S1P lyase cleaves S1P at the C_2-C_3 bond to produce hexadecanal (palmitaldehyde) and phosphoethanolamine (Van Veldhoven and Mannaerts, 1993). The enzyme requires the co-enzyme pyridoxal 5'phosphate and is specific for the D-*erythro* isomers of phosphorylated sphingoid bases including sphingosine-, DHS- and phytosphingosine1-phosphate. The enzyme is ubiquitously expressed in mammalian tissues with the notable exception of platelets where it is absent. A mammalian S1P lyase gene has recently been cloned (Zhou and Saba, 1998) and encodes a protein of 568 amino acids (63.7kDa). Hydropathy analysis indicates the presence of a single 20 amino acid transmembrane domain close to the N-terminus. Analysis of S1P lyase mRNA levels was consistent with previously reported S1P lyase activity levels, being most abundant in liver followed by kidney, lung, heart and brain. The mechanism of regulation of the enzyme remains to be determined.

3. APOPTOTIC EFFECTS OF SPHINGOSINE

Initiation of apoptosis in some cell types is dependent upon the deacylation of agonist-stimulated ceramide to sphingosine (Ohta *et al.*, 1994) although others suggest that the concentration of sphingosine required may not be attained under physiological conditions (Chmura *et al.*, 1996). Nevertheless, sphingosine (and DHS/sphinganine) induces apoptosis in many cell types (Jarvis *et al.*, 1996 and 1997; Ohta *et al.*, 1995; Sweeney *et al.*, 1996). Moreover, acylation of sphingosine to ceramide does not appear to be involved. In addition, myeloid leukaemia cells induced to terminally differentiate by phorbol esters exhibit a maturation-dependent increase in their ability to deacylate ceramide to sphingosine and, ultimately, undergo apoptosis (MacFarlane *et al.*, 1993; Ohta *et al.*, 1995). In addition, sphingosine sensitises cells to gamma-irradiation-induced apoptosis (Nava *et al.*, 2000) and increases the pro-apoptotic activity of ceramide (Jarvis *et al.*, 1996). The anti-neoplastic agent doxorubicin, which is known to increase cellular ceramide, has recently been shown to also increase sphingosine levels in human breast carcinoma MCF7 cells (Cuvillier *et al.*, 2001).

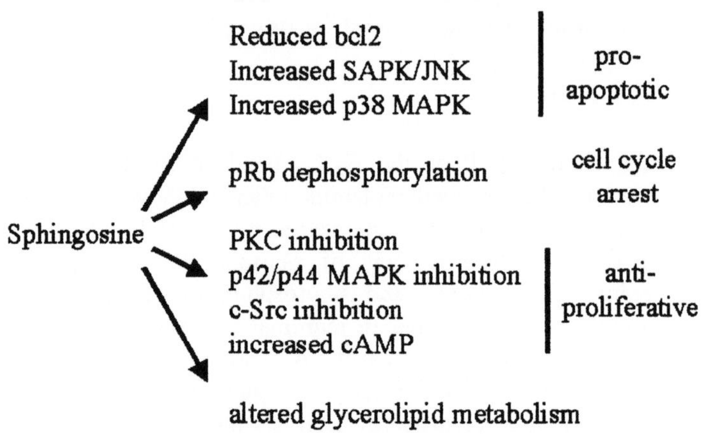

Figure 2 Actions of sphingosine

Many intracellular targets of sphingosine that may contribute to its apoptotic and anti-proliferative effects have been identified (Figure 2). These include proteins of the bcl-2 family, the retinoblastoma gene product, several enzymes and its ability to interact with DNA (reviewed by Alessenko, 2000). For example, sphingosine-induced apoptosis of HL60 cells correlated with reduced expression of the cell survival protein *bcl*-2 whereas the expression of *bcl*-XL and *bax* mRNA was unchanged (Sakakura *et al.*, 1996). In addition, sphingosine stimulated the dephosphorylation of the retinoblastoma gene product and promoted cell cycle arrest at G_0/G_1 (Pushkareva *et al.*, 1995). Both of these effects were independent of the well-established ability of sphingosine to inhibit conventional and novel isoforms of PKC (Hannun and Bell, 1989), which promote cell survival.

Sphingosine also inhibits agonist-stimulation of p42/p44 MAPK (Tolan *et al.*, 1996 and 1999). This effect may be due to the inhibition of PKC, an up-stream regulator of p42/p44 MAPK. Indeed, sphingosine analogs such as *threo*-DHS and *N,N*-DMS have a similar effect (Tolan *et al.*, 1999) and are also inhibitors of PKC (Hannun and Bell, 1989). In addition, sphingosine alters the cellular levels of the glycerolipid second messengers, diacylglycerol and phosphatidic acid. For example, sphingosine stimulates diacylglycerol kinase (Sakane *et al.*, 1989), phospholipase Cδ (Matecki and Pawelczyk, 1997) and phospholipase D (Lavie and Liscovitch, 1990) but inhibits phosphatidic acid phosphatase (Mullmann *et al.*, 1991). Indeed, these latter effects may contribute to the inhibition of the p42/p44 MAPK cascade by sphingosine (Tolan *et al.*, 1997). Another up-stream component of the p42/p44 MAPK cascade, the non-receptor tyrosine kinase c-Src, is also inhibited by sphingosine (Igarashi *et al.*, 1989).

Conversely, the SAPK/JNK cascade, which is closely associated with apoptosis, is activated by sphingosine (Pyne *et al.*, 1996). Exogenous sphingosine also activates caspase-7in a Bcl-x(L)-sensitive manner whereas caspase-8 was unaffected (Nava *et al.*, 2000; Cuvillier *et al.*, 2001). In addition, sphingosine stimulates p38 MAPK in osteoblast-like cells (Kozawa *et al.*, 2000) but not in oligodendrocytes where is produces lysis (Hida *et al.*, 1999). p38 MAPK is up-stream of MAPKAP kinase-2 and the phosphorylation of heat shock protein 27 (hsp27), which is produced in response to cellular stress.

The effect of sphingosine on other enzymes may also contribute to its apoptotic effect. These include the inhibition of calcium/calmodulin-requiring enzymes and DNA primase and the stimulation of casein kinase II and several unidentified kinases (Alessenko, 2000). In addition, sphingosine can increase the cellular concentration of cyclic AMP, which is inhibitory for proliferation in many cell types (Pyne and Pyne, 1996).

4. PROTECTIVE EFFECTS OF S1P: EXTRA- AND INTRACELLULAR ACTIONS?

Exogenously applied S1P binds to specific GPCRs to activate multiple signal transduction pathways. Additionally, some S1P could be taken up into cells. Furthermore, S1P produced intracellularly in response to receptor stimulation could act at an intracellular site or be released to act on S1P-specific GPCRs in an autocrine or paracrine manner. These possibilities are discussed below.

4.1 EDG receptors

To date, five closely related GPCRs of the EDG (endothelial differentiation gene) family (EDG1, EDG3, EDG5/AGR16/H218, EDG6 and EDG8/*nrg-1*) have been identified as high affinity S1P receptors (Hla and Macaig, 1990; Okazaki *et al.*, 1993; MacLennan *et al.*, 1994; Graler *et al.*, 1998; Yamazaki *et al.*, 2000; Im *et al*, 2000; Glickman *et al.*, 1999). They are integral membrane proteins that are probably glycosylated, predicted to have seven transmembrane domains and exhibit approximately 50% amino-acid sequence identity. The S1P-specific EDG GPCRs couple productively to different types of G protein α sub-units, βγ sub-units and Rho.

EDG1, which exhibits high affinity for S1P (K_d 8.1 nM, (Lee *et al.*, 1996); 13.2 nM (Kon *et al.*, 1999)) couples via G_{i1}, G_{i2}, G_{i3}, G_o and G_z (but not G_s, G_q, G_{12} or G_{13}) to a number of signaling pathways (see below) (Okamoto *et al.*, 1998; Zondag *et al.*, 1998; Windh *et al.*, 1999). EDG3 and EDG5 (which have K_d values for S1P of 23nM and 27nM, respectively (Van Brocklyn *et al.*, 1999) and 26.2nM and 20.4 nM, respectively (Kon *et al.*, 1999)) communicate not only via G_i but also through G_q, G_{12} and G_{13} (An *et al.*, 1998 and 1999; Ancellin and Hla, 1999). EDG6 (K_d of 12-19nM) couples *via* $G_{i/o}$ whereas a role for G_q, $G_{12/13}$ and/or Rho has yet to be addressed (Yamazaki *et al.*, 2000). EDG8 (K_d of 2nM for S1P) couples *via* $G_{i/o}$ but not through G_q (Im *et al.*, 2000).

Rho is also involved in EDG receptor signaling. For EDG3 and EDG5 receptors, coupling to $G_{12/13}$ may be the predominant route of Rho activation (Buhl *et al.,* 1995; Kozasa *et al.,* 1998). However, the activation of Rho via EDG1 is by another mechanism since EDG1 does not couple to G_{12} or G_{13}.

4.2 Inhibition of apoptosis by exogenous S1P

Numerous studies have demonstrated the protective effect of exogenously applied S1P. In many cases, the protective effects of S1P have been associated with the activation of signaling pathways mediated via EDG receptors (Figure 3). For example, S1P-dependent inhibition of serum-withdrawal-induced apoptosis of HTC4 hepatoma cells was mediated by transfected EDG3 or EDG5. S1P-induced protection involved EDG receptor-dependent activation of p42/p44 MAPK, inhibition of caspase-3 activity and expression of the AP-1 transcription factors, c-jun and c-fos (An *et al.*, 2000). EDG3 and EDG5 also mediated the S1P protection of T cells induced to apoptose by antibodies to Fas, CD2 and CD3 plus CD28. In this case, S1P suppressed cellular levels of the apoptosis-promoting protein Bax without altering the levels of Bcl-x_L or Bcl-2 (Goetzl *et al.*, 1999). In Jurkat T cells, Fas ligation or C_2-ceramide activated caspase-3/CPP32 and caspase-7/Mch3 followed by cleavage of poly(ADP-ribose) polymerase (PARP). These effects were blocked by exogenous S1P, which also prevented caspase-6/Mch2 activation. However, Fas-induced caspase-8/FLICE cleavage was not inhibited by S1P. Thus, S1P was suggested to act downstream of caspase-8/FLICE but upstream of caspases-3, -6 and –7 (Cuvillier *et al*, 1998). Another mechanism of protection from serum-deprived apoptosis of HUVECs involves the production of nitric oxide (NO) since S1P increased Ca^{2+}-sensitive endothelial nitric oxide synthase (eNOS) activity. An inhibitor of NOS reversed the effect whereas an inhibitor of guanylyl cyclase did not. A series of inhibitors were used to suggest that these effects of S1P were via EDG1- and EDG3-mediated activation of G_i, phospholipase C and Ca^{2+} mobilisation (Kwon *et al.*, 2001).

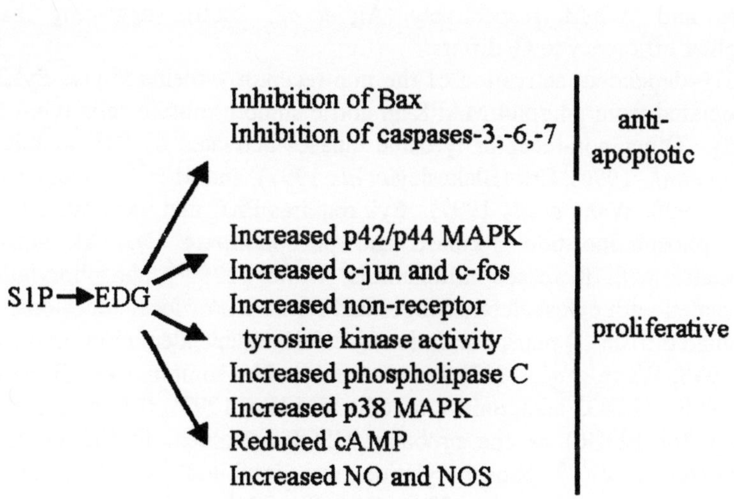

Figure 3 Actions of S1P

Many other signaling pathways are activated in response to S1P stimulation, depending on the specific cell type and the EDG receptors expressed (Pyne and Pyne, 2000). Examples include phospholipase C-catalysed inositol phosphates formation, Ca^{2+} mobilisation, translocation of protein kinase C isoforms, phospholipase D-catalysed phospholipid hydrolysis, non-receptor tyrosine kinases, muscarinic K^+ current and the activation and inhibition of adenylyl cyclase, dependent upon the cell type. In addition, S1P activates p42/p44 MAPK and stimulates the induction of AP1 transcription factors *c-jun* and *c-fos* (Pyne and Pyne, 2000).

S1P-stimulated activation of p42/p44 MAPK via EDG1 is pertussis toxin sensitive (Rakhit *et al.*, 1999; Lee *et al.*, 1996; Okamoto *et al.*, 1998; Zondag *et al.*, 1998). Signal intermediates include the non-receptor tyrosine kinase c-Src and phosphoinositide 3-kinase in airway smooth muscle (Rakhit *et al.*, 1999) and may require receptor internalisation (Lee *et al.*, 1996). However, activation of the p42/p44 MAPK pathway by S1P shows differential sensitivity to pertussis toxin (Guo *et al.*, 1998; Sato *et al.*, 1999). In C6 cells, the pertussis toxin-insensitive component was abolished by inhibition of PKC and was attributed to EDG5 linked to G_q/phospholipase C. In contrast, S1P induced p42/p44 MAPK activation occurs entirely through a $G_{i/o}$-dependent pathway in EDG5 transfected CHO cells (Gonda *et al.*, 1999) and HTC4 cells transfected with either EDG3 or EDG5 (An *et al.*, 2000). Interestingly, in the HTC4 cells, the concentration dependence for S1P activation of p42/p44 MAPK differs for EDG3 and EDG5, being maximal at 10nM and 100nM, respectively (An *et al.*, 2000), suggesting that the coupling efficiency to G_i differs.

S1P-dependent activation of the non-receptor tyrosine kinase Pyk2 was dissociated from p42/p44 MAPK in aortic smooth muscle cells (Guo *et al.*, 1998). Other non-receptor tyrosine kinases activated by S1P include Syk (Yang *et al.*, 1996), Crk (Blakesley *et al.*, 1997), and $p125^{FAK}$ (Van Brocklyn *et al.*, 1998; Wang *et al.*, 1997). Syk requires PKC and transmits signals to Ras, phosphoinositide 3-kinase and phospholipase Cγ. Crk activation correlates with increased mitogenesis, while $p125^{FAK}$ phosphorylation is associated with cytoskeletal changes and formation of focal adhesions. S1P-stimulated $p125^{FAK}$ phosphorylation is EDG1-independent (Van Brocklyn *et al.*, 1998; Wang *et al.*, 1997) and not inhibited by suramin (Van Brocklyn *et al.*, 1998) (EDG3 antagonist (Ancellin and Hla, 1999)) thereby implicating EDG5 (or EDG6) as the probable GPCR involved. Furthermore, S1P-stimulated $p125^{FAK}$ phosphorylation was inhibited by C3 exoenzyme, suggesting the involvement of Rho (Van Brocklyn *et al.*, 1998; Wang *et al.*, 1997).

S1P activates p38 MAPK in EDG5-transfected CHO cells (Gonda *et al.*, 1999). S1P induces hsp27 in A10 vascular smooth muscle cells and osteoblasts and amplifies inositol phosphates formation in response to vasopressin and prostaglandin $F_{2\alpha}$, respectively (Kozawa *et al.*, 1999a and b;

Kozawa *et al.*, 2000a and b). In both cell types, the p38 MAPK-specific inhibitor, SB203580, blocks S1P hsp27 induction and amplification of inositol phosphates. In osteoblastic cells, PGE stimulates cyclic AMP formation and inhibits apoptosis. This appears to be mediated by a cyclic AMP-dependent modulation of SPHK and S1P production (Machwate *et al.*, 1998)

4.3 Intracellular S1P

Exogenous S1P is also reported to protect cells from the apoptosis *via* an intracellular mechanism. For example, in PC12 cells, an intracellular role of S1P was suggested since protection correlated with S1P uptake but not with EDG1 expression (Edsall *et al.*, 1997; Van Brocklyn *et al.*, 1998). Furthermore, dihydrosphingosine 1-phosphate had no protective effect despite being able to bind to EDG1 (Van Brocklyn *et al.*, 1998). However, the potential role of other EDG receptors should also be considered. Indeed, dihydrosphingosine 1-phosphate competes with S1P for binding to EDG3 and EDG5 (Van Brocklyn *et al.*, 1999).

Evidence to support an intracellular role for S1P has been based mainly on the stimulation of SPHK by agonists, the use of SPHK inhibitors and, more recently, transfection of SPHK (Pyne and Pyne, 2000).

However, DNA synthesis in Swiss 3T3 cells was increased by microinjection of S1P (Van Brocklyn *et al.*, 1998). In addition, microinjection of S1P induced rapid, transient Ca^{2+} mobilisation, which was not antagonised by heparin, thereby excluding a role for inositol trisphosphate and its receptor (Meyer zu Heringdorf *et al.*, 1998). Similarly, heparin did not block S1P-induced mobilisation of Ca^{2+} in a microsomal preparation (Ghosh *et al.*, 1994). Furthermore, inhibition of SPHK blocks Ca^{2+} mobilisation to a variety of agonists. Most recently, the LPA-stimulated intracellular Ca^{2+} release in SH-SY5Y neuroblastoma cells was suggested to be mediated by S1P since the response was inhibited by SPHK inhibitors but not by down-regulation of inositol 1,4,5-trisphosphate receptors (Young *et al.*, 2000). However, intracellular release of caged DHS has also been shown to mobilise Ca^{2+} from ryanodine-sensitive intracellular stores in cultured neurones (Ayar *et al.*, 1998). Moreover, sphingosylphosphorylcholine (SPC) also releases microsomal Ca^{2+} (Ghosh *et al.*, 1994, Huang and Chueh, 1996), thereby questioning the specificity of the S1P effect. In addition, others have failed to detect S1P-induced mobilisation of intracellular calcium (Durieux *et al.*, 1993; Orlati *et al.*, 1997).

5. THE SPHINGOSINE: S1P RHEOSTAT

A signaling rheostat has been proposed where the relative level of S1P and ceramide determines cellular fate. This was based on the observation that ceramide, Fas and TNFα-mediated programmed cell death is suppressed by the addition of S1P (Cuvillier *et al.*, 1996). The model assumes that ceramide is a physiological mediator of apoptosis although this is subject to controversy (Hofmann and Dixit, 1998; Kolesnick and Hannun, 1999). Indeed, sphingosine is also implicated as an apoptotic agent (see above).

Quantification of sphingosine and S1P levels in a variety of tissue and cell types reveals that sphingosine is often in excess by ten-fold (Edsall and Spiegel, 1999). The notable exception to this is platelets, which lack S1P lyase activity. In fact, S1P is stored and released from platelets upon their activation (Igarashi and Yatomi, 1998). This contributes to the high serum concentration of S1P relative to sphingosine (Edsall and Spiegel, 1999). In addition, other blood cell types actively remove plasma sphingosine and convert it to S1P that is spontaneously released (Yang *et al.*, 1999). This may represent a specific protective mechanism for maintaining a low plasma sphingosine concentration.

The rheostat model in relation to sphingosine and S1P is supported in several ways. First, the addition of exogenous sphingosine induces apoptosis (see above) whereas apoptotic agents increase sphingosine and/or reduce S1P, e.g. gliotoxin markedly reduced the S1P content of bovine alveolar macrophages (Lutjohann *et al.*, 1998). Second, exogenous application of S1P to cells inhibits apoptosis (see above). Third, transfection of enzymes that regulate sphingosine and S1P levels affect cellular fate. Finally, SPHK is activated by agonists that induce proliferation or promote survival, whereas, inhibitors of SPHK induce apoptosis and prevent cell survival.

5.1 Transfection studies

The recently cloned SPHK and S1P phosphatase have been used as molecular tools to examine the physiological roles of sphingosine and S1P.

For example, Jurkat T cells apoptose in response to serum withdrawal, addition of exogenous cell-permeable ceramide or stimulation with the apoptotic agonist Fas. However, transfection with SPHK1, measured as a 400-600-fold increase in activity, reduced apoptosis and increased proliferation (Olivera *et al.*, 1999). These effects were correlated with an inhibition of the activity of caspase 3, an initiator caspase (Olivera *et al.*, 1999). A similar effect has recently been observed in rat pheochromocytoma PC12 cells (Edsall *et al.*, 2001). Trophic withdrawal led to the activation of initiator caspases, caspase-3 and caspase-7 and subsequent to activation of caspase-2, which is required for progression of apoptosis in this cell type. This effect was suppressed by SPHK1 overexpression as was trophic withdrawal-induced activation of SAPK/JNK. Surprisingly, there was no apparent increase in two known cell survival pathways, p42/p44 MAPK and Akt/PKB. A role for G_iPCRs was excluded since pertussis toxin had no effect. Furthermore, S1P release from PC12 cells could not be detected. In addition, transfection of SPHK1 into NIH3T3 cells or HEK 293 cells increased their ability to cross the G_1/S transition, to synthesize DNA and to proliferate (Olivera *et al.*, 1999). Similar observations on apoptosis and proliferation have yet to be made for SPHK2. These effects were correlated with a significant increase in S1P mass and decrease in sphingosine and ceramide (Kohama *et al.*, 1998; Olivera *et al.*, 1999). The net increase in S1P in these studies (0.95 and 0.04 pmol/nmol phospholipid, respectively) differs from the decrease in sphingosine (0.23 and 0.2 pmol/nmol total phospholipd, respectively). However, the most marked effect is a decrease in ceramide (6 and 10 pmol/nmol total phospholipid, respectively) (Kohama *et al.*, 1998; Edsall *et al.*, 2001). Therefore, the concomitant removal of ceramide upon SPHK over-expression may underlie or contribute to the prevention of apoptosis. Furthermore, in some cell types, over-expression of SPHK results in S1P release, which influences cellular function. For example, release of S1P from SPHK1-transfected HEK 293 cells contributes to their Rac-mediated migration in response to PDGF (Hobson *et al.*, 2001). A similar situation could contribute to cellular survival.

Transfection of mouse S1P phosphatase into HEK 293 cells, resulting in a 3-fold increase in membrane S1P phosphatase activity, caused a 50% decrease in S1P levels (reduced by 0.6 pmol/nmol phospholipid) and a 2 fold increase in ceramide (increased by 23 pmol/nmol phospholipid) whereas sphingosine levels were similar to vector controls (0.8 pmol/nmol phospholipid). S1P phosphatase transfected cells underwent apoptosis in response to serum withdrawal, C_2-ceramide, peroxide or doxorubicin with 2-3 fold higher frequency compared to vector-transfected control cells. Surprisingly, exogenously added S1P, which normally confers protection, increased apoptosis. This may be due to its metabolism to ceramide (Mandala *et al.*, 2000) although other factors may also be involved.

A similar use of gene over-expression and of deletion mutants in *Saccharomyces cerevisiae* has also provided insight into the physiological roles of sphingolipids in stress responses which are mediated by ceramide/sphingosine whereas phosphorylated sphingoid bases, including S1P, are survival signals.

5.1.1 Activation and inhibition of SPHK

An increase in SPHK activity has been shown to correlate with enhanced survival in a variety of cell types. For example, U937 monoblastic leukemia cells apoptose in response to TNFα but can be rescued by phorbol ester treatment. This correlates with an increase in SPHK activity and S1P levels (Cuvillier *et al.*, 1996). A similar observation was made in TNFα-stimulated C11 endothalial cells (Xia *et al.*, 1999). In contrast, TNFα fails to induce apoptosis in HUVECs where it increases SPHK activity (Xia *et al.*, 1999). Moreover, a dominant negative form of SPHK1 was used to show that TNFα-stimulation of p42/p44 MAPK requires S1P synthesis (Pitson *et al.*, 2000). However, in other cell types, a role for agonist-stimulated S1P production has been excluded from p42/p44 MAPK activation (Alemany *et al.*, 2000). Cyclic AMP-mediated protection from apoptosis has also been correlated with an increase in SPHK activity in periosteal cells (Machwate *et al.*, 1998). Increased endogenous SPHK expression also protects cells against apotosis. For example, PC12 cells induced to differentiate with nerve growth factor become more resistant to serum-withdrawal-mediated apoptosis as SPHK activity increases (Edsall *et al.*, 1997). A similar situation occurs in HL60 cells induced to differentiate with vitamin D_3 (Kleuser *et al.*, 1998).

In addition, endogenous SPHK activity correlates with adhesion of HL60 cells and their ability to apoptose. Thus, non-adherent HL60 cells have low SPHK activity and spontaneously apotose. In contrast, adherent cells, which are resistant to phorbol ester-induced apoptosis, have higher SPHK activity. These cells can be made apoptotic by the addition of the SPHK inhibitor, DMS (Nakamura *et al.*, 1998). This inhibitor has also been used to block the protective effect of various agonists including nerve growth factor in PC12 cells (Edsall *et al.*, 1997) and vitamin D_3 in HL60 cells (Kleuser *et al.*, 1998). DMS also sensitises cells to apoptosis. For example, HUVECs undergo apotosis in response to TNFα in the presence of DMS (Xia *et al.*, 1999). Additionally, DMS induces apoptosis by increasing intracellular sphingosine levels as demonstrated in platelets (Yatomi *et al.*, 1996). However, it should be noted that the SPHK inhibitors DMS and *threo*-DHS also inhibit PKC (Hannun and Bell, 1989), which may contribute to some of their observed effects (Tolan *et al.*, 1999).

6. EXCEPTIONS

Although the majority of current evidence supports a pro-apoptotic role for sphingosine and a survival/proliferative role for S1P, there are exceptions. For example, sphingosine stimulates p42/p44 MAPK and proliferation in glomerular mesangial cells (Coroneos *et al.*, 1996). Thus, the effect of sphingosine varies between cell types. This could, but does not necessarily, depend on its metabolism to S1P (Zhang *et al.*, 1991). Indeed, sphingosine has recently been shown to promote a marked increase in the autophosphorylation of 3-phosphoinositide-dependent kinase 1 (PDK1) and one of its substrates, Akt/PKB (King *et al.*, 2000). The phosphoinositide 3-kinase/Akt/PKB pathway is essential for cell survival where Akt/PKB phosphorylates Bad, thereby decreasing Bad/Bcl-xL binding at mitochondria and increasing Bad/14-3-3 binding in the cytosol. Akt/PKB has also been shown to inhibit activation of caspases –9 and –3 (Zhou *et al.*, 2000).

Moreover, S1P can also be anti-proliferative and even promote apoptosis. For instance, S1P inhibits proliferation of hepatic myofibroblasts via an acute activation of cyclooxygenase-2, the production of PGE_2 and subsequently cAMP. Prolonged treatment with S1P also induced cyclooxygenase-2 (Davaille *et al.*, 2000). S1P-induced apoptosis of hippocampal neurons involves the activation of protein phosphatase, possibly calcineurin and protein phosphatase 2A (or a related phosphatase) (Moore *et al.*, 1999) whereas S1P induced neurite retraction (one of the first steps in the apoptotic process) in N1E-115 cells via a GPCR- and Rho-dependent pathway that involves contraction of the actin cytoskeleton (Postma *et al.*, 1996). Recent studies have shown that S1P-induced neurite retraction involves EDG5 and, possibly, EDG3 (Van Brocklyn *et al.*, 1999). In contrast, S1P up-regulates the expression of Bax expression, a pro-apoptotic protein belonging to the Bcl-2 family, in human hepatoma cells. This was shown to be independent of p53-mediated transcription and growth arrest (Hung and Chuang, 1996).

7. CONCLUSION

S1P is involved in a variety of physiological and pathophysiological processes that involve proliferation and migration (Pyne and Pyne, 2000). These include tumor invasion, angiogenesis, atherosclerosis and differentiation. Anti-cancer therapies could include the use of EDG receptor sub-type-specific antagonists or EDG receptor knockout. The inhibition of SPHK is also a promising approach. For instance, DMS has been used to induce apoptosis of solid tumors (Park *et al.* 1994; Sakakura *et al.*, 1997). Novel SPHK inhibitors with higher potency (De Jonge *et al.*, 1999) are also being developed which may have improved clinical efficacy. Conversely, activation of SPHK or inhibition of S1P phosphatase and S1P lyase could be used to protect against apoptosis. This may be of particular relevance to neurological disorders, such as Alzheimer's and Parkinson's disease, which are associated with de-regulated apoptosis.

ACKNOWLEDGMENTS

This work is supported by The Wellcome Trust.

REFERENCES

Alderton, F.A., Darroch, P.I., Sambi, B., McKie, A., Pyne, N.J. and Pyne, S., 2001, G-protein-coupled receptor signaling to p42/p44 mitogen-activated protein kinase is attenuated by lipid phosphate phosphatases 1, 1a and 2 in HEK 293 cells, *J. Biol. Chem.* **276**: 13452-13460.

Alemany, R., Sichelschmidt, B., Meyer zu Heringdorf, D.M., Lass, H., Van Koppen, C.J. and Jakobs, K.H., 2000, Stimulation of sphingosine 1-phosphate formation by the $P2Y_2$ receptor in HL-60 cells: Ca^{2+} requirement and implication in receptor-mediated Ca^{2+} mobilization, but not MAP kinase activation, *Mol. Pharmacol.* **58**: 491-497.

Alessenko, A.V., 2000, The role of sphingomyelin cycle metabolites in transduction of signals of cell proliferation, differentiation and death, *Memb. Cell Biol.* **13**: 303-320.

An, S., Bleu, T. and Zheng, Y., 1999, Transduction of intracellular calcium signals through G protein- mediated activation of phospholipase C by recombinant sphingosine 1- phosphate receptors. *Mol. Pharmacol.* **55**: 787-794.

An, S., Goetzl, E.J. and Lee, H., 1998, Signaling mechanisms and molecular characteristics of G protein-coupled receptors for lysophosphatidic acid and sphingosine 1-phosphate. *J. Cell. Biochem.* **30/31**: 147-157.

An, S., Zheng, Y. and Bleu, T., 2000, Sphingosine 1-phosphate-induced cell proliferation, survival and related events mediated by G protein-coupled receptors EDG3 and EDG5. *J. Biol. Chem.* **275**: 288-296.

Ancellin, N. and Hla, T., 1999, Differential pharmacological properties and signal transduction of the sphingosine 1-phosphate receptors EDG-1, EDG-3, and EDG-5. *J. Biol. Chem.* **274**: 18997-19002.

Auge, N., Nikolova-Karakahian, M., Carpentier, S., Parthasarathy, S., Negre-Salvayre, A., Salvayre, A., Merrill, Jr., A.H. and Levade, T., 1999, Role of sphingosine 1-phosphate in the mitogenesis induced by oxidized low density lipoprotein in smooth muscle cells via activation of sphingomyelinase, ceramidase and sphingosine kinase, *J. Biol. Chem.* **274**: 21533-21538.

Ayar, A., Thatcher, N.M., Zehavi, U., Trentham, D.R. and Scott, R.H., 1998, Mobilization of intracellular calcium by intracellular flash photolysis of caged dihydrosphingosine in cultured neonatal rat sensory neurones, *Acta Biochim. Polon.* **45**: 311-326.

Blakesley, V.A., Beitner-Johnson, D., Van Brocklyn, J.R., Rani, S., Shen-Orr, Z., Stannard, B., Spiegel, S. and LeRoith, D., 1997, Sphingosine 1-phosphate stimulates tyrosine phosphorylation of Crk. *J. Biol. Chem.* **272**: 16211-16215.

Brindley, D.N. and Waggoner, D.W., 1998, Mammalian lipid phosphate phosphohydrolases, *J. Biol. Chem.* **273**: 24261-24284.

Buehrer, B.M., Bardes, E.S and Bell, R.M., 1996, Protein kinase C-dependent regulation of human erthyroleukemia (HEL) cell sphingosine kinase activity, *Biochim. Biophys. Acta* **1303**: 233-242.

Buhl, A.M., Johnson, N.L., Dhanasekaran, N. and Johnson, G.L., 1995, $G\alpha_{12}$ and $G\alpha_{13}$ stimulate Rho-dependent stress fibre formation and focal adhesion assembly. *J. Biol. Chem.* **270**: 24631-24634.

Chmura, S.J., Nodzenski, E., Crane, M.A., Virudachalam, S., Hallahan, D.E., Weichselbaum, R.R. and Quintans, J., 1996, Cross-talk between ceramide and PKC activity in the control of apoptosis in WEHI-231, *Adv. Exp. Med.* **406**: 39-55.

Choi, O.H., Kim, J.H. and Kinet, J.P., 1996, Calcium mobilisation via sphingosine kinase in signaling by the FcεRI antigen receptor, *Nature (London)* **380**: 634-636.

Coroneos, E., Martinez, M., McKenna, S. and Kester, M., 1995, Differential regulation of sphingomyelinase and ceramidase activities by growth factors and cytokines. Implications for cellular proliferation and differentiation, *J. Biol. Chem* **270**: 23305-23309.

Coroneos, E., Wang, Y., Panuska, J.R., Temepleton, D.J. and Kester, M., 1996, Sphingolipid metabolites differentially regulate extracellular signal-regulated kinase and stress-activated protein kinase cascades, *Biochem. J.* **316**: 13-17.

Cuvillier, O., Nava, V.E., Murthy, S.K., Edsall, L.C., Levade, T., Milstien, S. and Spiegel, S., 2001, Sphingosine generation, cytochrome c release, and activation of caspase-7 in doxorubicin-induced apoptosis of MCF7 breast adenocarcinoma cells, *Cell Death Differ* **8**: 162-171.

Cuvillier, O., Pirianov, G., Kleuser, B., Vanek, P.G., Coso, O.A., Gutkind, S. and Spiegel, S., 1996, Suppression of ceramide-mediated programmed cell death by sphingosine 1-phosphate, *Nature (London)* **381**: 800-803.

Cuvillier, O., Rosenthal, D.S., Smulson, M.E. and Spiegel, S., 1998, Sphingosine 1-phosphate inhibits activation of caspases that cleave poly(ADP-ribose)polymerase and lamins during Fas- and ceramide-mediated apoptosis in jurkat T lymphocytes, *J. Biol. Chem.* **273**: 2910-2916.

Davaille, J., Gallois, C., Habib, A., Li, L., Mallat, A., Tao, J., Levade, T. and Lotersztajn, S., 2000, Antiproliferative properties of sphingosine 1-phosphate in hepatic myofibroblasts: a cyclooxygenase-2-mediated pathway, *J. Biol. Chem.* **275**: 34628-34633.

De Cuester, P., Mannaerts, G.P. and Van Veldhoven, P.P., 1995, Identification and sub-cellular localization of sphinganine phosphatases in rat liver, *Biochem. J.* **311**: 139-146.

De Jonge, S., Van Overmeire, I., Poulton, S., Hendrix, J., Busson, R., Van Calenbergh, S., De Keukeleire, D., Spiegel, S. & Herdewijn, P., 1999, Structure-activity relationship of short-chain sphingoid bases as inhibitors of sphingosine kinase, *Bioorg. Med. Chem. Lett.* **9**: 3175-3180.

Durieux, M.E., Carlisle, S.J., Salafranca, M.N. and Lynch, K.R., 1993, Responses to sphingosine-1-phosphate in *X. laevis* oocytes: Similarities with lysophosphatidic acid signaling, *Am. J. Physiol.* **264:** C1360-C1364.

Edsall, L.C. and Spiegel. S., 1999, Enzymatic measurement of sphingosine 1-phosphate, Anal. Biochem. **272:** 80-86.

Edsall, L.C., Cuvillier, O., Twitty, S., Spiegel, S. and Milstien, S., 2001, Sphingosine kinase expression regulates apoptosis and caspase activation in PC12 cells, *J. Neurochem.* **76**: 1573-1584.

Edsall, L.C., Pirianov, G.G. and Spiegel, S., 1997, Involvement of sphingosine 1-phosphate in nerve growth factor-mediated neuronal survival and differentiation, *J. Neurosci.* **17**: 6952-6960.

El Bawab, S., Birbes, H., Roddy, P., Szulc, Z.M., Bielawska, A. and Hannun, Y.A., 2001, Biochemical characterization of the reverse activity of rat brain ceramidase. a CoA-independent and fumonisin b1-insensitive ceramide synthase, *J. Biol. Chem.* **276**: 16758-16766.

El Bawab, S., Roddy, P., Qian, T., Bielawska, A., Lemasters, J.J. and Hannun, Y.A., 2000, Molecular cloning and characterization of a human mitochondrial ceramidase. *J. Biol. Chem.* **275**: 21508-21513.

Ghosh, T.K., Bian, J. and Gill, D.L., 1994, Sphingosine 1-phosphate generated in the endoplasmic reticulum membrane activates release of stored calcium, J. Biol. Chem. **269**: 22628-22635.

Glickman, M., Malek, R.L., Kwitek-Black, A.E., Jacob, H.J. and Lee, N.H., 1999, Molecular cloning, tissue-specific expression, and chromosomal localization of a novel nerve growth factor-regulated G-protein-coupled receptor, *nrg-1*, *Mol. Cell. Neurosci.* **14**: 141-152.

Goetzl, E.J., Kong, Y. and Mei, B., 1999, Lysophosphatidic acid and sphingosine 1-phosphate protection of T cells from apoptosis in association with suppression of Bax, *J. Immunol.* **162**: 2049-2056.

Gonda, K., Okamoto, H., Takuwa, N., Yatomi, Y., Okazaki, H., Sakura, T., Kimura, S., Sillard, R., Harii, K. and Takuwa, Y., 1999, The novel sphingosine 1-phosphate receptor AGR16 is coupled via pertussis toxin-sensitive and -insensitive G-proteins to multiple signaling pathways. *Biochem. J.* **337**: 67-75.

Graler, M.H., Bernhardt, G. and Lipp, M., 1998, A lymphoid tissue-specific receptor, EDG6, with potential immune modulatory functions mediated by extracellular lysophospholipids, *Genomics* 53: 164-169.

Guo, C., Zheng, C., Martin-Padura, I., Bian, Z.C. and Guan J-L., 1998, Differential stimulation of proline-rich tyrosine kinase 2 and mitogen-activated protein kinase by sphingosine 1-phosphate. *Eur. J. Biochem.* **257**: 403-408.

Hannun, Y.A. and Bell, R.M., 1989, Functions of sphingolipids and sphingolipid breakdown products in cellular regulation, *Science* **243**: 500-507.

Hida, H., Nagano, S., Takeda, M. and Soliven, B., 1999, Regulation of mitogen-activated protein kinases by sphingolipid products in oligodendrocytes, *J. Neurosci.* **19**: 7458-7467.

Hla, T. and Macaig, T., 1990, An abundant transcript induced in differentiating human endothelial cells encodes a polypeptide with structural similarities to G-protein-coupled receptors, *J. Biol. Chem.* **265**: 9308-9313.

Hobson, J.P., Rosenfeldt, H.M., Barak, L.S., Olivera, A., Poulton, S., Caron, M., Milstien, S. and Spiegel, S., 2001, Role of sphingosine 1-phosphate receptor EDG1 in PDGF-induced cell motility, *Science* **291**: 1800-1803.

Hofmann, K. and Dixit, V.M., 1998, Ceramide in apoptosis - Does it really matter?, *Trends Biochem. Sci.* **23**: 374-377.

Huang, W-C. and Chueh, S-H., 1996, Calcium mobilization from the intracellular mitochondrial and non-mitochondrial stores of the rat cerebellum, *Brain Res.* **18**: 151-158.

Hung, W-C. and Chuang, L-Y., 1996, Induction of apoptosis by sphingosine 1-phosphate in human hepatoma cells is associated with increased expression of bax gene product. *Biochim. Biophys. Res. Comm.* **229**: 11-15.

Igarashi, Y. and Yatomi, Y., 1998, Sphingosine 1-phosphate is a blood constituent released from activated platelets, possible playing a variety of physiological and pathophysiological roles. *Acta Biochim. Pol.* **45**: 299-309.

Igarashi, Y., Hakomori, S., Toyokuni, T., Dean, B., Fujita, S., Sugimoto, M., Ogawa, T., El-Ghendy, K. and Racker E., 1989, Effect of chemically well-defined sphingosine and its

N-methyl derivatives on protein kinase C and src kinase activities, *Biochem.* **28**: 6796-6800.

Im, D-S., Heise, C.E., Ancellin. N., O'Dowd, B.F., Shei, G-J., Heavens, R.P., Rigby, M.R., Hla, T., Mandala, S., McAllister, G., George, S.R. and Lynch, K.R., 2000, Characterization of a novel sphingosine 1-phosphate receptor, EDG8, *J. Biol. Chem.* **275**: 14281-14286.

Jarvis, W.D., Fornari Jr, F.A., Auer, K.L., Freemerman, A.J., Szabo, E., Birrer, M.J., Johnson, C.R., Barbour, S.E., Dent, P. and Grant, S., 1997, Coordinate regulation of stress- and mitogens-activated protein kinases in the apoptotic actions of ceramide and sphingosine, *Mol. Pharmacol.* **52**: 935-947.

Jarvis, W.D., Fornari Jr, F.A., Traylor, R.S., Martin, H.A., Kramer, L.B., Erukulla. R.K., Bittman, R. and Grant, S., 1996, Induction of apoptosis and potentiation of ceramide-mediated cytotoxicity by sphingoid bases in human myeloid leukaemia cells, *J. Biol. Chem.* **271**: 8275-8284.

King, C.C., Zenke, F.T., Dawson, P.E., Dutil, E.M., Newton, A.C., Hemmings, B.A. and Bokoch, G.M., 2000, Sphingosine is a novel activator of 3-phosphoinositide-dependent kinase 1, *J. Biol. Chem.* **275**: 18108-18113.

Kleuser, B., Cuvillier, O. and Spiegel, S., 1998, 1α,25-Dihydroxyvitamin D_3 inhibits programmed cell death in HL60 cells by activation of sphingosine kinase, *Cancer Res.* **58**: 1817-1824.

Koch, J., Gartner, S., Li, C.M., Quintern, L.E., Bernardo, K., Levran, O., Schnabel, D., Desnick, R.J., Schuchman, E.H. and Sandhoff, K., 1996, Molecular cloning and characterization of a full-length complementary DNA encoding human acid ceramidase. Identification of the first molecular lesion causing Farber disease, *J. Biol. Chem.* **271**: 33110-33115.

Kohama, T., Olivera, A., Edsall, L., Nagiec, M.M., Dickson, R. and Spiegel, S., 1998, Molecular cloning and functional characterization of murine sphingosine kinase, *J. Biol. Chem.* **273**: 23722-23728.

Kolesnick, R. and Hannun ,Y.A., 1999, Ceramide and apoptosis, *Trends Biochem. Sci.* **24**: 224-225.

Kon, J., Sato, K., Watanabe, T., Tomura, H., Kuwabara, A., Kimura, T., Tamama, K-I., Ishizuka, T., Murata, N., Kanda, T., Kobayashi, I., Ohta, H., Ui, M. and Okajima, F., 1999, Comparison of intrinsic activities of the putative sphingosine 1-phosphate receptor sub-types to regulate several signaling pathways in their cDNA-transfected Chinese hamster ovary cells. *J. Biol. Chem.* **274**: 23940-23947.

Kozasa, T., Jiang, X., Hart, M.J., Sternweiss, P.M., Singer, W.D., Gilman, A.G., Bollag, G. and Sternweis, P.C., 1998, p115 RhoGEF, a GTPase activating protein for $G\alpha_{12}$ and $G\alpha_{13}$. *Science* **280**: 2109-2111.

Kozawa, O., Kawamura, H, and Uematsu, T., 2000a, Sphingosine 1-phosphate amplifies phosphoinositide hydrolysis stimulated by prostaglandin F2α in osteoblasts: involvement of p38 MAP kinase, *Prostaglandins Leukot. Essent. Fatty Acids* **62**: 355-359.

Kozawa, O., Kawamura, H., Matsuno, H. and Uematsu, T., 2000b, p38 MAP kinase is involved in the signaling of sphingosine in osteoblasts: sphingosine inhibits prostaglandin F2α-induced phoshoinositide hydrolysis, *Cell Signal* **12**: 447-450.

Kozawa, O., Niwa, M., Matsuno, H., Tokuda, H., Miwa, M., Ito, H., Kato, K. & Uematsu, T., 1999a, Sphingosine 1-phosphate induces heat shock protein 27 via p38 mitogen-activated protein kinase activation in osteoblasts, *J. Bone. Min. Res.* **14**: 1761-1767.

Kozawa, O., Tanabe, K., Ito, H., Matsuno, H., Niwa, M., Kato, K. and Uematsu, T., 1999b, Sphingosine 1-phosphate regulates heat shock protein 27 induction by a p38 MAP kinase-dependent mechanism in aortic smooth muscle cells, *Exp. Cell Res.* **250**: 376-380.

Kozawa, O., Yamamoto, T., tanabe, K., Akamatsu, S., Dohi, S. and Uematsu, T., 2000, Enhancement by sphingosine 1-phosphate in vasopressin-induced phosphoinositide hydrolysis in aortic smooth muscle cell: involvement of p38 MAP kinase, *J. Cell. Biochem.* **18**: 46-52.

Kwon, Y-G., Min, J-K., Kim, K-M., Lee, D-J., Billiar, T.R. and Kim, Y-M., 2001, Sphingosine 1-phosphate protects human umbilical vein endothelial cells from serum-deprived apoptosis by nitric oxide production, *J. Biol. Chem.* **276**: 10627-10633.

Lavie, Y. and Liscovitch, M., 1990, Activation of phospholipase D by sphingoid bases in NG108-15 neural-derived cells, *J. Biol. Chem.* **265**: 3868-3872.

Lee, M-J., Van Brocklyn, J.R., Thangada, S., Liu, C.H., Hand, A.R., Menzeleev, R., Spiegel, S. and Hla, T., 1996, Sphingosine-1-phosphate as a ligand for the G protein-coupled receptor EDG-1. *Science* **279**: 1552-1555.

Liu, H., Sugiura, M., Nava, V.E., Edsall, L.C., Kono, K., Poulton, S., Milstien, S., Kohama, T. and Spiegel, S., 2000, Molecular cloning and functional characterization of a novel mammalian sphingosine kinase type 2 isoform, *J. Biol. Chem.* **275**: 19513-19520.

Lutjohann, J., Spiess, A.N. and Gercken, G., 1998, Agonist-stimulated alveolar macrophages: apoptosis and phospholipid signaling, *Toxicol. Lett.* **96-97**: 59-67.

Macfarlane, D.E. and O'Donnell, P.S., 1993, Phorbol ester induces apoptosis in HL-60 promyelocytic leukemia cells but not in HL-60 PET mutant, *Leukemia* 7: 1846-1851.

Machwate, M., Rodan, S.B., Rodan, G.A. and Harada, S.-I., 1998, Sphingosine kinase mediates cyclic AMP suppression of apoptosis in rat periosteal cells, *Mol. Pharmacol.* **54**: 70-77.

MacLennan, A.J., Browe, C.S., Gaskin, A.A., Lado, D.C. and Shaw, G., 1994, Cloning and characterization of a putative G-protein coupled receptor potentially involved in development. *Mol. Cell. Neurosci.* **5**: 201-209.

Makamura, H., Oda, T., hamada, K., Hirano, T., Shimizu, N., Utiyama, H., 1998, Survival by Mac-1-mediated adherence and anoikis in phorbol ester-treated HL-60 cells, *J. Biol. Chem.* **273**: 15234-15351.

Mandala, S.M., 2001, Sphingosine 1-phosphate phosphatases, *Prostaglandins Lipid Mediators* **64**: 143-156.

Mandala, S.M., Thornton, R., Galve-Roperh, I., Poulton, S., Peterson, C., Olivera, A., Bergstrom, J., Kurtz, M.B. and Spiegel, S., 2000, Molecular cloning and characterization of a novel lipid phosphohydrolase that degrades sphingosine 1-phosphate and induces cell death, *Proc. Nat. Acad. Sci. U.S.A.* **97**: 7859-7864.

Mao, C., Xu, R., Szulc, Z.M., Bielawska, A., Galadari, S.H. and Obeid, L.M., 2001, Cloning and characterization of a novel human alkaline ceramidase: a mammalian enzyme that hydrolyzes phytoceramide, *J. Biol. Chem.* (in press).

Marshall, C.J., 1995, Specificity of receptor tyrosine kinase signaling: transient versus sustained extracellular signal-regulated kinase activation, *Cell* **80**: 179-185.

Matecki, A. and Pawelczyk, T., 1997, Regulation of phospholipase C delta1 by sphingosine, *Biochim. Biophys. Acta* **1325**: 287-296.

Melendez, A., Floto, R.A., Gillooly, D.J., Harnett, M.M and Allen, J.M., 1998, FcγRI coupling to phospholipase D initiates sphingosine kinase-mediated calcium mobilization and vesicular trafficking, *J. Biol. Chem.* **273**: 9393-9402.

Meyer zu Heringdorf, D.M., Lass, H., Alemany, R., Laser, K.T., Neumann, E., Zhang, C., Schmidt, M., Rauen, U., Jakobs, K.H. and Van Koppen, C.J., 1998, Sphingosine kinase-mediated Ca^{2+} signaling by G-protein-coupled receptors, *EMBO J.* **17**: 2830-2837.

Meyer zu Heringdorf, D.M., Lass, H., Kuchar, I., Alemany, R., Guo, Y., Schmidt, M. and Jakobs, K.H., 1999, Role of sphingosine kinase in Ca^{2+} signaling by epidermal growth factor receptor, *FEBS Lett.* **461**: 217-222.

Moore, A.N., Kampfl, A.W., Zhao, X., Hayes, R.L. and Dash, P.K., 1999, Sphingosine 1-phosphate induces apoptosis of cultured hippocampal neuronsthat requires protein phosphatases and activator protein-1 complexes, *Neurosci.* **94**: 405-415.

Mullmann, T.J., Siegel, M.I., Egan, R.W. and Billah, M.M., 1991, Sphingosine inhibits phosphatidate phosphohydrolase in human neutrophils by a protein kinase C-independent mechanism, *J. Biol. Chem.* **266**: 2013-2016.

Nakamura, H., Oda, T., Hamada, K., Hirano, T., Shimizu, N. and Utiyama, H., 1998, Survival by Mac-1-mediated adherence and anoikis in phorbol ester-treated HL-60 cells, *J. Biol. Chem.* **273**: 15345-15351.

Nava, V.E., Lacana, E., Poulton, S., Liu, H., Sugiura, M., Kono, K., Milstien, S., Kohama, T. and Spiegel, S., 2000, Functional characterization of human sphingosine kinase-1, *FEBS Lett.* **473**: 81-84.

Nikolova-Karakashian, M, Morgan, E.T., Alexander, C., Liotta, D.C. and Merrill Jr., A.H., 1997, Bimodal regulation of ceramidase by interleukin-1beta. Implications for the regulation of cytochrome P450 2C11 (CYP2C11), *J. Biol. Chem.* **272**: 18718-18724.

Ohta, H., Sweeney, E.A., Masamune, A., Yatomi, Y., Hakomori, S-I. and Igarashi, Y., 1995, Induction of apoptosis by sphingosine in human leukemic HL-60 cells: a possible endogenous modulator of apoptotic DNA fragmentation occurring during phorbol ester-induced differentiation, *Cancer Res.* **55**: 691-697.

Ohta, H., Yatomi, Y., Sweeney, E.A., Hakomori, S-I. and Igarashi, Y., 1994, A possible role of sphingosine in induction of apoptosis by tumor necrosis factor-alpha in human neutrophils, *FEBS Lett.* **355**: 267-270.

Okamoto, H., Takuwa, N., Gonda, K., Okazaki, H., Chang, K., Yatomi, Y., Shigematsu, H. and Takuwa, Y., 1998, EDG1 is a functional sphingosine-1-phosphate receptor that is linked via a G(i/o) to multiple signaling pathways, including phospholipase C activation, Ca2+ mobilization, ras-mitogen-activated protein kinase activation and adenylate cyclase inhibition. *J. Biol. Chem.* **273**: 27104-27110.

Okazaki, H., Ishizaka, N., Sakurai, T., Kurokawa , K., Goto, K., Kumada, M. and Takuwa, Y., 1993, Molecular cloning of a novel putative G protein-coupled receptor expressed in the cardiovascular system, *Biochem. Biophys. Res. Comm.* **190**: 1104-1109.

Olivera, A. and Spiegel, S., 1993, Sphingosine 1-phosphate as a second messenger in cell proliferation induced by PDGF and FCS mitogens, *Nature (London)* **365**: 557-560.

Olivera, A., Edsall, L., Poulton, S., Kazlauskas, A. and Spiegel, S., 1999, Platelet-derived growth factor-induced activation of sphingosine kinase requires phosphorylation of the PDGF receptor tyrosine residue responsible for binding of PLCγ, *FASEB J.* **13**: 1593-1600.

Olivera, A., Kohama, T., Edsall, L., Nava, V., Cuvillier, O., Poulton, S. and Spiegel, S., 1999, Sphingosine kinase expression increases intracellular sphingosine 1-phosphate and promotes cell growth and survival, *J. Cell Biol.* **147**: 545-557.

Orlati, S., Hrelia, S. and Rugolo, M., 1997, Pertussis toxin- and PMA-insensitive calcium mobilization by sphingosine in CFPAC-1 cells: Evidence for a phosphatidic acid-dependent mechanism, *Biochim. Biophys. Acta* **1358**: 93-102.

Park, Y.S., Hakomori, S-I., Kawa, S., Ruan, F. & Igarashi Y., 1994, Liposomal N,N,N-trimethylsphingosine (TMS) as an inhibitor of B16 melanomacell growth and metastasis with reduced toxicity and enhanced drug efficacy compared to free TMS: Cell membrane signaling as a target in cancer therapy III, *Cancer Res.* **54**: 2213-2217.

Pitson, S., Moretti, P.A.B., Zebol, J.R., Xia, P., Gamble, J.R., Vadas, M.A., D'Andrea, R.J. and Wattenberg, B.W., 2000, expression of a catalytically inactive sphingosine kinase mutant blocks agonis-induced sphingosine kinase activation. A dominant negative sphingosine kinase, *J. Biol. Chem.* **275**: 33945-33950.

Postma, F.R., Jalink, K., Hengeveld, T. and Moolenaar, W.H., 1996, Sphingosine 1-phosphate rapidly induces Rho-dependent neurite retraction: action through a specific cell surface receptor. *EMBO J.* **15**: 2388-2392.

Pushkareva, M., Chao, R., Bielawska, A., Merrill Jr, A.H., Crane, H.M., Lagu, B., Liotta, D. and Hannun, Y.A., 1995, Stereoselectivity of induction of the retinoblastoma gene product (pRb) dephosphorylation by D-erythro-sphingosine supports a role for pRb in growth suppression by sphingosine, *Biochem.* **34**: 1885-1892.

Pyne, S. and Pyne, N.J., 1996, The differential regulation of cyclic AMP by sphingomyelin-derived lipids and the modulation of sphingolipid-stimulated extracellular signal regulated kinase-2 in airway smooth muscle. *Biochem. J.* **315**: 917-923.

Pyne, S. and Pyne, N.J., 2000, Sphingosine 1-phosphate signaling in mammalian cells, *Biochem. J.* **349**: 385-402.

Pyne, S., Chapman, J., Steele, L. and Pyne, N.J., 1996, Sphingomyelin-derived lipids differentially regulate the extracellular signal-regulated kinase 2 (ERK-2) and c-Jun N-terminal kinase (JNK) signal cascades in airway smooth muscle, *Eur. J. Biochem.* **237**: 819-826.

Rakhit, S., Conway, A-M., Tate, R., Bower, T., Pyne, N.J. and Pyne, S., 1999, Sphingosine 1-phosphate stimulation of the p42/p44 mitogen-activated protein kinase pathway in airway smooth muscle: role of endothelial differentiation gene-1, c-Src tyrosine kinase and phosphoinositide 3-kinase. *Biochem. J.* **338**: 643-649.

Sakakura, C., Sweeney, E.A., Shirahama, T., Hakomori, S-I. and Igarashi, Y., 1996, Suppression of Bcl-2 gene expression by sphingosine in the apoptosis of human leukemic HL-60 cells during phorbol ester-induced terminal differentiation, *FEBS Lett.* **379**: 177-180.

Sakakura, C., Sweeney, E.A., Shirahama, T., Ruan, F., Solca, F., Kohno, M., Hakomori, S-I., Fischer, E.H. & Igarashi, Y., 1997, Inhibition of MAP kinase by sphingosine and its methylated derivative, N,N-dimethylsphingosine: A correlation with induction of apoptosis in solid tumor cells, *Intl. J. Oncol.* **11**: 31-39.

Sakane, F., Yamada, K. and Kanoh H., 1989, Different effects of sphingosine, R59022 and anionic amphiphiles on two diacylglycerol kinase isozymes purified from porcine thymus cytosol, *FEBS Lett.* **255**: 409-413.

Sato, K, Tomura, H., Igarashi, Y., Ui, M. and Okajima, F., 1999, Possible involvement of cell surface receptors in sphingosine 1-phosphate-induced activation of extracellular signal-regulated kinase in C6 glioma cells. *Mol. Pharmacol.* **55**: 126-133.

Strasser, A., O'Connor, L. and Dixit, V.M., 2000, Apoptosis signaling, *Ann. Rev. Biochem.* **69**: 217-245.

Sweeney, E.A., Sakakura, C., Shirahama, T., Masamune, A., Ohta, H., Hakomori, S-I. and Igarashi, Y., 1996, Sphingosine and its methylated derivative N,N-dimethylsphingosine (DMS) induce apoptosis in a variety of human cancer cell lines, *Int. J. Cancer* 66: 358-366.

Tani, M., Okino, N., Mori, K., Tanigawa, T., Izu, H. and Ito, M., 2000, Molecular cloning of the full-length cDNA encoding mouse neutral ceramidase. A novel but highly conserved gene family of neutral/alkaline ceramidases, *J. Biol. Chem.* **275**: 11229-11234.

Tolan, D., Conway, A-M., Pyne, N.J. and Pyne, S., 1997, Sphingosine prevents diacylglycerol signaling to mitogen-activated protein kinase in airway smooth muscle. *Am. J. Physiol.* **273** (*Cell. Physiol* **42**): C928-C936.

Tolan, D., Conway, A-M., Rakhit, S., Pyne, N.J. and Pyne, S., 1999, Assessment of the extracellular and intracellular actions of sphingosine 1-phosphate by using the p42/p44 mitogen-activated protein kinase cascade as a model. *Cell. Signal.* **11**: 349-354.

Van Brocklyn, J.R., Lee, M-J., Menzeleev, R., Olivera, A., Edsall, L., Cuvillier, O., Thomas, D.M., Coopman, P.J.P., Thangada, S., Liu, C.H., Hla, T. and Spiegel, S., 1998, Dual actions of sphingosine 1-phosphate: extracellular through the G(i)-coupled receptor EDG1 and intracellular to regulate proliferation and survival, *J. Cell Biol.* **142**: 229-240.

Van Brocklyn, J.R., Tu, Z., Edsall, L., Schmidt, R.R. and Spiegel, S., 1999, Sphingosine 1-phosphate-induced cell rounding and neurite retraction are mediated by the G protein-coupled receptor H218, *J. Biol. Chem.* **274**: 4626-4632.

Van Veldhoven, P.P. and Mannaerts, G.P., 1993, Sphingosine-phosphate lyase, *Adv. Lipid Res.* **26**: 69-98.

Van Veldhoven, P.P. and Mannaerts, G.P., 1994, Sphinanine 1-phosphate metabolism in cultured skin fibroblasts: evidence for the existence of a sphingosine phosphatase, *Biochem. J.* **299**: 597-601.

Verheij, M., Bose, R., Lin, X.H., Yao, B., Jarvis, W.D., Grant, S., Birrer, M.J., Szabo, E., Zon, L.I., Kyriakis, J.M., Haimovitz-Friedman, A., Fuks, Z. and Kolesnick, R.N., 1996, Requirement for ceramide-initiated SAPK/JNK signaling in stress-induced apoptosis, *Nature* **380**: 75-79.

Wang, F., Nobes, C.D., Hall, A. and Spiegel, S., 1997, Sphingosine 1-phosphate stimulates Rho-mediated tyrosine phosphorylation of focal adhesion kinase and paxillin in Swiss 3T3 fibroblasts. *Biochem. J.* **324**: 481-488.

Windh, R.T., Lee, M-J., Hla, T., An, S., Barr, A.J. and Manning, D.R., 1999, Differential coupling of the sphingosine 1-phosphate receptors EDG1, EDG3 and H218/EDG5 to the G_i, G_q, and G_{12} families of heterotrimeric G proteins. *J. Biol. Chem.* **274**: 27351-27358.

Xia, P., Wang, L., Gamble, J.R. and Vadas, M.A., 1999, Activation of sphingosine kinase by tumor necrosis factor-a inhibits apoptosis in human endothelial cells, *J. Biol. Chem.* **274**: 34499-34505.

Xu, J., Love, L.M., Singh, I., Zhang, Q-X., Dewald, J., Wang, D-A., Fischer, D.J., Tigyi, G., Berthiaume, L.G., Waggoner, D.W. and Brindley, D.N., 2000, Lipid phosphate phosphatase-1 and Ca2+ control lysophosphatidate signaling through EDG-2 receptors, *J. Biol. Chem.* **275**: 27520-27530.

Yamazaki, Y., Kon, J., Sato, K., Tomura, H., Sato, M., Yoneya, T., Okazaki, H., Okajima, F. and Ohta, H., 2000, EDG-6 as a putative sphingosine 1-phosphate receptor coupling to Ca2+ signaling pathway. *Biochem. Biophys. Res. Comm.* **268**: 583-589.

Yang, L., Yatomi, Y., Hisano, N., Qi, R., Asazuma, N., Satoh, K., Igarashi, Y., Ozaki, Y. and Kume, S., 1996, Activation of protein-tyrosine kinase Syk in human platelets stimulated with lysophosphatidic acid or sphingosine 1-phosphate. *Biochem. Biophys. Res. Comm.* **229**: 440-444.

Yang, L., Yatomi, Y., Miura, Y., Satoh, K. and Ozaki, Y., 1999, Metabolism and functional effects of sphingolipids in blood cells, *Br. J. Hemeatol.* **107**: 282-293.

Yatomi, Y., Ruan, F., Megidish, T., Toyokumi, T., Hakomori, S. and Igarashi, Y., 1996, N,N-dimethylsphingosine inhibition of sphingosine kinase and sphingosine 1-phosphate activity in human platelets, *Biochem. J.* **35**: 626-633.
Young, K.W., Bootman, M.D., Channing, D.R., Lipp, P., Maycox, P.R., Meakin, J., Challis, J.A. and Nahorski, S.R., 2000, Lysophosphatidic acid-induced Ca^{2+} mobilisation requires intracellular sphingosine 1-phospahte production, *J. Biol. Chem.* **275**: 38532-38539.
Zhang, H., Desai, N.N., Olivera, A., Seki, T., Brooker, G. and Spiegel, S., 1991, Sphingosine-1-phosphate, a novel lipid, involved in cellular proliferation. *J. Cell Biol.* **114**: 155-167.
Zhou, H., Li, X-M., Meinkoth, J. and Pittman, R.N., 2000, Akt regulates cell survival and apoptosis at a postmitochondrial level, *J. Cell Biol.* **151**: 483-494.
Zhou, J. and Saba, J., 1998, Identification of the first mammalian sphingosine phosphate lyase gene and its functional expression in yeast, *Biochim. Biophys. Res. Comm.* **242**: 502-507.
Zondag, G.C., Postma, F.R., Van Etten, A., Verlaan, I. and Moolenaar, W.H., 1998, Sphingosine 1-phosphate signaling through the G-protein-coupled receptor Edg-1. *Biochem. J.* **330**: 605-609.

Chapter 14

Ceramide in Regulation of Apoptosis

Implication in multitoxicant resistance

Jean-Pierre Jaffrézou, Guy Laurent, and Thierry Levade
INSERM E9910, Institut Claudius Régaud, 20 rue du Pont St Pierre, 31052 Toulouse, France and INSERM U466,CHU Rangueil, 31403 Toulouse, France

1. INTRODUCTION

Studying cell death signaling pathways may provide a few answers in the field of anticancer pharmacology and may be useful for tailoring new therapies. Indeed, while extensive studies of the biochemical and molecular pharmacology of chemotherapeutic drugs have revealed intimate drug-target interactions, how and why such interactions should bring about cell death has been hitherto unclear. In fact, in many cases cellular damage actually caused by active doses of these agents (intercalators, topoisomerase inhibitors, antimitotics, etc.) is not sufficient to explain the observed toxicity (Chabner and Myers, 1989; Dive and Hickman, 1991).

There is mounting evidence that anticancer agents kill by activating a metabolic pathway, rather than by crippling cellular metabolism (Fisher, 1994). The identification of apoptosis as a common response of neoplastic cells suggests that the cytotoxic action of chemotherapeutic drugs requires the active participation of the target cell (Dive and Hickman, 1991; Hickman, 1992; Fisher, 1994; Martin and Green, 1994). Although it remains to be determined if this participation occurs before or after effector induced cell damage; this observation has forced a reconsideration of the mechanism(s) whereby tumor cells respond to cytotoxic agents.

Until recently, anticancer research has been, for the most part, more empirical that rational. In our endeavour to combat neoplastic diseases, researchers have developed a battery of treatments, most based on a simple

Subcellular Biochemistry, Volume 36: Phospholipid Metabolism in Apoptosis.
Edited by Quinn and Kagan. Kluwer Academic/Plenum Publishers, New York, 2002

strategy, to kill the assailant cell by intercepting a specific cellular target. However, both clinicians and pharmacologists concede that little gain has been made in the therapeutic outcome of cancer patients in the past 60 years. Indeed, neoplastic cells have the uncanny aptitude (inherently or acquired) of resistance to a diverse panel or cytotoxic agents (multitoxicant resistance or MTR). This standpoint has forced investigators to explore a new concept in cell response to cytotoxic stress: apoptosis signaling. The recent identification of a ceramide (CER) mediated apoptotic signaling pathway triggered by antitumor agents offers a new perspective for the treatment of neoplastic cells. In this chapter, we have attempted to delineate how cellular regulation of stress-induced CER generation may contribute to cell resistance.

2. CERAMIDE IN APOPTOSIS SIGNALING

2.1 Stress-induced ceramide generation

Ceramide (CER) has emerged as a critical mediator of apoptosis. Hannun and co-workers were the first to report that TNFα could stimulate a neutral sphingomyelinase (SMase) responsible for sphingomyelin (SM) hydrolysis and subsequent generation of CER in the human myeloid cell line U937. These authors also demonstrated that a synthetic cell-permeable CER analog (C_2-CER), induced apoptosis in these cells (Obeid *et al.*, 1993). These results strongly supported the implication of the SM-CER pathway in apoptosis induced by TNFα, and, perhaps, by other cytotoxic molecules. For example, the anthracycline daunorubicin (DNR) and the nucleoside analog 1-β-D-arabinofuranosylcytosine (Ara-C) induce apoptosis in HL-60 and U937 cells at concentrations which are clinically relevant (Quillet-Mary *et al.*, 1996; Bezombes *et al.*, 2001). DNR cytotoxicity is generally thought to be the result of drug-induced damage to DNA resulting from free radicals stemming from the quinone-generated redox activity, by intercalation-induced distortion of the double helix, or by stabilization of the cleavable complex formed between DNA and topoisomerase II (Chabner and Myers, 1989). Ara-C is, for its part, metabolized into Ara-CTP and incorporates into and inhibits DNA synthesis resulting in strand breakage and apoptosis (Fram and Kufe, 1982; Gunji *et al.*, 1991). Although specific biochemical lesions have been associated with DNR and Ara-C mediated cytotoxicity, the pathways leading to the induction of apoptosis remain unknown.

DNR- and Ara-C-triggered apoptosis has been correlated with CER generation (Jaffrézou *et al.*, 1996; Mansat *et al.*, 1997a; Mansat *et al.*, 1997b; Bezombes *et al.*, 2001). Several other chemotherapeutic agents have also been shown to activate a SMase leading to SM hydrolysis and CER generation, e.g., vincristine (Zhang *et al.*, 1996), dexamethasone (Quintans *et al.*, 1994), and mitoxantrone (Bettaïeb *et al.*, 1999). Finally, resistance to apoptosis triggered by several toxicants (including DNR, TNFα and ionizing radiation) has also been correlated with the inability to generate CER (Bettaïeb *et al.*, 1996; Santana *et al.*, 1996; Wright *et al.*, 1996; Bruno *et al.*, 1998).

Central to the concept of the SM-CER pathway is the role played by effector-target interactions. For example, chemotherapeutic drugs, which possess nuclear targets, undergo at least three main translocation processes. One can imagine that the activation of a SMase can occur in any of these steps. The first one is the passage through the plasma membrane, where there are a number of cell "sensors". The second is intracytoplasmic, from the cell membrane to the perikaryon. The third intracellular movement corresponds to nuclear translocation. Of course, there is also a continuous exchange of molecules between the cytoplasm and the nucleus through the nuclear envelope, the nuclear pore complex and the nuclear lamina. Moreover, the SMase involved in CER generation triggered by cytotoxic agents has yet to be unambiguously identified (Levade and Jaffrézou, 1999). It also appears that there may exist distinct pools of SM implicated in CER generation, situated in the plasma membrane inner leaflet (Linardic and Hannun, 1994; Andrieu *et al.*, 1996), in caveolae (Liu and Anderson, 1995), and in endosomal vesicles (Wiegmann *et al.*, 1994). Furthermore, due to its hydrophobicity, natural CER is not readily distributed within the cell so that CER signaling is likely to be confined to its site of production (Chatelut *et al.*, 1998). Of course, a signaling cascade originating from the plasmic membrane and leading to DNA fragmentation implies an intricate pathway. A simpler model would assume that such a signaling pathway would be generated at or near the nucleus (closer to the ergotic drug/DNA target interaction site). Indeed, neutral SMase activity has been found in isolated cell nuclei (Tamiya-Koizumi *et al.*, 1989), as well as significant amounts of SM (Spangler *et al.*, 1975; Cocco *et al.*, 1980). Moreover, in the particular case of ionizing radiation triggered apoptosis in myeloid cells, this nuclear enzyme has been shown to be activated (Jaffrézou *et al.*, 2001).

Effector-induced SMase activation plays a highly pivotal role in apoptosis signaling since it has been described that the absence of CER generation, due to a defect in SMase stimulation, leads to cell resistance (for review see Levade and Jaffrézou, 1999). However, the spacio-temporal organization of these events is unclear. SM, a major constituent of mammalian cell membranes, is found essentially within the plasma membrane and concentrated mostly in its outer leaflet. However, we and others have demonstrated that in response to cytotoxic effectors such as

TNF-α, SM is hydrolyzed within the plasma membrane inner leaflet (Linardic and Hannun, 1994; Andrieu *et al.* 1996). This apparent discrepancy was partially resolved by the study of van Blitterswijk and co-workers which demonstrated that the CER generated from the plasma membrane SM (initially located in the outer leaflet), gains access to a cytosolic SMase by flipping to the inner leaflet due to phospholipid scrambling (Tepper *et al.*, 2000).

2.2 Ceramide regulation and resistance

The implication of the SM-CER signaling pathway in MTR opens intriguing avenues of research. Indeed, it has been suggested, for example, that apoptosis-resistant cells are deficient in SMase activation or SM hydrolysable pools (Santana *et al.*, 1996; Bettaïeb *et al.*, 1996; Wright *et al.*, 1996). In fact, we have shown in some resistant myeloid leukemia cells that the incapacity to generate CER in the presence of effector was in relation to a deficit of (hydrolysable) SM pool (Bettaïeb *et al.*, 1996). Thus, it is possible that, in some AML cells, the activation of a yet undefined SM translocase results not only in modification of SM transverse asymmetry but also in reduced SM hydrolysable pool, decreased CER production, and inhibition of apoptosis (Bezombes *et al.* 1998). Hence one can conclude from these studies that SMase activation is substrate dependent and therefore it is conceivable that spacial SM organisation is essential in the initiation of a SMase dependent apoptotic pathway. If this hypothesis were valid, one could concede that by modifying SM transverse distribution it would be possible to restore an apoptotic pathway in MTR cells.

A very recent report demonstrated that human multidrug resistant (MDR)1 and mouse mdr1a P-glycoprotein (P-gp) can translocate short-chain lipids (including short-chain SM analogs) at the cell surface across the plasma membrane, and that P-gp inhibitors blocked lipid translocation across the membrane (van Helvoort *et al.*, 1996). From these important observations, it has been proposed that MDR1 P-gp may function as a lipid translocase of broad specificity (van Helvoort *et al.*, 1996; van Helvoort *et al.*, 1997). A good model to study the potential role of P-gp in resistance to apoptosis is the myeloid leukemia cell line KG1a. As an example, KG1a cells are completely resistant to the cytotoxic (and therefore apoptotic) effect of TNFα (Munker *et al.*, 1987; Higuchi and Aggarwal, 1994). These cells also express functional P-gp (Bailly *et al.*, 1995). If P-gp may potentially be a SM translocator, we hypothesized that P-gp inhibition may limit SM movement from the inner to the outer leaflet of the plasma membrane leading to an increase of hydrolysable SM pool, and thus sensitize KG1a cells to TNFα. Indeed, we have shown that the cyclosporin derivative

PSC833, a potent and specific P-gp blocker (Twentyman and Bleehen, 1991), sensitizes KG1a cells to TNFα-mediated apoptosis (Bezombes *et al.*, 1998). Furthermore, we presented evidence for the translocation of a cytosolic SMase to the inner leaflet plasma membrane upon SM enrichment.

2.3 Origin of ceramide production

Since the subcellular location of drug-induced SMase activation remains unclear, we focused on the possible compartmentalisation of SM signaling pools. A model for membrane structure that describes the organisation of the lipid microdomains as platforms, the rafts, has been proposed. It has been demonstrated that proteins can selectively be included or excluded from these microdomains which can serve as relay stations in intracellular signaling (Simons and Ikonen, 1997). The main forces driving the formation of rafts are lipid-lipid interactions, dependent on the biophysical characteristics of the lipid components. Detergent-insoluble rafts (and caveolae) have been suggested to be the plasma membrane compartments producing CER, since they are enriched in SM (and neutral SMase [Brown and Rose, 1992; Liu and Anderson, 1995; Liu *et al.* 1997; Bilderback *et al.*, 1997; Veldman *et al.*, 2001]), which confers a low-buoyant density in sucrose. Such microdomains have been described in several hematopoietic cells and other cell types, and contain selected membrane glycoproteins, such as CD36 a surface receptor molecule involved in signal transduction in both platelets and monocytes (Yuan *et al.*, 1995). To address whether rafts are implicated in drug-induced apoptosis signaling, we investigated, using subcellular fractionation, the trafficking of the Lyn-SMase cascade in Ara-C-treated myeloid leukemic cells (U937).

Induction of apoptosis by DNA damaging agents such as Ara-C includes the activation of Lyn protein tyrosine kinase (Huang *et al.*, 1991). We have previously established that Ara-C-induced activation of Lyn results in its binding to a neutral SMase and is requisite for its stimulation and the induction of apoptosis in U937 cells. (Bezombes *et al.*, 2001). However, the cellular origin of these events was unclear. Recently, we observed that part of the total cellular neutral SMase activity is sequestered in SM-enriched plasma membrane microdomains (rafts) (unpublished data). Under Ara-C treatment, Lyn is rapidly activated and translocated into rafts. The compartmentalisation of Lyn (as well as neutral SMase activation) induced by Ara-C was blocked by the tyrosine kinase inhibitor herbimycin A. In conclusion, we observed that DNA damaging agents such as Ara-C rapidly induce Lyn activation and its translocation into membrane rafts. This in turn leads to neutral SMase activation and raft-associated SM hydrolysis with the concomitant generation of CER. These findings may have important implications, since it is now conceivable that the lack of apoptosis signaling in drug-resistant cells is possibly linked to the presence and/or constitution of rafts. Further investigations are necessary to determine to what extent the compartmentalisation of these signaling processes are modified/absent in drug-resistant cells.

2.4 Cellular modulators of ceramide production

2.4.1 Protein Kinase C

In previous studies it has been demonstrated that not only protein kinase C (PKC) activators including phorbol esters (TPA) and diacylglycerol (DAG) inhibited the ability of cell-permeant CER to induce apoptosis (Jarvis *et al.* 1996), but also that PKC inhibitors enhanced CER-induced apoptosis (Jarvis *et al.*, 1994). From these studies it has been suggested that PKC is a key negative regulator of CER-induced apoptosis.

Little is known about the role of PKC in the regulation of CER production. It has been recently reported that PKC inhibitors triggered neutral SMase activation, suggesting that PKC may also play an important role by regulating basal SMase activity (Chmura *et al.*, 1996). These findings are consistent with those reported by two other groups, who observed that TPA inhibited CER generation and apoptosis in irradiated bovine endothelial cells and in TNF-α treated leukaemia cells (Haimovitz-Friedmann *et al.* 1994; Obeid *et al.*, 1993). We have shown that PKC activators TPA and phosphatidylserine inhibited DNR-induced neutral SMase stimulation, SM hydrolysis, CER generation and apoptosis (Mansat *et al.*, 1997a). Only the classic (α, β, γ) and novel (δ, ε, θ, μ) PKC isoforms present a TPA binding site whereas all of the PKCs have an amino-terminal regulatory domain containing a phosphatidylserine-binding site (Nishizuka, 1995). Therefore, DNR-induced neutral SMase stimulation can be negatively manipulated by one or several classic and/or novel PKC isoforms but likely not by atypical (ζ, λ, ι) isoforms. The mechanism by which PKC regulates SMase activity remains to be elucidated. The fact that PKC activation had no influence on basal SM activity suggests that PKC regulates signaling events triggered by cytotoxic effectors upstream SMase stimulation.

2.4.2 Serine Proteases

The observation that serine prctease inhibitors (serpins) can block several forms of apoptosis (Higuchi *et al.*, 1995, Nicholson *et al*, 1995, Martin *et al.*, 1996, Lahti *et al.*, 1996: Dubrez *et al.*, 1996) has led to the central hypothesis that the activation of serine proteases are responsible for mediating cell death (Kumar and Harvey, 1995). Such proteolytic events which occur during apoptosis, are believed to be a relatively late event in the apoptotic pathway. Moreover, several studies have suggested that activation of these proteases occurs after (and is perhaps dependent on) CER generation (Machleidt *et al.*, 1994; Smyth *et al.*, 1996; Wright *et al.*, 1996; Chinnaiyan *et al.*, 1996). We have previously reported that DNR-induced CER generation and apoptosis could be blocked by serine protease inhibitors (Mansat *et al.*, 1997b). Moreover, several other studies have shown that CER generation triggered by various cytotoxic effectors could be negatively regulated by some tri- or tetrapeptide inhibitors as well as by cowpox virus CrmA (Pronk *et al.*, 1996; Gamard *et al.*, 1997; Dbaido *et al.*, 1997), which suggested a role for proteases belonging to the caspase-1 or initiator caspase family. Since CrmA blocked TNF and Fas agonist induced CER generation and apoptosis, this suggested that caspase-8 was a probable candidate (Dbaido *et al.*, 1997; Wesselborg *et al.*, 1999)

We and others have shown that serpins could block DNR- and TNFα–triggered apoptosis (Mansat *et al.*, 1997b; Dbaido *et al.*, 1997). This effect was due to the inhibition of neutral SMase activation. This observation is critical since it situates a signal transducing protease activity upstream of CER generation. The identity of this protease in unknown, but one could speculate for an ICE family member. Indeed, inhibition of CER generation by serpins has also been described in apoptosis during drosophila embryo development and it was suggested that an ICE-like protease may be responsible for both CER generation and cell death (Pronk *et al.*, 1996).

2.4.3 Diacylglycerol

It stands out that DAG plays a central role in the negative control of drug- and ionizing radiation-induced apoptosis mediated by CER by potentially intervening upstream of its generation (by blocking SMase activity) and downstream (by blocking the apoptotic effect of CER) (Hannun and Obeid, 1995). In this respect, it is advisable to recall that most antitumor agents susceptible in triggering the SM-CER pathway are equally capable of activating in parallel the production of DAG (Posada *et al.*, 1989; Strum *et al.*, 1994; Rubin *et al.*, 1992; Nishio *et al.*, 1992). For example, DNR and ionizing radiation are capable of inducing in a dose-dependent manner the production of DAG by hydrolysis from phosphatidylcholine through the activation of phosphatidylcholine phospholipase C or D (Bettaïeb *et al.*, 1999; Avila *et al.*, 1993). Thus, for antitumor agents, it is possible that the cellular response is the result between a noxious pathway mediated by CER and a protection pathway mediated by DAG, the balance of these two pathways conditioning the amplitude of the cytotoxic effect (Jaffrézou *et al.*, 1998). This interpretation opens many perspectives of pharmacological manipulations aimed at increasing the production of CER and/or diminishing the liberation of DAG by using inhibitors of the phospholipases concerned (phosphatidylcholine phospholipase C and D).

2.5 Ceramide metabolism

The intracellular CER concentration results from the equilibrium between CER production and CER metabolism. It now appears that CER metabolism plays a critical role in regulating intracellular CER concentration and therefore effector-induced cytotoxicity. The four major CER metabolic pathways include its reconversion into SM, its catabolism to sphingosine (followed by sphingosine 1-phosphate) and its metabolism to glucosylceramide or galactosylceramide. All of which may contribute to limiting stress-induced CER elevation, except for the galactosylceramide pathway which has of yet not been described.

SM resynthesis requires that CER be transferred to a phosphorylcholine group to generate SM and DAG by a SM synthase. It is likely that this metabolic pathway is activated after SM hydrolysis during the SM-CER cycle. This enzyme therefore has the important ability to directly regulate, in opposite directions, CER and DAG levels within the cells (Luberto and Hannun, 1998). However, despite the great biological potential of SM synthase, very little is known about location, distribution, and regulation of this enzyme (Hampton and Morand, 1989; Andrieu-Abadie *et al.*, 1998). Whether SM synthase plays a role in drug-induced cytotoxicity remains to be investigated.

CER can also be catabolized by a ceramidase into sphingosine which, in turn, can be converted into sphingosine-1-phosphate through sphingosine-1-kinase. Studies performed in Spiegel's laboratory, have shown that sphingosine-1-phosphate inhibits apoptosis induced by CER and other effectors, perhaps by inhibiting CER-induced JNK stimulation and caspase activities (Cuvillier *et al.*, 1996; Cuvillier *et al.*, 1998; Kleuser *et al.*, 1998; Perez *et al.*, 1997). Therefore, sphingosine-1-phosphate and sphingosine-1-kinase might play an important role in regulating effector-induced apoptosis and therefore contribute to MTR. However, we have reported that effector-induced CER production and apoptosis can be observed in cells derived from Farber's disease (which are genetically deficient for lysosomal ceramidase, the major component of cellular ceramidase activity) (Ségui *et al.*, 2000). Although we can not totally exclude the implication of extralysosomal ceramidases, this result suggests that CER conversion to sphingosine may not be a critical metabolic pathway for CER generated by cytotoxic effectors.

Finally, CER can be transformed to glucosylceramide by a glucosylceramide synthase. Previous reports have indicated that glucosylceramide has no cytotoxic property or even may stimulate cell proliferation (Shayman *et al.*, 1991), and that glucosylceramide synthase inhibitors have been found to display some antitumor activity (Abe *et al.*, 1995; Rani *et al.*, 1995). These results suggest that glucosylceramide synthase plays an important role in cellular protection. In fact, cells transduced by a glucosylceramide synthase gene were highly resistant to anthracyclines (Liu *et al.*, 1999). Furthermore, it appears that this enzymatic activity is significantly boosted in MDR cells (Lavie *et al.*, 1997). Conversely, transfection of glucosylceramide synthase antisense reverses adriamycin resistance (Liu *et al.*, 2000). Therefore, one can speculate that in drug-treated MDR cells, CER is rapidly converted to glucosylceramide (Lavie *et al.*, 1996; Lucci *et al.*, 1998), which makes this enzyme an attractive target for MDR reversal. Supporting this hypothesis is the observation that the inhibition of CER glycosylation pathway increases MDR cell sensitivity to cytotoxics (Lucci *et al.*, 1999a), and that most MDR modulators, including cyclosporin A, tamoxifen, and verapamil are potent glucosylceramide synthase inhibitors whereas the cyclosporin A analog PSC 833 (Valspodar) increases intracellular CER concentration by stimulating CER synthase (Lavie *et al.*, 1997; Lucci *et al.*, 1999b; Cabot *et al.*, 1999). These results suggest that these agents may exert their chemosensitizing effect not only through their P-gp binding capacity, as previously postulated, but also by facilitating CER accumulation.

3. CONCLUSION

CER generation and its metabolism play a critical role in cell response to cytotoxic stress. Several cellular metabolic pathways have been described in regulating CER levels and it is apparent that these biochemical events are complex (Figure 1). It is also of importance to discriminate between early CER generation (pre-mitochondrial damage) responsible for the initiation of apoptosis and late CER generation (post-mitochondrial damage) potentially implicated in the execution phase of apoptosis. It is also essential to delineate among resistance mechanisms, such as those which lead to decreased effector-target interactions and those which interfere with cell death commitment. Finally, the reinterpretation of the action of well-established therapeutic agents in the light of recent advances in apoptosis signaling should allow cellular pharmacology of anti-neoplastic agents to continue gathering momentum. The tantalizing implication of these findings is that by pharmacologically manipulating key steps in the apoptosis signaling cascade, one could possibly increase chemosensitivity of neoplastic cells (Hunnun *et al.*, 1997; Laurent and Jaffrézou, 2001).

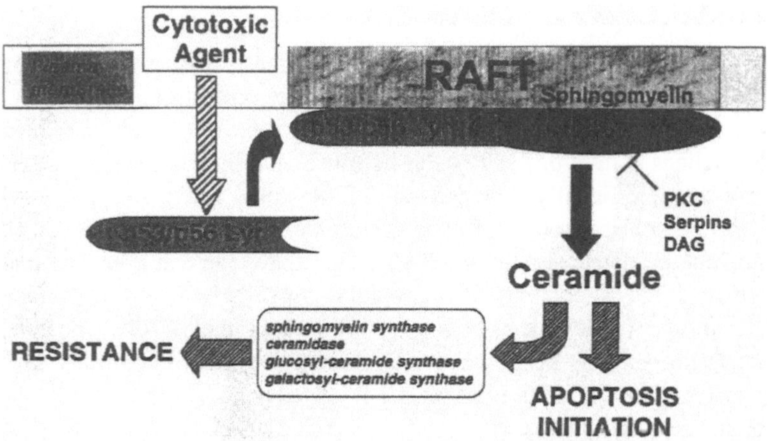

Figure 1. Ceramide mediated apoptosis signaling. Recent investigations hypothesize that, at least for drug induced apoptosis, the src kinase Lyn is activated and recruited to plasma membrane rafts. There it binds to and activates a neutral sphingomyelinase (SMase) within minutes. This initial rise in ceramide levels initiates an apoptotic cascade. However several cellular events can inhibit ceramide production (by blocking sphingomyelinase activation or clearing out ceramide accumulation through increased metabolisation) and therefore contribute to cellular resistance/survival.

ACKNOWLEDGMENTS

This work was supported by la Ligue Nationale Contre le Cancer and les Comités Départementaux du Gers, de l'Aveyron, du Lot et de la Haute-Garonne (J.P.J.), and in part by l'Association pour la Recherche sur le Cancer grants 5526 (G.L.), La Faculté de Médecine Toulouse-Rangueil (G.L), and INSERM (G.L. and T.L).

REFERENCES

Abe, A., Radin, N. S., Shayman, J. A., Wotring, L. L., Zipkin, R. E., Sivakumar, R., Ruggieri, J. M., Carson, K. G., and Ganem, B., 1995, Structural and stereochemical studies of potent inhibitors of glucosylceramide synthase and tumor cell growth. *J. Lipid Res.* **36**: 611-621.

Andrieu, N., Salvayre, R., and Levade, T., 1996, Comparative study of the metabolic pools of sphingomyelin and phosphatidylcholine sensitive to tumor necrosis factor. *Eur. J. Biochem.* **236**: 738-745.

Andrieu-Abadie, N., Carpentier, S., Salvayre, R., and Levade,T., 1998, The tumour necrosis factor-sensitive pool of sphingomyelin is resynthesized in a distinct compartment of the plasma membrane. *Biochem J.* **333**: 91-97.

Avila, M. A., Otero, G., Cansado, J., Dritschilo, A., Velasco, J. A., and Notario, V., 1993, Activation of phospholipase D participates in signal transduction pathways responsive to γ-radiation. *Cancer Res.* **53**: 4474-4476.

Bailly, J. D., Muller, C., Jaffrézou, J. P., Demur, C., Cassar, G., Bordier, C., and Laurent, G., 1995, Lack of correlation between expression and function of P-glycoprotein in acute myeloid leukemia cells. *Leukemia* **9**: 799-807.

Bettaïeb, A., Record, M., Côme, M. G., Bras, A. C., Chap, H., Laurent, G., and Jaffrézou, J. P., 1996, Opposite effects of tumor necrosis factor α on the sphingomyelin-ceramide pathway in two myeloid leukemia cell lines: Role of transverse sphingomyelin distribution in the plasma membrane. *Blood* **88**: 1465-1472.

Bettaïeb, A., Plo, I., Mansat-De Mas, V., Quillet-Mary, A., Levade, T., Laurent, G., and Jaffrézou, J. P., 1999, Daunorubicin and mitoxantrone-triggered phosphatidylcholine hydrolysis: Implication in drug-induced ceramide generation and apoptosis. *Mol. Pharmacol.* **55**: 118-125.

Bezombes, C., Maestre, N., Laurent, G., Levade, T., Bettaïeb, A., and Jaffrézou, J. P., 1998, Restoration of TNF-α-induced ceramide generation and apoptosis in resistant human leukemia KG1a cells by the P-glycoprotein blocker PSC833. *FASEB J.* **12**: 101-109.

Bezombes, C., Mansat-De Mas, V., Plo, I ., Quillet-Mary, A., Nègre-Salvayre, A., Laurent, G. and Jaffrézou, J. P., 2001, Oxidative stress-induced activation of Lyn recruits sphingomyelinase and is requisite for its stimulation by Ara-C *FASEB J.* **15**: 1583-1585.

Bilderback, T. R., Grigsby, R. J., and Dobrowsky, R. T., 1997, Association of p75(NTR) with caveolin and localization of neurotrophin-induced sphingomyelin hydrolysis to caveolae. *J. Biol. Chem.* **272**: 10922-10927.

Brown, D. A., and Rose, J. K., 1992, Sorting of GPI-anchored proteins to glycolipid-enriched membrane subdomains during transport to the apical cell surface. *Cell* **68**: 533-544.

Bruno, A. P. , Laurent, G., Blaise, R., Averbeck, D., Demur, C., Bonnet, J., Bettaïeb, A., Levade, T., and Jaffrézou, J.P., 1998, Lack of ceramide generation in TF-1 human myeloid leukemic cells resistant to ionizing radiation. *Cell Death Diff.* 5: 172-182.

Cabot, M. C., Giulano, A. E., Han, T-Y., and Liu, Y-Y., 1999, SDZ PSC 833, the cyclosporin A analog and multidrug resistance modulator, activates ceramide synthesis and increases vinblastine sensitivity in drug-sensitive and drug-resistant cancer cells. *Cancer Res.* **59**: 880-885.

Chabner, B. A, and Myers, C. E., 1989, Clinical pharmacology of cancer chemotherapy, in: *Cancer: Principles and Practice of Oncology,* (V. T. DeVitta, J., Hellman, and S., Rosenberg, S. A.) J. B. Lippencott Co., Philadelphia, PA. pp. 349-395,

Chatelut, M., Leruth, M., Harzer, K., Dagan, A., Marchesini, S., Gatt, S., Salvayre, R., Courtoy, P., and Levade T., 1998, Natural ceramide is unable to escape the lysosome, in contrast to a florescent analog. *FEBS Lett.* **426**: 102-106.

Chinnaiyan, A. M., Tepper, C. G., Seldin, M. F., O'Rourke, K., Kischkel, F. C., Hellbardt, S., Krammer, P. H., Peter, M. E., and Dixit, V. M., 1996, FADD/MORT1 is a common mediator of CD95 (Fas/APO-1) and tumor necrosis factor receptor-induced apoptosis. *J. Biol. Chem.* **271**: 4961-4965.

Chmura, S. J., Nodzenski, E., Weichselbaum, R. R., and Quintans, J., 1996, Protein kinase C inhibition induces apoptosis and ceramide production through activation of a neutral sphingomyelinase. *Cancer Res.* **56**: 2711-2714.

Cocco, L., Maraldi, N. M., and Manzoli, F. A., 1980, Phospholipid interactions in rat liver nuclear matrix. *Biochem. Biophys. Res. Commun.* **96**: 890-898.

Cuvillier, O., Pirianov, G., Kleuser, B., Vanek, P. G., Coso, O. A., Gutkind, J. S., and Spiegel, S., 1996 Suppression of ceramide-mediated programmed cell death by sphingosine-1-phosphate. *Nature* **381**: 800-803.

Cuvillier, O., Rosenthal, D. S., Smulson, M. E., and Spiegel, S., 1998, Sphingosine 1-phosphate inhibits activation of caspases that cleave poly(ADP-ribose) polymerase and lamins during Fas- and ceramide-mediated apoptosis in Jurkat T lymphocytes. *J. Biol. Chem.* **273**: 2910-2916.

Dbaido, G. S., Perry, D. K., Gamard, C. J., Platt, R., Poirier, G. G., Obeid, L. M., and Hannun, Y. A., 1997, Cytokine response modifier A (CrmA) inhibits ceramide formation in response to tumor necrosis factor (TNF)-a: CrmA and Bcl2 target distinct components in the apoptotic pathway. *J. Exp. Med.* 185: 481-490.

Dive, C., and Hickman, J. A., 1991, Drug-target interactions: only the first step in the commitment to a programmed cell death? *Br. J. Cancer* **64**: 192-196.

Dubrez, L., Savoy, I., Hamman, A., and Solary, E., 1996, Pivotal role of a DEVD-sensitive step in etoposide-induced and Fas-mediated apoptotic pathways. *EMBO J.* **15**: 5504-5512.

Fisher, D. E., 1994, Apoptosis in cancer therapy: crossing the threshold. *Cell* **78**: 539-542.

Fram, R. J., and Kufe, D., 1982, DNA strand breaks caused by inhibitors of DNA synthesis: 1-beta-D-arabinofuranosylcytosine and aphidicolin.*Cancer Res.* **42**: 4050-4053.

Gamard, C.J.L., Dbaibo, G.S., Liu, B., Obeid, L.M. and Hannun, Y.A., 1997, Selective involvement of ceramide in cytokine-induced apoptosis. Ceramide inhibits phorbol ester activation of nuclear factor kappaB. *J. Biol Chem.*, **272**: 16474-16481.

Gunji, H., Kharbanda, S., and Kufe, D., 1991, Induction of internucleosomal DNA fragmentation in human myeloid leukemia cells by 1-beta-D-arabinofuranosylcytosine. *Cancer Res.* **51**: 741-743.

Haimovitz-Friedman, A., Kan, C. C., Ehleiter, D., Persaud, R. S., McLoughlin, M., Fuks, Z., and Kolesnick, R. N., 1994, Ionizing radiation acts on cellular membranes to generate ceramide and initiate apoptosis. *J. Exp. Med.* **180**: 525-535.

Hampton, R. Y., Morand, O. H., 1989, Sphingomyelin synthase and PKC activation. *Science.* **246**: 1050.

Hannun, Y. A., and Obeid, L. M., 1995, Ceramide: an intracellular signal for apoptosis. *Trends Biochem. Sci.* **20**: 73-77.

Hannun, Y. A., 1997, Apoptosis and the dilemma of cancer chemotherapy. *Blood* **89**: 1845-1853.

Hickman, J. A., 1992, Apoptosis induced by anticancer drugs. *Cancer and Metastasis Rev.* **11**: 121-139.

Higuchi, M., and Aggarwal, B. B., 1994, Differential roles of two types of the TNF receptor in TNF-induced cytotoxicity, DNA fragmentation, and differentiation. *J. Immunol.* **152**: 4017-4022.

Higuchi, M., Singh, S., Chan, H., and Aggarwal, B. B., 1995, Protease inhibitors differentially regulate tumor necrosis factor-induced apoptosis, nuclear factor -κB activation, cytotoxicity, and differentiation. *Blood* **86**: 2248-2256.

Huang, M. M., Bolen, J. B., Barnwell, J. W., Shattil, S. J., and Brugge, J. S., 1991, Membrane glycoprotein IV (CD36) is physically associated with the Fyn, Lyn, and Yes protein-tyrosine kinases in human platelets. *Proc. Natl. Acad. Sci. USA* **88**: 7844-7848.

Jaffrézou, J. P., Levade, T., Bettaïeb, A., Andrieu, N., Bezombes, C., Maestre, N., Vermeersch, S., Rousse, A., and Laurent, G., 1996, Daunorubicin-induced apoptosis: triggering of ceramide generation through sphingomyelin hydrolysis. *EMBO J.* **15**: 2417-2424.

Jaffrézou, J. P., Bettaïeb, A., Levade, T., and Laurent, G., 1998, Antitumor agent-Induced apoptosis in myeloid leukemia cells: a controlled suicide. *Leukemia Lymphoma* **29**: 453-463

Jaffrézou, J. P., Bruno, A. P., Moisand, A., Levade, T., and Laurent, G., 2001, Activation of a nuclear sphingomyelinase in radiation induced apoptosis. *FASEB J.* **15**: 123-133.

Jarvis, W. D., Fornari, F. A., Browning, J. L., Gewirtz, D. A., Kolesnick, R. N., and Grant, S. G., 1994, Attenuation of ceramide-induced apoptosis by diglyceride in human myeloid leukemia cells. *J. Biol. Chem.* **269**: 31685-31692.

Jarvis, W. D., Fornani, F. A., Traylor, R. S., Martin, H. A., Kramer, L. B., Erukulla, R. K., Bittman, R., and Grant, S., 1996, Induction of apoptosis and potentiation of ceramide-mediated cytotoxicity by sphingoid bases in human myeloid leukemia cells. *J. Biol. Chem.* **271**: 8275-8284.

Kleuser, B., Cuvillier, O., and Spiegel, S., 1998, 1α,25-dihydroxyvitamin D3 inhibits programmed cell death in HL-60 cells by activation of sphingosine kinase. *Cancer Res.* **58**: 1817-1824.

Kumar, S., and Harvey, N. L., 1995, Role of multiple cellular proteases in the execution of programmed cell death. *FEBS Lett.* **375**: 169-173.

Lahti, J. M., Xiang, J., Heath, L. S., Campana, D., and Kidd, V.J., 1996, PITSLRE protein kinase activity is associated with apoptosis. *Mol. Cell. Biol.* **15**: 1-11.

Laurent, G., and Jaffrézou, J. P., 2001, Signaling pathways activated by daunorubicin. *Blood*, **15**: in press.

Lavie, Y., Cao, H-t., Bursten, S.L., Giulano, A.E. and Cabot, M.C., 1996, Accumulation of glucosylceramide in multidrug-resistant cancer cells. *J. Biol. Chem.*, **271**: 19530-19536.

Lavie, Y., Cao, H-t., Volner, A., Lucci, A., Han, T. Y., Geffen, V., Giuliano, A. E., and Cabot, M. C., 1997, Agents that reverse multidrug resistance, tamoxifen, verapamil, and cyclosporin A, block glycosphingolipid metabolism by inhibiting ceramide glycosylation in human cancer cells. *J. Biol. Chem.* **272**: 1682-1687.

Levade, T., and Jaffrézou, J. P., 1999, Signaling sphingomyelinase: which, where, how and why. *Biochim. Biophys. Acta,* **1438**: 1-7.

Linardic, C. M., and Hannun, Y. A., 1994, Identification of a distinct pool of sphingomyelin involved in the sphingomyelin cycle. *J. Biol. Chem.* **269**: 23530-23537.

Liu, P., and Anderson, R. G. W., 1995, Compartmentalized production of ceramide at the cell surface. *J. Biol. Chem.* **270**: 27179-27185.

Liu, J., Oh, P., Horner, T., Rogers, R. A., and Schnitzer, J. E., 1997, Organized endothelial cell surface signal transduction in caveolae distinct fromglycosylphosphatidylinositol-anchored protein microdomains. *J. Biol. Chem.* **272**: 7211-7222.

Liu, Y. Y., Han, T. Y., Giulano, A. E., and Cabot, M. C., 1999, Expression of glucosylceramide synthase, converting ceramide to glucosylceramide, confers adriamycin resistance in human breast cancer cells. *J. Biol. Chem.* **274**: 1140-1146.

Liu, Y-Y., Han, T-Y., Giulano, A. E., Hansen, N., and Cabot, M. C., 2000, Uncoupling ceramide glycosylation by transfection of glucosylceramide synthase antisens reverses adriamycin resistance. *J. Biol. Chem.* **275**: 7138-7143.

Luberto, C., and Hannun, Y. A., 1998, Sphingomyelin synthase, a potential regulator of intracellular levels of ceramide and diacylglycerol during SV40 transformation. Does sphingomyelin synthase account for the putative phosphatidylcholine-specific phospholipase C? *J. Biol. Chem.* **273**: 14550-14559.

Lucci, A., Cho, W.I., Han, T.Y., Giulano, A.E. and Cabot, M.C., 1998, Glucosylceramide: a marker for multdrug-resistant cancers. *Anticancer Res.*, **18**: 475-480.

Lucci, A., Han, T. Y., Liu, Y. Y., Giuliano, A. E., and Cabot, M. C., 1999a, Modification of ceramide metabolism increases cancer cell sensitivity to cytotoxins. Int. J. Oncol., **15**: 541-546.

Lucci, A., Han, T. Y., Liu, Y. Y., Giuliano, A. E., and Cabot, M. C., 1999b, Multidrug resistance modulators and doxorubicin synergize to elevate ceramide levels and elicit apoptosis in drug-resistant cancer cells. *Cancer* **86**: 299-310.

Machleidt, T., Wiegmann, K., Henkel, T., Schütze, S., Baeuerle, P., and Krönke, M., 1994, Sphingomyelinase activates proteolytic IκB-α degradation in a cell-free system. *J. Biol. Chem.* **269**: 13760-13765.

Mansat, V., Laurent, G. Levade, T., Bettaïeb, A., and Jaffrézou, J.P., 1997a, The protein kinase C inhibitors phorbol ester and phosphatidylserine block neutral sphingomyelinase activation, ceramide generation and apoptosis triggered by daunorubicin. *Cancer Res.* **57**: 5300-5304.

Mansat, V., Bettaïeb, A., Levade, T., Laurent, G., and Jaffrézou, J.P., 1997b, Serine-protease inhibitors block neutral sphingomyelinase activation, ceramide generation and apoptosis triggered by daunorubicin. *FASEB J.* **11**: 695-702.

Martin, S. J., and Green, D. R., 1994, Apoptosis as a goal of cancer therapy. *Curr. Opin. Oncol.* **6**: 616-621.

Martin, S. J., Amarante-Mendes, G. P., Shi, I., Chuang, T. H., Casiano, C. A., O'Brien, G. A., Fitzgerald, P., Tan, E. M., Bokoch, G. M., Greenberg, A. H., and Green, D. R., 1996, The cytotoxic cell protease granzyme B initiates apoptosis in a cell-free system by proteolytic processing and activation of thez ICE/CED-3 family protease, CPP32, via novel two-step mechanism. *EMBO J.* **15**: 2407-2416.

Munker, R., DiPersio, J., and Koeffler, H. P., 1987, Tumor necrosis factor: receptors on hematopoietic cells. *Blood* **70**: 1730-1734.

Nicholson, D.W., Ali, A., Thornberre, N.A., Vaillancourt, J.P., Ding, C.K., Gallant, M., Gareau, Y., Griffin, P.R., Labelle, M., Lazenbnik, Y.A., Munday, N.A., Raju, S.M., Smulson, M.E., Yamin, T.T., Yu, V.L. and Miller, D.K., 1995, Identification and inhibition of the ICE/CED-3 protease necessary for mammalian apoptosis. *Nature*, **376**: 37-43.

Nishio, K., Sugimoto, Y., Fujiwara, Y., Ohmori, T., Morikage, T., Takeda, Y., and Saijo, N., 1992, Phospholipase C-mediated hydrolysis of phosphotidylcholine is activated by cis-diamminedichloroplatinum (II). *J. Clin. Invest.* 89: 1622-1628.

Nishizuka, Y., 1995, Protein kinase C and lipid signaling for sustained cellular responses. *FASEB J.* **9**: 484-496.

Obeid, L. M., Linardic, C. M., Karolak, L. A., and Hannun Y. A., 1993, Programmed cell death induced by ceramide. *Science.* **259**: 1769-1771.

Perez, G. I., Knudson, C. M., Leykin, L ., Korsmeyer, S. J., and Tilly, J. L., 1997, Apoptosis-associated signaling pathways are required for chemotherapy-mediated female germ cell destruction. *Nature Med.* **3**: 1228-1232.

Posada, J., Vichi, P., and Tritton, T. R., 1989, Protein kinase C in adriamycin action and resistance in mouse sarcoma 180 cells. *Cancer Res.* **49**: 6634-6639.

Pronk, G. J., Ramer, K., Amiri, P., and Williams, L. T., 1996, Requirement of an ICE-like protease for induction of apoptosis and ceramide generation by REAPER. *Science* **271**: 808-810.

Quillet-Mary, A., Mansat, V., Duchayne, E., Côme, M. G., Allouche, M., Bailly, J. D., Bordier, C., and Laurent, G., 1996, Daunorubicin-induced internucleosomal DNA fragmentation in acute myeloid cell lines. *Leukemia* **10**: 417-425.

Quintans, J., Kilkus, J., McShan, C. L., Gottschalk, A. R., and Dawson, G., 1994, Ceramide mediates the apoptotic response of WEHI 231 cells to anti-immunoglobulin, corticosteroids and irradiation. *Biochem. Biophys. Res. Commun.* **202**: 710-714.

Rani, C. S., Abe, A., Chang, Y., Rosenzweig, N., Saltiel, A. R., Radin, N. S., and Shayman, J. A., 1995, Cell cycle arrest induced by an inhibitor of glucosylceramide synthase. Correlation with cyclin-dependent kinases. *J. Biol. Chem.* **270**: 2859-2867.

Rubin, E., Kharbanda, S., Gunji, H., Weichselbaum, R., and Kufe, D., 1992, Cis-diamminedichloroplatinum (II) induces c-jun expression in human myeloid leukemia cells: potential involvement of a PKC-dependent signaling pathway. *Cancer Res.* **52**: 878-882.

Santana, P., Peña, L. A., Haimovitz-Friedman, A., Martin, S., Green, D. R., McLoughlin, M., Cordon-Cardo, C., Schuchman, E. H., Fuks, Z., and Kolesnick, R., 1996, Acid sphingomyelinase-deficient human lymphoblasts and mice are defective in radiation-induced apoptosis. *Cell* **86**: 189-199.

Ségui, B., Bezombes, C., Uro-Costes, E., Medin, J. A., Andrieu-Abadie, N., Auge, N., Brouchet, A., Laurent, G., Salvayre, R., Jaffrezou, J. P., and Levade, T., 2000, Stress-induced apoptosis is not mediated by endolysosomal ceramide. *FASEB J.* **14**: 36-47.

Shayman, J. A., Deshmukh, G. D., Mahdiyoun, S., Thomas, T. P., Wu, D., Barcelon, F. S., and Radin, N. S., 1991, Modulation of renal epithelial cell growth by glucosylceramide. Association with protein kinase C, sphingosine, and diacylglycerol. *J. Biol. Chem.* **266**: 22968-22974.

Simons, K., and Ikonen, E., 1997, Functional rafts in cell membranes.*Nature* **387**: 569-572.

Spangler, M., Coetzee, M. L., Katyal, S. L., Morris, H. P., and Ove, P., 1975, Some biochemical characteristics of rat liver and morris hepatoma nuclei and nuclear membranes. *Cancer Res.* **35**: 3131-3145.

Strum, J. C., Small, G. W., Pauig, S. B., and Daniel, L. W., 1994, 1-β-D-arabinofuranosyl-cytosine stimulates ceramide and diglyceride formation in HL-60 cells. *J. Biol. Chem.* **269**: 15493-15497.

Smyth, M. J., Perry, D. K., Zhang, J., Poirier, G. G., Hannun, Y. A., and Obeid, L. M., 1996, prICE: a downstream target for ceramide-induced apoptosis and for the inhibitory action of Bcl-2. Bio*chem. J.* **316**: 25-28.

Tamiya-Koizumi, K., Umekawa, H., Yoshida, S., and Kojima, K., 1989, Existence of Mg^{2+}-dependent, neutral sphingomyelinase in nuclei of rat ascites hepatoma cells. *J. Biochem.* **106**: 593-598.

Tepper, A. D., Ruurs, P., Wiedmer, T., Sims, P. J., Borst, J., and van Blitterswijk, W. J., 2000, Sphingomyelin hydrolysis to ceramide during the execution phase of apoptosis results from phospholipid scrambling and alters cell-surface morphology. *J. Cell Biol.* **150**: 155-164.

van Helvoort, A., Smith, A., Sprong, H., Fritzsche, I., Schinkel, A. H., Borst, P., van Meer, G., 1996, MDR1 P-glycoprotein is a lipid translocase of broad specificity, while MDR3 P-glycoprotein specifically translocates phosphatidylcholine. *Cell* **87**: 507-517.

Twentyman, P. R., and Bleehen, N. M., 1991, Resistance modification by PSC-833, a novel nonimmunosuppressive cyclosporin. *Eur. J. Cancer* **27**: 1639-1642.

van Helvoort, A., Giudici, M. L., Thielemans, M., and van Meer, G., 1997, Transport of sphingomyelin to the cell surface is inhibited by brefeldin A and in mitosis, where C6-NBD-sphingomyelin is translocated across the plasma membrane by a multidrug transporter activity. *J. Cell Sci.* **110**: 75-83.

Veldman, R. J., Maestre, N., Aduib, O. M., Medin, J. A., Salvayre, R., and Levade, T., 2001, A neutral sphingomyelinase resides in sphingolipid-enriched microdomains and is inhibited by the caveolin-scaffolding domain: potential implications in tumor necrosis factor signaling. *Biochem. J.* **355**: 859-868.

Wesselborg, S., Engels, I.H., Rossmann, E., Los, M. and Schulze-Osthoff, K., 1999, Anticancer drugs induce caspase-8/FLICE activation and apoptosis in the absence of CD95 receptor/ligand interaction. *Blood*, **93**: 3053-3063.

Wiegmann, K., Schütze, S., Machleidt, T., Witte, D., and Krönke, M., 1994, Functional dichotomy of neutral and acidic sphingomyelinases in tumor necrosis factor signaling. *Cell* **78**: 1005-1008.

Wright, S. C., Zheng, H., and Zhong, J., 1996, Tumor cell resistance to apoptosis due to a defect in the activation of sphingomyelinase and the 34 kDa apoptotic protease (AP24). *FASEB J.* **326**: 325-332.

Yuan, Z. M., S. Kharbanda, and D. Kufe., 1995, 1-beta-D-arabinofuranosylcytosine activates tyrosine phosphorylation of p34cdc2 and its association with the src-like p56/p53lyn kinase in human myeloid leukemia cells. *Biochemistry.* **34**: 1058-1063.

Zhang, J., Alter, N., Reed, J. C., Borner, C., Obeid, L. M., and Hannun Y. A., 1996, Bcl-2 interrupts the ceramide-mediated pathway odf cell death. *Proc. Natl. Acad. Sci. USA* **93**: 5325-5328.

Chapter 15

Lipid Signaling in CD95-mediated Apoptosis

Alessandra Rufini and Roberto Testi
Department of Experimental Medicine and Biochemical Science, University of Rome "Tor Vergata", Italy

1. SIGNALING VIA THE DEATH RECEPTORS

1.1 A family of death receptors

CD95 (or Fas/Apo-1) is a glycosylated cell surface receptor of approximately 42 to 52 kDa, expressed on a wide variety of tissues and cell lines. It is an integral membrane type I glycoprotein that can also be produced in a soluble form when the transmembrane portion is spliced out by an alternative splicing. It belongs to the superfamily of the Tumor Necrosis Factor Receptor (TNF-R), whose members are characterized by two to five copies of a cysteine-rich domain in their extracellular portion. Among them, receptors such as CD95, TNF-R1, Death Receptor 3 (DR3 or wsl-1/APO-3/LARD/TRAMP), DR4 (or TRAIL-R1) and DR5 (or TRAIL-R2/APO-2/TRICK/KILLER) form the Death Receptors subfamily (Ashkenazi and Dixit, 1998). All the members of this subfamily possess an intracellular domain of approximately 80 aminoacids, called Death Domain (DD), which confers on them the ability to transduce an apoptotic signal inside the cells. The integrity of this domain is strictly required for the induction of apoptosis. Death receptors can trigger apoptosis when crosslinked by their natural ligands or by specific agonistic antibodies (Peter and Krammer, 1998). However, members of the death receptor family can mediate also a number of different cellular processes, such as proliferation and differentiation.

Subcellular Biochemistry, Volume 36: Phospholipid Metabolism in Apoptosis.
Edited by Quinn and Kagan. Kluwer Academic/Plenum Publishers, New York, 2002

Most of the ligands for the death receptors have been identified and they constitute a corresponding family of proteins, the death ligands family, the prototype being the Tumor Necrosis Factor (TNF), which most likely co-evolved with the death receptors family. The death ligands are usually synthesized as type II transmembrane proteins that can be subsequently cleaved by specific proteases to allow the release of the soluble form of the proteins. Specific metalloproteases have been identified for the CD95 ligand (CD95L) and TNF. They share a beta-sheet receptor binding structure arranged in a beta-jelly roll topology. It seems that all members of the TNF family act in the form of trimers, and usually homotrimers, to induce the clustering of the receptors monomers on the cell surface (Nagata, 1999). It was however recently shown that CD95 and TNF-R1 exist in a pre-assembled form (Chan *et al.*, 2000; Siegel *et al.*, 2000; Papoff *et al.*, 1999), an interaction mediated by the "pre-ligand assembly domain" (PLAD) and necessary for the induction of the death signal. It was also proposed that CD95 can be induced to oligomerize in a ligand-independent fashion when cells are exposed to UV radiation (Rehemtulla *et al.*, 1997).

The gene coding for human CD95 is localized on chromosome 10q23, while the mouse homologue was mapped to chromosome 19. Both the human and mouse CD95L genes were mapped to chromosome 1. Analysis of the effects of spontaneously occurring or artificially induced genetic alterations in these genes allowed the understanding of their physiological role. Loss of CD95 function results mainly in abnormal expansion of lymphoid organs, indicating a predominant role of this receptor in the regulation of lymphocytes homeostasis (Krammer, 2000). Several mutations in the mouse genes coding for CD95 and CD95L have been identified and were shown to cause similar complex disorders involving mostly the immune system, such as lymphoadenopathy and autoimmune diseases. The *lpr* (lymphoproliferation) mutation is responsible for a recessive disease similar to systemic lupus erythematosus (SLE). The defect can be either caused by a retroviral early transposable element insertion into the second intron of CD95 gene, causing a premature termination of the transcription and an extremely reduced level of expression of the protein, or by a single point mutation substituting an isoleucine to an asparagine within the intracellular death domain of CD95 and impairing the transduction of the death signal. On the contrary, the *gld* (generalised lymphoproliferative disease) mutation affects the gene coding for CD95L, introducing a point mutation that leads to its inability to interact properly with the receptor. In both cases the phenotype is characterized by lymphoadenopathy, abnormal accumulation of T cells and autoimmune disorders (Nagata and Suda, 1995). A similar disorder, called autoimmune lymphoproliferative syndrome (ALPS) (Lenardo *et al.*, 1999), has been identified in humans, in which mutations in the genes coding for CD95 or for some of the molecules involved in the downstream transduction of the death signal, such as caspase-10 (Wang *et al.*, 1999), have been described.

1.2 The CD95 death-inducing signaling complex

Signaling through the death receptors may result in many different outcomes, from cell death to cell proliferation or differentiation and is differently regulated depending on the cell type and the stage of development. Different regions of the receptors may be required for the different functions. Several death receptors share common elements in their signal transduction pathways, however some relevant differences account for the specificity and pleiotropy of different biological outcomes (Wallach *et al.*, 1999).

Most studies have focused on the generation of the death signal, which was widely shown to be dependent on the integrity of the intracellular death domain. The structure of the death domain of CD95 has recently been determined by NMR analysis and consists of six amphipatic alpha helices arranged in an antiparallel orientation (Huang *et al.*, 1996). Signal generation through the death domain requires multimerization of the receptor molecules, which is probably induced by the trimeric conformation of the ligand. This would also induce a conformational change within the intracellular portion of the receptor, which is required to trigger the apoptotic signal.

Differently from the well-studied growth factors receptors, death receptors do not possess an intrinsic enzymatic activity, but the cross-linking of the receptor results in the recruitment of adaptor proteins that mediate the downstream signaling. Multiple screening approaches, including yeast two hybrid system and co-immunoprecipitation of receptor-associated molecules followed by two dimensional SDS-PAGE and nanoelectrospray tandem mass spectrometry, led to the identification of a number of proteins interacting with the intracellular portion of CD95. Many of these molecules interact with each other and with the receptor itself through homotypic interactions mediated by modular domains.

The molecules recruited to the receptor form a complex of proteins often referred to as the DISC (death-inducing signaling complex) (Medema *et al.*, 1997). In order to prevent spontaneous and undesired activation of the apoptotic cascade, proteins involved in the propagation of the signal are usually maintained in an inactive conformation and must undergo conformational changes, which require the recruitment to the DISC, in order to become active. FADD is an adaptor protein that binds to the receptor death domain through its own death domain (Berglund *et al.*, 2000;

Chinnaiyan *et al.*, 1995) and is in turn responsible for the recruitment of cysteine protease caspase-8 (or FLICE/MACH/Mch5) (Boldin *et al.*, 1996; Muzio *et al.*, 1996) via its death effector domain (DED). DEDs are homotypic protein-protein interaction domains, with a similar overall folding compared to the death domains, yet with different interacting surfaces, likely involving hydrophobic residues (Eberstadt *et al.*, 1998). Caspase-8 contains two repeated DEDs in its N-terminal prodomain which interact with the DED of FADD. This event leads to the activation of caspase-8 and triggers all the subsequent biochemical cascade. Activation of the caspase cascade is a crucial event in CD95-induced apoptosis since preventing caspase-8 activation with specific inhibitors completely blocks cell death. Moreover, cells derived from mice deficient in either caspase-8 (Varfolomeev *et al.*, 1998) or FADD (Zhang *et al.*, 1998) are unable to respond to CD95-induced apoptosis.

Another protein recruited to the DISC cloned by several groups is c-FLIP (or CASH/Casper/Mrit/I-Flice/FLAME/CLARP/usurpin) (Tschopp *et al.*, 1998). Initially identified by homology with the viral protein v-FLIP, which is able to prevent CD95-induced apoptosis, c-FLIP seems to have multiple roles in the DISC (Yeh *et al.*, 2000). It shares homology with casapse-8 and caspase-10 sequences, with two repeated N-terminal DEDs and a caspase domain. It lacks however some crucial residues involved in the catalysis and it is thought to have little proteolytic activity. However its recruitment to the DISC can lead to the displacement of caspase-8 or prevent the rapid turnover of caspase-8 molecules at the DISC and therefore reduce caspase activation (Scaffidi *et al.*, 1999). Recently, other roles have been proposed for FLIP that are apparently mediated by the DEDs, which involve the activation of JNK and NF-kB (Hu *et al.*, 2000). Similar roles have been also proposed for the DEDs contained in caspase-8 (Chaudhary *et al.*, 2000).

Recently Imai *et al* (1999) cloned a novel protein, FLASH, that contains a region with structural similarity to Apaf-1 and a duplicated domain similar to the DEDs of caspase-8. This region, designated DRD (DED-recruiting domain), would allow its interaction with caspase-8 and FADD and its recruitment to the DISC, where it is thought to participate in caspase-8 activation.

1.3 The caspase cascade

Caspases are a family of evolutionary conserved cysteine proteases that cleave their substrates after an aspartic acid residue within a defined consensus sequence (Alnemri *et al.*, 1996). They play a crucial role in the apoptotic program, as most of the characteristic features observed in apoptosis are attributable to their action (Earnshaw *et al.*, 1999). They are homologous to the nematode ced-3 protein, whose targeted deletion impairs the apoptotic cell death observed during development (Meier *et al.*, 2000). So far 14 members of the caspase family have been identified in mammals and more are likely to be discovered in the future. They can be classified into subgroups based on their sequence similarity, substrate specificity or functional role. Following the latter criterion, caspases can be divided into "initiator" and "executor" caspases, with caspase-8, -10, -9 and -2 considered as initiators, while caspase-3, -6, -7 included in the executor subgroup (Stennicke and Salvesen, 2000).

Caspases are synthesized as inactive zymogens and their activation proceeds through subsequent proteolytic events mediated by other caspases (Budihardjo *et al.*, 1999). They are constituted by a prodomain followed by a large subunit and a small subunit. The activation leads to the release of the prodomain and the formation of a tetrameric structure consisting of two heterodimers formed by a large and a small subunit. The active site QACR/QG is strictly conserved and is located in the large subunit, although residues from both subunits are involved in the catalysis, so that two catalytic centers are present in each tetramer (Grutter, 2000).

Initiator caspases are characterized by a long prodomain in their N-terminal portion, which contains an oligomerization motif that confers them the ability to self-activate. Different sets of experiments clearly demonstrated that forced oligomerization, or even dimerization, of caspase-8 molecules is sufficient to trigger its autoprocessing and activation, with the release of the small subunit and then of the prodomain and the formation of the active tetramer. This is probably due to the mild proteolytic activity contained in the zymogen form of the initiator caspases and led to the definition of the so-called "induced proximity model" (Salvesen and Dixit, 1999). *In vivo,* the oligomerization of caspase-8 molecules is mediated by the DEDs and occurs upon its interaction with FADD and its recruitment to the DISC (Medema *et al.*, 1997). Caspase-10 is thought to follow a similar mechanism of activation. Caspase-9 is characterized by a different domain in its prodomain called CARD (for caspase-activation recruitment domain) and is activated when cytochrome c and ATP, which serve as cofactors for its activation, are released from damaged mitochondria. The CARD of caspase-9 interacts, via a similar domain, with an adaptor molecule called Apaf-1, a multidomain protein homologous to the nematode ced-4. Apaf-1 itself oligomerizes, undergoes conformational changes, promoted by cytochrome c and ATP, and forms a sort of platform where caspase-9 is recruited and is activated through self-processing (Hu *et al.*, 1999). The CARD domain displays a folding similar to the death domain and the death effector domain, but the homotypic interaction is here mediated by opposed charged residues (Qin *et al.*, 1999).

Once the activator caspases are activated they can in turn cleave and activate other caspases, the executioner caspases, which are responsible for all the morphological and biochemical features that distinguish the apoptotic form of cell death. The number of caspase substrates identified is growing exponentially, with new ones identified every month (Nicholson, 1999). They include structural proteins like fodrin, lamin, gelsolin, actin, plectin, whose cleavage by caspases accounts for changes in the cell shape observed in apoptosis and the nuclear condensation. Cleavage by caspases, however, not only leads to a loss of function of the substrate protein but can indeed result in the activation of some signaling molecules, like PAK2 (p21-activated kinase), which once activated participate in the execution of the apoptotic program, likely through the activation of JNK. Some other signaling proteins are also cleaved by caspases, such as the sterol regulatory element binding proteins SREBP-1 and SREBP-2, protein kinase C δ, the PISTLRE kinase, MEKK1 and the cytosolic phospholipaseA_2 (cPLA2). Furthermore proteins involved in DNA metabolism and repair are often substrate for caspases, like PARP, DNA-PK and the oncogenes MDM2 and Rb. Importantly, the caspase-activated DNase (CAD) was identified as the nuclease responsible for DNA degradation into internucleosomal fragments, a hallmark of apoptotic cell death (Nagata, 2000), and was found to be negatively regulated by an inhibitor, ICAD, which is cleaved by caspase-3 during apoptosis, leading to activation of CAD (Enari, 1998).

Another important event is the cleavage of the Bcl-2 family member BID, mediated by caspase-8 (Li *et al.*, 1998; Luo *et al.*, 1998). The truncated C-terminal fragment generated by caspase cleavage, tBID, translocates to mitochondria and initiates mitochondrial changes characteristic of apoptosis, such as loss of the transmembrane potential, permeability transition and release of apoptogenic factors such as cytochrome c and AIF, which will then contribute to the final phase of the apoptotic program, being responsible for post-mitochondrial caspase activation and DNA fragmentation, respectively.

1.4 Additional signaling pathways

Other molecules are known to bind the intracellular portion of CD95 and regulate both positively or negatively the cell death signal. FAP-1 is a protein tyrosine phosphatase, which is highly expressed in CD95-resistant tumors and that was shown to bind to the C-terminal part of CD95. This stretch of 15 aminoacids downstream of the death domain is known to have an inhibitory regulatory role, and such inhibition is mediated by FAP-1 (Sato *et al.*, 1995).

DAXX is an adaptor that is recruited to the CD95 death domain upon receptor clustering, via its N-teminal portion. DAXX in turn binds and activates the ASK1 kinase, an initiator of the stress-activated protein kinase (SAPK)/Jun N-terminal kinase (JNK)/p38 cascades, and positively cooperates in the propagation of the apoptotic signal (Yang *et al.*, 1997). Intriguingly, DAXX was also found to interact with the PML oncogene in nuclear bodies, where it could contribute to transcriptional activation of pro-apopotic genes (Zhong *et al.*, 2000).

CD95 clustering has been shown to trigger the localized hydrolysis of membrane phospholipids and the generation of diffusible lipid molecules, which act as signal mediators. Lipid mediators generated by the action of specific phospolipases are in fact crucial components of the apoptotic machinery. Unfortunately, the difficulties encountered in studying lipids, the lack of availability of utilitarian techniques and the enormous boost that protein science received with the advent of proteomics technologies made lipid signaling a field of study afforded by few laboratories. Lipid molecules of the sphingolipid and glycosphingolipid class are considered evolutionarily-conserved mediators of the stress response in all eukaryotes, and evidence of their role in signaling can be traced to yeast (Hannun and Luberto, 2000). Ceramide was shown to be a mediator of the apoptotic response in different cell types (Obeid *et al.*, 1993) and Cifone *et al* (1995, 1994) demonstrated that an acidic sphingomyelinase (ASM) is rapidly activated upon CD95 cross-linking in lymphoid cells, leading to the accumulation of ceramide within intracellular acidic compartments. CD95-induced ASM activation requires diacylglycerol (DAG) released by the action of phosphatidylcholine-specific phospholipase C (PC-PLC), since blocking PC-PLC prevented CD95-induced ASM activation. This early PC-PLC/ASM pathway is strictly dependent on the integrity of the death domain and the activation of early caspases. (Brenner *et al.*, 1998; De Maria *et al.*, 1997).

The accumulating evidence for a role of ceramides in apoptosis signaling has generated interest in cellular sphingomyelinases, as emerging key components of the lipid pathway to programmed cell death.

2. A LIPID PATHWAY TO PROGRAMMED CELL DEATH

2.1 More than one sphingomyelinase

Sphingomyelinases represent a widely distributed and heterogeneous family of type C phospholipase. They catalyse the hydrolysis of sphingomyelin to yield ceramide and phosphorylcholine. Sphingomyelin is a phospholipid located mainly in the outer leaflet of the plasma membrane. Its breakdown is considered to be the main source of ceramide that contributes to cell signaling. Sphingomyelin hydrolysis can be induced by a number of stimuli including stress-inducing stimuli, such as heat shock, UV and ionizing radiation, oxidative stress, some chemotherapeutics agents, like daunorubicin and vincristine, and cytokines like TNFα and FasL (Levade and Jaffrezou, 1999).

Different sphingomyelinases can be identified. They can be classified depending on their pH optimum and cofactors requirement (Samet and Barenholz, 1999). The acidic sphingomyelinase (ASM) displays a pH optimum around 4.5. It is localized in acidic compartments of the cells, such as endosomes, lysosomes and caveolae, and is associated with the outer leaflet of the membranes. It is synthesized as a 75 kDa precursor protein, which is then converted by subsequent processing events in the 52 kDa mature form of the enzyme (Bartelsen *et al.*, 1998). It is a highly glycosylated protein (Ferlinz *et al.*, 1997) and displays a sphingolipid-activator protein (SAP)-like domain, characteristic of several lipid binding proteins (Ferlinz *et al.*, 1999). Additionally, a secreted form of Zn^{2+}-dependent acidic sphingomyelinase was identified in fetal bovine serum and was also shown to be secreted by human endothelial cells upon IL-1β or IFNγ stimulation (Marathe *et al.*, 1998). The neutral sphingomyelinases (NSM) require a neutral pH for optimal activity. They can occur as membrane-bound, Mg^{2+}-dependent forms or cytosolic Mg^{2+}-independent forms. A putative Mg^{2+}-dependent NSM, was recently cloned (Tomiuk *et al.*, 2000) and consists of an approximately 47 kDa polypeptide with two predicted transmembrane domains.

The relative contribution of individual sphingomyelinases to the propagation the apoptotic program has been recently clarified. The identification of CD95-resistant clones of the T cell lymphoma cell line HuT78 provided the first clue (Cifone *et al.*, 1995). A HuT78 clone expressing a differently spliced form of CD95, lacking exon 8 and devoid of death domain (Cascino *et al.*, 1996), was selected as unable to undergo CD95-induced apoptosis. It retained the ability to activate NSM, the MAP kinase ERK2 and PLA2 in response to CD95 cross-linking but could not induce ASM activation, pointing towards a prominent role of ASM in the propagation of the cell death signal. Furthermore, cells from Niemann-Pick disease (NPD) patients provided a unique model to assess the role of ASM in CD95-induced apoptosis. This autosomal recessive disorder is characterized by a genetic defect in the gene coding for ASM resulting in a deficiency in functional ASM, sphingomyelin accumulation with consequent spleen and liver hyperplasia and the presence of characteristic storage cells in the bone marrow. The execution of the apoptotic program is strongly delayed in lymphoblastoid cell lines derived from NPD patients (De Maria *et al.*, 1998). These cells are unable to induce early ASM activation in response to CD95 cross-linking, although other components of the apoptotic machinery are present and activation of NSM could still be induced. Adoptive replacement of the missing ASM allowed NPD cells to fully recover an efficient CD95-induced apoptotic program.

ASM knockout mice display a phenotype similar to the Niemann-Pick disorder, showing defects in the nervous system development and death around the sixth month (Otterbach and Stoffel, 1995). They provided genetic support for the crucial role of this enzyme in CD95-induced apoptosis both *in vivo* in peripheral lymphocytes and hepatocytes (Kirschnek *et al.*, 2000) and in cells derived from the knockout mice (Lozano *et al.*, 2001). In these cells, indeed, signals requiring ceramide for progression, like those generated by TNFα, CD95L, radiation and serum withdrawal, resulted impaired and could be restored by exogenous addition of ceramide.

There are other reports indicating however a role for the NSM-derived ceramide in the death pathway. A cytosolic NSM is infact activated late in the CD95-induced apoptotic program (Tepper *et al.*, 1995). A unifying model has been proposed in which ceramide generated by the hydrolysis of sphingomyelin mediated by ASM in the early phases of CD95 response can induce mitochondrial damage and subsequent activation of the cytosolic NSM, which in turn leads to a late and prolonged wave of ceramide accumulation in the following hours (Jaffrezou *et al.*, 1998).

An additional source for apoptogenic ceramide is its neosynthesis. This occurs through the condensation of serine and palmitoyl-CoA to yield ketosphinganine, subsequently reduced to dihydrosphingosine. The enzyme ceramide synthase then mediates the acylation of dihydrosphingosine to give dihydroceramide, which is then converted to ceramide by dihydroceramide reductase. These reactions are thought to occur in the cytosolic face of the endoplasmic reticulum (Michel and van Echten-Deckert, 1997). Bose *et al.* (1995) showed that daunorubicin-induced ceramide elevation was mediated by ceramide synthase and that treatment with fumonisin B1, a fungal toxin that blocks the enzyme responsible for the synthesis of ceramide prevented ceramide accumulation and apoptosis in response to drug treatment.

However it was later shown that fumonisin B1 treatment does not impair receptor- or drug-induced apoptosis (Jaffrezou *et al.*, 1996), strenghtening the notion that sphingomyelin breakdown plays the major part in the generation of the biologically relevant ceramide.

2.2 Ceramide biology, an issue of compartimentalization

Cellular compartimentalization of diffusible signal mediators generation is crucial in determining different functional outcomes.

Ceramide, or N-acyl-sphingosine, is a known mediator of many cellular processes, such as cell proliferation, differentiation, cell cycle arrest, cell senescence and apoptosis (Kolesnick and Kronke, 1998; Perry and Hannun, 1998). The pleiotropy of effects that has been attributed to ceramide suggests that regulated and specific subcellular compartmentalisation couples ceramide to a number of different downstream effectors (Kolesnick *et al.*, 2000; Testi, 1996). The identification of targets that can mediate ceramide action has supported its prominent role in signal transduction. Ceramide is known to activate serine/threonine phosphatases of the PP1 and PP2A families. Ceramide-activated serine/threonine phosphatase (CAPP) of the PP2A family was recently reported to mediate the dephosphorylation and consequent inactivation of the PKCα (Lee *et al.*, 1996) and of Bcl-2 (Ruvolo *et al.*, 1999). Moreover, ceramide is known to activate a membrane-bound, proline-directed serine/threonine kinase, the ceramide-activated protein kinase (CAPK), which is thought to phosphorylate Raf-1. Raf-1 in turn phosphorylates and activates MEK, leading to the activation of the MAP kinase cascade. CAPK activity was recently shown to be in fact the previously identified kinase suppressor of ras (KSR) (Zhang *et al.*, 1997). PKCζ is another target kinase for ceramide, which might be involved in NF-kB activation. Ceramide was also reported to interfere with the cell cycle regulation through the dephosphorylation of the oncogene Rb, mediated by a PP1 phosphatase (Kishikawa *et al.*, 1999), the down regulation of c-myc and the activation of the transcription factor c-Jun (Reyes *et al.*, 1996).

Importantly, ceramide can serve as a precursor for the biosynthesis of more complex sphingolipids and glycosphingolipids in the Golgi apparatus, through the conversion to glucosylceramide and the stepwise addition of sugar units. In an attempt to identify downstream effectors of the apoptotic signal mediated by ceramide, changes in glycosphingolipid and ganglioside metabolism and expression pattern were investigated. De Maria *et al.* (1997) found that ceramide is rapidly and transiently converted into the ganglioside GD3 in lymphoid cells undergoing apoptosis after CD95 cross-linking. This event is required for the progression of the apoptotic program. Ceramides generated by different sphingomyelinases are biochemically indistinguishable, however it was clearly shown (De Maria *et al.*, 1998) that only ceramide generated in the early phase of the apoptotic program by the hydrolysis of sphingomyelin by ASM can be converted into GD3, since cells from Niemann-Pick disease patients fail to accumulate GD3 in response to CD95 triggering. Ceramide generated by the membrane-associated NSM cannot be recruited for ganglioside biosynthesis that leads to the early accumulation of GD3. This is most likely due to a different localization inside the cells that makes it unavailable as a substrate for the enzymes responsible for its conversion into gangliosides.

3. GANGLIOSIDES AS INTRACELLULAR MEDIATORS OF APOPTOSIS

3.1 Structural heterogeneity of gangliosides

Gangliosides are a class of glycosphingolipid characterized by the presence of one or more sialic acid residues. They are located in the cell membranes, particularly in the plasma membrane, of almost all cell types. They are anchored in the membranes through their ceramide moiety and have the polar sugar-containing group in the extracytoplasmic side. They are particularly abundant in the nervous system where their expression pattern is modulated during development.

The first step along the biosynthetic pathway of gangliosides is the conversion of ceramide into glucosylceramide, which is mediated by a glucosyltransferase located in early compartments of the proximal Golgi. Once glucosylceramide is generated, it is flipped into the lumen of these compartments where it is used for the biosynthesis of more complex glycolipids and gangliosides along the Golgi compartments and the trans Golgi network (TGN). The addition of a galactose unit, mediated by a galactosyltransferase, converts it into lactosylceramide. The action of different sialyltransferases then converts lactosylceramide into the gangliosides GM3, GD3 and GT3, through stepwise addition of sialic acid residues to the galactose unit of lactosylceramide. GM3, GD3 and GT3 can in turn generate the a-, b-, and c-series of gangliosides, respectively, through the sequential addition of N-acetylgalactosamine, galactose and more sialic acid residues to the galactose residue. The o-series of gangliosides is generated directly from lactosylceramide (Huwiler *et al.*, 2000) (Figure 1).

Most of the glycosyltransferases involved in the biosynthesis of gangliosides are integral transmembrane proteins and catalyse the transfer of a monosaccharide group from a sugar nucleotide donor to the sphingolipid acceptor, which is also inserted into membranes. Ceramide galactosyltransferase is a type I membrane protein located in the endoplasmic reticulum (ER), with its catalytic site facing the lumen of the ER (Sprong *et al.*, 1998). On the contrary, enzymes that catalyse downstream glycosylation and addition of sialic acid residues are generally type II transmembrane protein with a domain organization consisting of a short cytoplasmic N-terminal tail, a transmembrane domain containing the localization signal, and a luminal part with a stem region and a large catalytic domain at the C-terminus (Haraguchi *et al.*, 1994).

Glycolipid glycosyltransferase activity can possibly be modulated by post-translational modification as many putative phosphorylation, N-glycosylation and N-myristoilation consensus sites have been identified in their primary sequences. Recent data indicate that N-glycosylation of sialyltransferases is critical for enzyme stability, catalytic activity and proper sorting from the ER to the Golgi complex (Maccioni *et al.*, 1999). Site directed removal of N-glycosylation sites in the N-acetylgalactosaminyltransferase generates a catalytically inactive enzyme that retains its proper Golgi location (Haraguchi *et al.*, 1995). Moreover, inhibition of GD3 synthase N-glycosylation, results in the generation of an unstable and inactive form of the enzyme that is retained in the ER (Martina *et al.*, 1998). Although direct *in vivo* evidence for the phosphorylation of glycosyltransferases is still lacking, there are data reporting a role for protein kinase C and A in the regulation of many sialyltransferases (Bieberich *et al.*, 1998).

3.2 GD3 is a killer ganglioside

Ceramide is a mediator of cell death signaling in lymphoid cells and is rapidly accumulated upon CD95 cross-linking by the action of ASM. De Maria and colleagues (1997) found that ceramide requires its conversion into the ganglioside GD3 for the progression of the apoptotic signal. GD3 accumulation is rapid and transient and peaks 15 minutes after CD95 stimulation in hemopoietic cells. Similarly GD3 accumulation could be induced by treating the cells with the cell-permeable ceramide analog C2-ceramide. GD3 was shown to be a potent mediator of cell death, since its exogenous addition was able to induce mitochondrial damage, with consequent dissipation of transmembrane potential, DNA fragmentation and apoptosis. None of these effects could be induced by other gangliosides, like GD1a or GM1. Interestingly, CD95-resistant cell clones, expressing a death domain-defective variant of CD95, failed to accumulate GD3 in response to CD95 stimulation, but could still be induced to undergo apoptosis upon exposure to GD3. Furthermore, CD95-induced GD3 accumulation was shown to be dependent on upstream activation of caspases, since it could be prevented by treatment with the tetrapeptide caspase inhibitors.

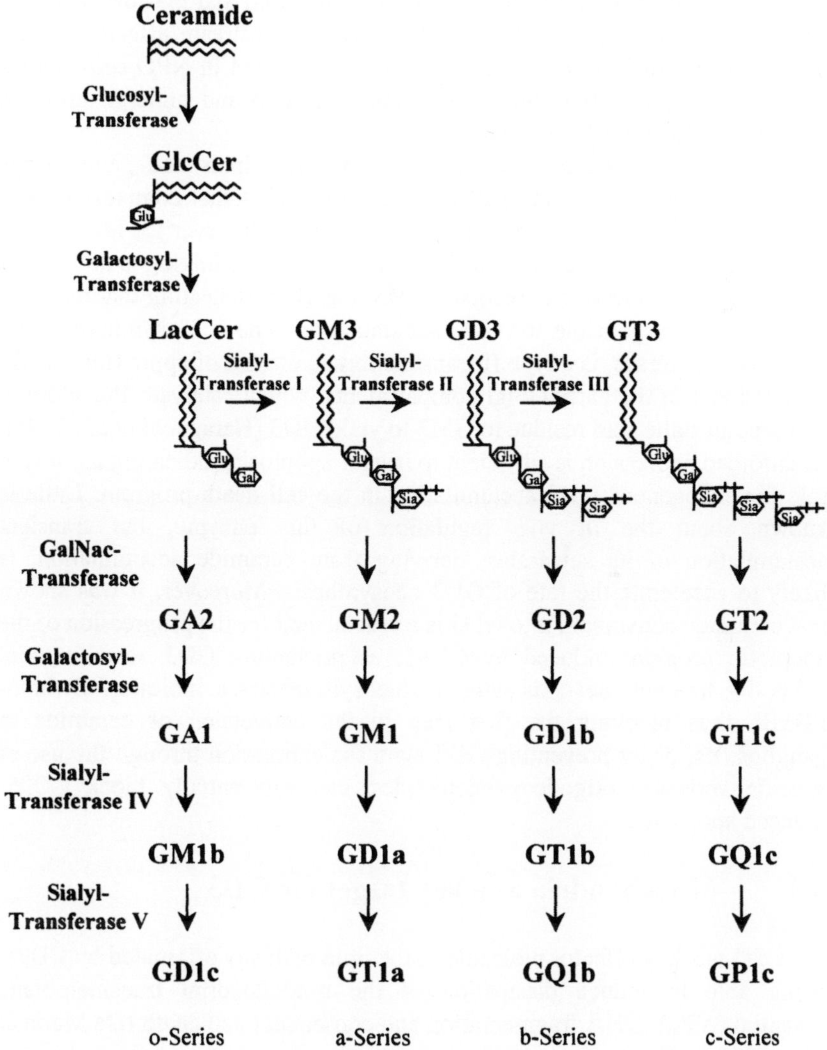

Figure 1. Ganglioside biosynthesis

It was also demonstrated that ceramides derived from sphingomyelin breakdown mediated by ASM are used for ganglioside biosynthesis, up to the generation of GD3. Indeed, lymphoblastoid cells derived from Niemann-

Pick disease patients, which show a substantial delay in the execution of the apoptotic program induced by CD95, fail to accumulate GD3 in response to CD95 (De Maria *et al.*, 1998). However they retain sensitivity to ceramide, which is still capable of inducing GD3 accumulation, suggesting that signals downstream of ceramide generation are not impaired and that the defect in ASM activity uncouples CD95 signaling with downstream effectors of the lipid pathway. In fact, adoptive reconstitution of ASM in NPD cells allows the cells to recover the ability to accumulate GD3 and undergo efficient apoptosis following CD95 clustering.

In addition to the *de novo* synthesis of glycolipids and gangliosides described above, a recycling pathway from the endosomal compartments to the Golgi has been described (Gillard *et al.*, 1998). However De Maria *et al.* (1997) showed that GD3 accumulated after CD95 cross-linking is derived by *de novo* synthesis from its precursor GM3, therefore indicating that the GD3 synthase was responsible for GD3 accumulation. The GD3 synthase, or α-2,8-sialyltransferase, is a type II transmembrane protein of approximately 40 kDa resident in the early Golgi compartments, which catalyses the addition of a second sialic acid residue to GM3 to yield GD3 (Haraguchi *et al.*, 1994). Its enforced expression is sufficient to trigger apoptosis, indicating an *in vivo* role for endogenous GD3 accumulation in the cell death program. Little is known about the *in vivo* regulation of this enzyme, but transient accumulation of its substrates, deriving from ceramide accumulation, is likely to accelerate the rate of GD3 neosynthesis. Moreover, it was shown that ceramide conversion into GD3 is a crucial step for the progression of the apoptotic program induced by CD95, as preventing GD3 accumulation, either by treating the cells with a glucosyltransferase inhibitor, D-threo-PDMP, thus blocking the first step in the conversion of ceramide to gangliosides, or by preventing GD3 synthase expression through the use of specific antisense oligodeoxynucleotides, can substantially block CD95-induced apoptosis.

3.3 Mitochondria as a key target for GD3

GD3 is a key effector molecule in the lipid pathway originated by CD95, being able to induce dissipation of the mitochondrial transmembrane potential ($\Delta\Psi_m$), DNA fragmentation and consequent cell death (De Maria *et al.*, 1997). However its direct molecular targets remain unknown. Mitochondria constitute key elements in the propagation and amplification of the apoptotic signal (Kroemer and Reed, 2000). Dissipation of mitochondrial transmembrane potential is a critical step in the apoptotic program. It is thought to occur through the opening of the permeability transition pore complex (PTPC), located in the contact sites between the inner and outer mitochondrial membranes. The PTPC is composed by a complex of proteins including the voltage-dependent anion channel (VDAC), the adenine nucleotide transporter (ANT) and cyclophilin-D (Crompton *et al.*, 1998). Its opening allows the efflux of molecules with a molecular mass below 1500 Da and leads to the dissipation of the proton gradient, with consequent inactivation of the oxidative phosphorylation and efflux of calcium from mitochondria into the cytoplasm (Kristal and Brown, 1999a). These events are usually associated with mitochondria swelling and rupture of the outer membrane, with consequent release of apoptogenic factors, such as cytochrome c, ATP, AIF (Susin *et al.*, 1999) and mitochondrial caspases (Crompton, 1999). Once released in the cytosol, cytochrome c and ATP serve as cofactors for the Apaf1-mediated activation of caspase-9 (Saleh *et al.*, 1999), while AIF is able to drive nuclear changes characteristic of apoptosis, such as chromatin condensation and large fragment DNA fragmentation (Ferri and Kroemer, 2000).

Recent reports (Garcia-Ruiz *et al.*, 2000; Inoki *et al.*, 2000; Rippo *et al.*, 2000; Kristal and Brown, 1999a; Scorrano *et al.*, 1999) indicate mitochondria as immediate downstream targets of GD3. Rippo *et al.* (2000) recently showed that GD3 directly interacts with mitochondria and is capable of inducing mitochondrial loss of $\Delta\Psi_m$, with consequent release of cytochrome c, AIF and caspase-9. GD3 is able to induce swelling of isolated rat liver mitochondria within minutes, while other gangliosides, like GD1a or GM3, have no effect. Interestingly, ceramide is not able to directly induce mitochondrial swelling, further indicating the importance of its conversion into GD3 for signal progression. GD3-induced loss of $\Delta\Psi_m$ is inhibited by cyclosporin A (Rippo *et al.*, 2000), ADP, trifluoperazine and Mg^{2+} (Kristal and Brown, 1999b), all known inhibitors of the PTPC opening, indicating that GD3 is actually acting at the pore level. However, whether this occurs through interaction with an as yet unidentified target in the PTPC or through perturbation of mitochondrial membranes, which might affect physical and chemical parameters critical for the PTPC function, remains to be established.

It is noteworthy that, unlike the previously described inducers of PTPC opening, GD3 is able to exert its effect on mitochondria even in the presence of submicromolar calcium concentration (Kristal and Brown, 1999a), a property shared only by the most potent PT inducers, such as the thiol cross-linking agent phenylarsine oxide and by the peptide mastoparan. This low calcium requirement is important because intracellular calcium concentration during CD95-mediated apoptosis was not described to rise above 0.4 μM (Oshimi and Miyazaki, 1995).

Importantly, apoptogenic factors released from mitochondria after GD3 treatment are able to induce DNA fragmentation in isolated nuclei in a cell-free system. On the contrary, GD3 itself is unable to drive nuclear changes and DNA fragmentation, arguing against a generic role for GD3 in perturbing cellular membranes. Furthermore, endogenously accumulated GD3, upon ceramide exposure, relocates to mitochondria and induces mitochondrial changes *in vivo.* As described above, GD3 is generated in the early compartments of the Golgi complex (EG) and is thought to reach mitochondria through physical continuity between ER/EG (Rizzuto *et al.*, 1998; Rusinol *et al.*, 1994) and mitochondrial membranes. However an active transport mechanism cannot be excluded.

The opening of the PTPC was widely shown to be under the control of the Bcl-2 family members. Indeed the proapoptotic proteins Bax and Bak interact with ANT (Marzo *et al.*, 1998) and VDAC (Griffiths *et al.*, 1999) cooperating to the opening of the PTPC. Bax was also shown to translocate to mitochondria during apoptosis and to form BH3-mediated oligomeric aggregates, which might directly affect mitochondrial permeability transition (Crompton, 2000). Moreover the truncated form of Bid, tBid, produced by a caspase 8-mediated cleavage, is found associated with mitochondria in apoptotic cells, where contributes to the release of cytochrome c (Gross *et al.*, 1999). On the contrary, anti-apoptotic proteins, like Bcl-2 and Bcl-XL, are known to exert a protective effect on mitochondrial function (Shimizu *et al.*, 1999). In line with this evidence, the role of Bcl-2 in GD3-induced mitochondrial damage was investigated by Rippo and colleagues (2000). Lymphoid cells stably overexpressing Bcl-2 showed a substantial resistance to GD3-induced apoptosis and showed a marked delay in GD3-induced mitochondrial changes (for a comprehensive model of GD3 action within CD95 pathway see Figure 2)

4. CONCLUDING REMARKS

The role of GD3 as a key effector molecule of the apoptotic signal in the CD95 pathway has been originally elucidated in lymphoid cells. However, accumulating data point towards a more general role for GD3 as a mediator of cell death. Ceramide accumulates rapidly in brain tissue after ischemic injury (Herr *et al.*, 1999) and contributes to neuronal loss, which occurs after reperfusion. A crucial role for ASM in the generation of apoptogenic ceramide following brain ischemia has been recently shown (Yu *et al.*, 2000). It is likely that, similarly to hemopoietic cells, ASM-derived ceramide requires conversion into GD3 to exert its apoptogenic effects, since preventing GD3 accumulation significantly suppresses cell death in both *in vitro* and *in vivo* ischemia/reperfusion models (Martin-Villalba *et al.*, submitted). Moreover, the sensitivity to GD3 displayed by neurons and oligodendrocytes (unpublished), together with the evidence for GD3 accumulation in pathological tissues derived from Alzheimer's, multiple sclerosis and Creutzfeldt-Jacob's patients, strongly suggests a role for apoptogenic GD3 in neurodegeneration.

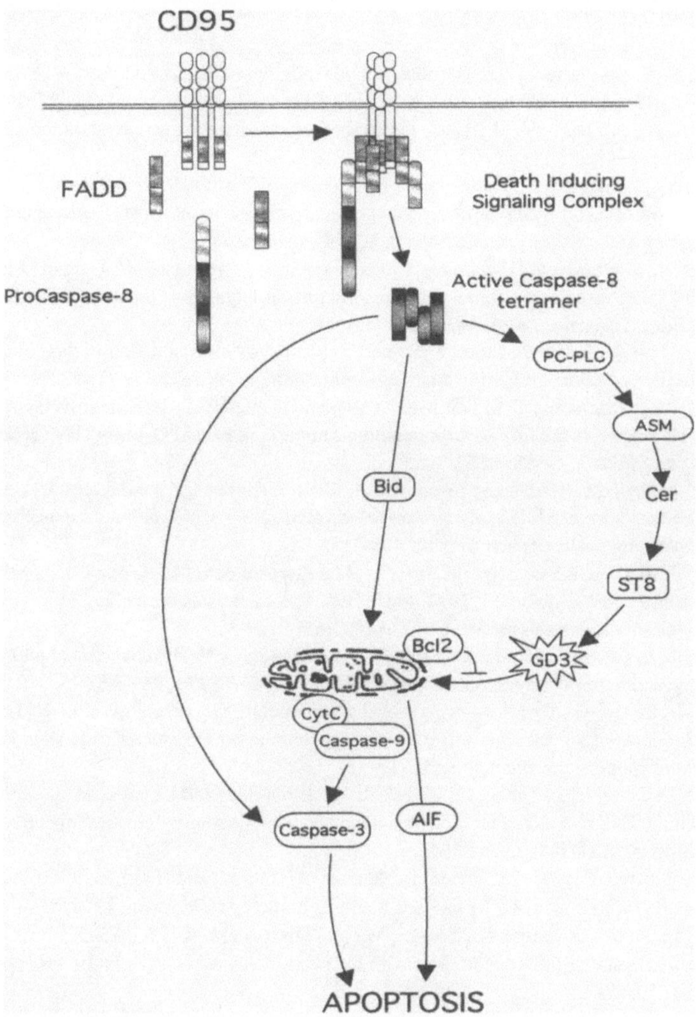

Figure 2. CD95 signaling pathways

ACKNOWLEDGMENTS

Our work is supported by Associazione Italiana Ricerca sul Cancro and Agenzia Spaziale Italiana.

REFERENCES

Alnemri, E. S., Livingston, D. J., Nicholson, D. W., Salvesen, G., Thornberry, N. A., Wong, W. W., and Yuan, J., 1996, Human ICE/CED-3 protease nomenclature. *Cell* **87**: 171.

Ashkenazi, A., and Dixit, V. M., 1998, Death Receptors: Signaling and Modulation. *Science* **281**: 1305–1308.

Bartelsen, O., Lansmann, S., Nettersheim, M., Lemm, T., Ferlinz, K., and Sandhoff, K., 1998, Expression of recombinant human acid sphingomyelinase in insect Sf21 cells: purification, processing and enzymatic characterization. *J Biotechnol* **63**: 29-40.

Berglund, H., Olerenshaw, D., Sankar, A., Federwisch, M., McDonald, N. Q., and Driscoll, P. C., 2000, The three-dimensional solution structure and dynamic properties of the human FADD death domain. *J Mol Biol* **302**: 171-188.

Bieberich, E., Freischutz, B., Liour, S. S., and Yu, R. K., 1998, Regulation of ganglioside metabolism by phosphorylation and dephosphorylation. *J Neurochem* **71**: 972-979.

Boldin, M. P., Goncharov, T. M., Goltsev, Y. V., and Wallach, D., 1996, Involvement of MACH, a novel MORT1/FADD-interacting protease, in Fas/APO-1- and TNF receptor induced cell death. *Cell* **85**: 803-815.

Bose, R., Verheij, M., Haimovitz-Friedman, A., Scotto, K., Fuks, Z., and Kolesnick, R., 1995, Ceramide synthase mediates daunorubicin-induced apoptosis: an alternative mechanism for generating death signals. *Cell* **82**: 405-414.

Brenner, B., Ferlinz, K., Grassme, H., Weller, M., Koppenhoefer, U., Dichgans, J., Sandhoff, K., Lang, F., and Gulbins, E., 1998, Fas/CD95/Apo-I activates the acidic sphingomyelinase via caspases. *Cell Death Differ* **5**: 29-37.

Budihardjo, I., Oliver, H., Lutter, M., Luo, X., and Wang, X., 1999, Biochemical pathways of caspase activation during apoptosis. *Annu Rev Cell Dev Biol* **15**: 269-290.

Cascino, I., Papoff, G., De Maria, R., Testi, R., and Ruberti, G., 1996, Fas/APO-1/CD95 receptor lacking the intracytoplasmic signaling domain protects tumor cells from Fas-mediated apoptosis. *J. Immunol.* **156**: 13-17.

Chan, F. K., Chun, H. J., Zheng, L., Siegel, R. M., Bui, K. L., and Lenardo, M. J., 2000, A domain in TNF receptors that mediates ligand-independent receptor assembly and signaling. *Science* **288**: 2351-2354.

Chaudhary, P. M., Eby, M. T., Jasmin, A., Kumar, A., Liu, L., and Hood, L., 2000, Activation of the NF-kappaB pathway by caspase 8 and its homologs. *Oncogene* **19**: 4451-4460.

Chinnaiyan, A. M., O'Rourke, K., Tewari, M., and Dixit, V. M., 1995, FADD, a novel death domain-containing protein, interacts with the death domain of Fas and initiates apoptosis. *Cell* **81**: 505-512.

Cifone, M. G., De Maria, R., Roncaioli, P., Rippo, M. R., Azuma, M., Lanier, L. L., Santoni, A., and Testi, R., 1994, Apoptotic signaling through CD95 (Fas/APO-1) activates an acidic sphingomyelinase. *J. Exp. Med.* **180**: 1547-1552.

Cifone, M. G., Roncaioli, P., De Maria, R., Camarda, G., Santoni, A., Ruberti, G., and Testi, R., 1995, Multiple signaling originate at the Fas/Apo-1 (CD95) receptor: sequential involvement of phosphatidylcholine-specific phospholipase C and acidic sphingomyelinase in the propagation of the apoptotic signal. *EMBO J.* **14**: 5859-5868.

Crompton, M., 1999, The mitochondrial permeability transition pore and its role in cell death. *Biochem J* **341**: 233-249.

Crompton, M., 2000, Bax, Bid and the permeabilization of the mitochondrial outer membrane in apoptosis. *Curr Opin Cell Biol* **12**: 414-419.

Crompton, M., Virji, S., and Ward, J. M., 1998, Cyclophilin-D binds strongly to complexes of the voltage-dependent anion channel and the adenine nucleotide translocase to form the permeability transition pore. *Eur J Biochem* **258**: 729-735.

De Maria, R., Lenti, L., Malisan, F., d'Agostino, F., Tomassini, B., Zeuner, A., Rippo, M. R., and Testi, R., 1997, Requirement for GD3 ganglioside in CD95- and ceramide-induced apoptosis. *Science* **277**: 1652-1655.

De Maria, R., Rippo, M. R., Schuchman, E. H., and Testi, R., 1998, Acidic sphingomyelinase (ASM) is necessary for Fas-induced GD3 ganglioside accumulation and efficient apoptosis of lymphoid cells. *J Exp Med* **187**: 897-902.

Earnshaw, W. C., Martins, L. M., and Kaufmann, S. H., 1999, Mammalian caspases: structure, activation, substrates, and functions during apoptosis. *Annu Rev Biochem* **68**: 383-424.

Eberstadt, M., Huang, B., Chen, Z., Meadows, R. P., Ng, S. C., Zheng, L., Lenardo, M. J., and Fesik, S. W., 1998, NMR structure and mutagenesis of the FADD (Mort1) death-effector domain. *Nature* **392**: 941-945.

Enari, M., Sakahira, H., Yokoyama, H., Okawa, K., Iwamatsu, A., Nagata, S., 1998, A caspase-activated DNase that degrades DNA during apoptosis, and its inhibitor ICAD. *Nature* **391**: 43-50.

Ferlinz, K., Hurwitz, R., Moczall, H., Lansmann, S., Schuchman, E. H., and Sandhoff, K., 1997, Functional characterization of the N-glycosylation sites of human acid sphingomyelinase by site-directed mutagenesis. *Eur J Biochem* **243**: 511-517.

Ferlinz, K., Linke, T., Bartelsen, O., Weiler, M., and Sandhoff, K., 1999, Stimulation of lysosomal sphingomyelin degradation by sphingolipid activator proteins. *Chem Phys Lipids* **102**: 35-43.

Ferri, K. F., and Kroemer, G., 2000, Control of apoptotic DNA degradation. *Nat Cell Biol* **2**: E63-64.

Garcia-Ruiz, C., Colell, A., Paris, R., and Fernandez-Checa, J. C., 2000, Direct interaction of GD3 ganglioside with mitochondria generates reactive oxygen species followed by mitochondrial permeability transition, cytochrome c release, and caspase activation. *Faseb J* **14**: 847-858.

Gillard, B. K., Clement, R. G., and Marcus, D. M., 1998, Variations among cell lines in the synthesis of sphingolipids in de novo and recycling pathways. *Glycobiology* **8**: 885-890.

Griffiths, G. J., Dubrez, L., Morgan, C. P., Jones, N. A., Whitehouse, J., Corfe, B. M., Dive, C., and Hickman, J. A., 1999, Cell damage-induced conformational changes of the pro-apoptotic protein Bak *in vivo* precede the onset of apoptosis. *J Cell Biol* **144**: 903-914.

Gross, A., Yin, X. M., Wang, K., Wei, M. C., Jockel, J., Milliman, C., Erdjument-Bromage, H., Tempst, P., and Korsmeyer, S. J., 1999, Caspase cleaved BID targets mitochondria and is required for cytochrome c release, while BCL-XL prevents this release but not tumor necrosis factor-R1/Fas death. *J Biol Chem* **274**: 1156-1163.

Grutter, M. G., 2000, Caspases: key players in programmed cell death. *Curr Opin Struct Biol* **10**: 649-655.

Hannun, Y. A., and Luberto, C., 2000, Ceramide in the eukaryotic stress response. *Trends Cell Biol.* **10**: 73-80.

Haraguchi, M., Yamashiro, S., Yamamoto, A., Furukawa, K., Takamiya, K., Lloyd, K. O., Shiku, H., and Furukawa, K., 1994, Isolation of GD3 synthase gene by expression cloning of GM3 α-2,8-sialyltransferase cDNA using anti-GD2 monoclonal antibody. *Proc. Natl. Acad. Sci. USA* **91**: 10455-10459.

Haraguchi, M., Yamashiro, S., Furukawa, K., Takamiya, K., and Shiku, H., 1995, The effects of the site-directed removal of N-glycosylation sites from beta-1,4-N-acetylgalactosaminyltransferase on its function. *Biochem J* **312**: 273-280.

Herr, I., Martin-Villalba, A., Kurz, E., Roncaioli, P., Schenkel, J., Cifone, M. G., and Debatin, K. M., 1999, FK506 prevents stroke-induced generation of ceramide and apoptosis signaling. *Brain Res* **826**: 210-219.

Hu, W. H., Johnson, H., and Shu, H. B., 2000, Activation of NF-kappaB by FADD, Casper, and caspase-8. *J Biol Chem* **275**: 10838-10844.

Hu, Y., Benedict, M. A., Ding, L., and Nunez, G., 1999, Role of cytochrome c and dATP/ATP hydrolysis in Apaf-1-mediated caspase-9 activation and apoptosis. *Embo J* **18**: 3586-3595.

Huang, B., Eberstadt, M., Olejniczak, E. T., Meadows, R. P., and Fesik, S. W., 1996, NMR structure and mutagenesis of the Fas (APO-1/CD95) death domain. *Nature* **384**: 638-641.

Huwiler, A., Kolterb, T., Pfeilschiftera, J., and Sandhoffb, K., 2000, Physiology and pathophysiology of sphingolipid metabolism and signaling. *Biochim Biophys Acta* **1485**: 63-99.

Imai, Y., Kimura, T., Murakami, A., Yajima, N., Sakamaki, K., and Yonehara, S., 1999, The CED-4-homologous protein FLASH is involved in Fas-mediated activation of caspase-8 during apoptosis. *Nature* **398**: 777-785.

Inoki, Y., Miura, T., Kajimoto, T., Kawase, M., Kawase, Y., Yoshida, Y., Tsuji, S., Kinouchi, T., Endo, H., Kagawa, Y., and Hamamoto, T., 2000, Ganglioside GD3 and its mimetics induce cytochrome c release from mitochondria. *Biochem Biophys Res Commun* **276**: 1210-1216.

Jaffrezou, J. P., Levade, T., Bettaieb, A., Andrieu, N., Bezombes, C., Maestre, N., Vermeersch, S., Rousse, A., and Laurent, G., 1996, Daunorubicin-induced apoptosis: triggering of ceramide generation through sphingomyelin hydrolysis. *Embo J* **15**: 2417-2424.

Jaffrezou, J. P., Maestre, N., de Mas-Mansat, V., Bezombes, C., Levade, T., and Laurent, G., 1998, Positive feedback control of neutral sphingomyelinase activity by ceramide. *Faseb J* **12**: 999-1006.

Kirschnek, S., Paris, F., Weller, M., Grassme, H., Ferlinz, K., Riehle, A., Fuks, Z., Kolesnick, R., and Gulbins, E., 2000, CD95-mediated apoptosis *in vivo* involves acid sphingomyelinase. *J Biol Chem* **275**: 27316-27323.

Kishikawa, K., Chalfant, C. E., Perry, D. K., Bielawska, A., and Hannun, Y. A., 1999, Phosphatidic acid is a potent and selective inhibitor of protein phosphatase 1 and an inhibitor of ceramide-mediated responses. *J Biol Chem* **274**: 21335-21341.

Kolesnick, R., and Kronke, M., 1998, Regulation of ceramide production and apoptosis. *Annu. Rev. Physiol.* **60**: 643-665.

Kolesnick, R. N., Goni, F. M., and Alonso, A., 2000, Compartmentalization of ceramide signaling: physical foundations and biological effects. *J Cell Physiol* **184**: 285-300.

Krammer, P. H., 2000, CD95's deadly mission in the immune system. *Nature* **407**: 789-795.

Kristal, B. S., and Brown, A. M., 1999a, Apoptogenic ganglioside GD3 directly induces the mitochondrial permeabilitytransition. *J Biol Chem* **274**: 23169-23175.

Kristal, B. S., and Brown, A. M., 1999b, Ganglioside GD3, the mitochondrial permeability transition, and apoptosis. *Ann N Y Acad Sci* **893**: 321-324.

Kroemer, G., and Reed, J. C., 2000, Mitochondrial control of cell death. *Nat Med* **6**: 513-519.
Lee, J. Y., Hannun, Y. A., and Obeid, L. M., 1996, Ceramide inactivates cellular protein kinase Calpha. *J Biol Chem* **271**: 13169-13174.
Lenardo, M., Chan, K. M., Hornung, F., McFarland, H., Siegel, R., Wang, J., and Zheng, L., 1999, Mature T lymphocyte apoptosis--immune regulation in a dynamic and unpredictable antigenic environment. *Annu Rev Immunol* **17**: 221-253.
Levade, T., and Jaffrezou, J. P., 1999, Signaling sphingomyelinases: which, where, how and why? *Biochim Biophys Acta* **1438**: 1-17.
Li, H., Zhu, H., Xu, C. J., and Yuan, J., 1998, Cleavage of BID by caspase 8 mediates the mitochondrial damage in the Fas pathway of apoptosis. *Cell* **94**: 491-501.
Lozano, J., Menendez, S., Morales, A., Ehleiter, D., Liao, W. C., Wagman, R., Haimovitz-Friedman, A., Fuks, Z., and Kolesnick, R., 2001, Cell Autonomous Apoptosis Defects in Acid Sphingomyelinase Knockout Fibroblasts. *J Biol Chem* **276**: 442-448.
Luo, X., Budihardjo, I., Zou, H., Slaughter, C., and Wang, X., 1998, Bid, a Bcl2 interacting protein, mediates cytochrome c release from mitochondria in response to activation of cell surface death receptors. *Cell* **94**: 481-490.
Maccioni, H. J., Daniotti, J. L., and Martina, J. A., 1999, Organization of ganglioside synthesis in the Golgi apparatus. *Biochim Biophys Acta* **1437**: 101-118.
Marathe, S., Schissel, S. L., Yellin, M. J., Beatini, N., Mintzer, R., Williams, K. J., and Tabas, I., 1998, Human vascular endothelial cells are a rich and regulatable source of secretory sphingomyelinase. Implications for early atherogenesis and ceramide-mediated cell signaling. *J Biol Chem* **273**: 4081-4088.
Martina, J. A., Daniotti, J. L., and Maccioni, H. J., 1998, Influence of N-glycosylation and N-glycan trimming on the activity and intracellular traffic of GD3 synthase. *J Biol Chem* **273**: 3725-3731.
Marzo, I., Brenner, C., Zamzami, N., Susin, S. A., Beutner, G., Brdiczka, D., Remy, R., Xie, Z. H., Reed, J. C., and Kroemer, G., 1998, Bax and adenine nucleotide translocator cooperate in the mitochondrial control of apoptosis. *Science* **281**: 2027-2031.
Medema, J. P., Scaffidi, C., Kischkel, F. C., Shevchenko, A., Mann, M., Krammer, P. H., and Peter, M. E., 1997, FLICE is activated by association with the CD95 death-inducing signaling complex (DISC). *Embo J* **16**: 2794-2804.
Meier, P., Finch, A., and Evan, G., 2000, Apoptosis in development. *Nature* **407**: 796-801.
Michel, C., and van Echten-Deckert, G., 1997, Conversion of dihydroceramide to ceramide occurs at the cytosolic face of the endoplasmic reticulum. *FEBS Lett* **416**: 153-155.
Muzio, M., Chinnaiyan, A. M., Kischkel, F. C., O'Rourke, K., Shevchenko, A., Ni, J., Scaffidi, C., Bretz, J. D., Zhang, M., Gentz, R., Mann, M., Krammer, P. H., Peter, M. E., and Dixit, V. M., 1996, FLICE, a novel FADD-homologous ICE/CED-3-like protease, is recruited to the CD95 (Fas/APO-1) death-inducing signaling complex. *Cell* **85**: 817-827.
Nagata, S., 1999, Fas ligand-induced apoptosis. *Annu Rev Genet* **33**: 29-55.
Nagata, S., 2000, Apoptotic DNA fragmentation. *Exp Cell Res* **256**: 12-18.
Nagata, S., and Suda, T., 1995, Fas and Fas ligand: lpr and gld mutations. *Immunol. Today* **16**: 39-43.
Nicholson, D. W., 1999, Caspase structure, proteolytic substrates, and function during apoptotic cell death. *Cell Death Differ* **6**: 1028-1042.
Obeid, L. M., Linardic, C. M., Karolak, L. A., and Hannun, Y. A., 1993, Programmed cell death induced by ceramide. *Science* **259**: 1769-1771.
Oshimi, Y., and Miyazaki, S., 1995, Fas antigen-mediated DNA fragmentation and apoptotic morphologic changes are regulated by elevated cytosolic Ca^{2+} level. *J Immunol* 154: 599-609.

Otterbach, B., and Stoffel, W., 1995, Acid sphingomyelinase-deficient mice mimic the neurovisceral form of human lysosomal storage disease (Niemann-Pick). *Cell* **81**: 1053-1061.

Papoff, G., Hausler, P., Eramo, A., Pagano, M. G., Di Leve, G., Signore, A., and Ruberti, G., 1999, Identification and characterization of a ligand-independent oligomerization domain in the extracellular region of the CD95 death receptor. *J Biol Chem* **274**: 38241-38250.

Perry, D., and Hannun, Y., 1998, The role of ceramide in cell signaling. *Biochim. Biophys. Acta* **1436**: 233-243.

Peter, M. E., and Krammer, P. H., 1998, Mechanisms of CD95 (APO-1/Fas)-mediated apoptosis. *Curr Opin Immunol* **10**: 545-551.

Qin, H., Srinivasula, S. M., Wu, G., Fernandes-Alnemri, T., Alnemri, E. S., and Shi, Y., 1999, Structural basis of procaspase-9 recruitment by the apoptotic protease-activating factor 1. *Nature* **399**: 549-557.

Rehemtulla, A., Hamilton, C. A., Chinnaiyan, A. M., and Dixit, V. M., 1997, Ultraviolet radiation-induced apoptosis is mediated by activation of CD-95 (Fas/APO-1). *J Biol Chem* **272**: 25783-25786.

Reyes, J. G., Robayna, I. G., Delgado, P. S., Gonzalez, I. H., Aguiar, J. Q., Rosas, F. E., Fanjul, L. F., and Galarreta, C. M., 1996, c-Jun is a downstream target for ceramide-activated protein phosphatase in A431 cells. *J Biol Chem* **271**: 21375-21380.

Rippo, M. R., Malisan, F., Ravagnan, L., Tomassini, B., Condo', I., Costantini, P., Susin, S. A., Rufini, A., Todaro, M., Kroemer, G., and Testi, R., 2000, GD3 ganglioside directly targets mitochondria in a bcl-2-controlled fashion. *Faseb J* **14**: 2047-2054.

Rizzuto, R., Pinton, P., Carrington, W., Fay, F. S., Fogarty, K. E., Lifshitz, L. M., Tuft, R. A., and Pozzan, T., 1998, Close contacts with the endoplasmic reticulum as determinants of mitochondrial Ca^{2+} responses. *Science* 280: 1763-1766.

Rusinol, A. E., Cui, Z., Chen, M. H., and Vance, J. E., 1994, A unique mitochondria-associated membrane fraction from rat liver has a high capacity for lipid synthesis and contains pre-Golgi secretory proteins including nascent lipoproteins. *J Biol Chem* **269**: 27494-27502.

Ruvolo, P. P., Deng, X., Ito, T., Carr, B. K., and May, W. S., 1999, Ceramide induces Bcl2 dephosphorylation via a mechanism involving mitochondrial PP2A. *J Biol Chem* **274**: 20296-20300.

Saleh, A., Srinivasula, S. M., Acharya, S., Fishel, R., and Alnemri, E. S., 1999, Cytochrome c and dATP-mediated oligomerization of Apaf-1 is a prerequisite for procaspase-9 activation. *J Biol Chem* **274**: 17941-17945.

Salvesen, G. S., and Dixit, V. M., 1999, Caspase activation: the induced-proximity model. *Proc Natl Acad Sci U S A* **96**: 10964-10967.

Samet, D., and Barenholz, Y., 1999, Characterization of acidic and neutral sphingomyelinase activities in crude extracts of HL-60 cells. *Chem Phys Lipids* **102**: 65-77.

Sato, T., Irie, S., Kitada, S., and Reed, J. C., 1995, FAP-1: a protein tyrosine phosphatase that associate with Fas. *Science* **268**: 411-415.

Scaffidi, C., Schmitz, I., Krammer, P., and Peter, M., 1999, The role of c-FLIP in modulation of CD95-induced apoptosis. *J Biol Chem.* 274(3): 1541-1548.

Scorrano, L., Petronilli, V., Di Lisa, F., and Bernardi, P., 1999, Commitment to apoptosis by GD3 ganglioside depends on opening of the mitochondrial permeability transition pore. *J Biol Chem* **274**: 22581-22585.

Shimizu, S., Narita, M., and Tsujimoto, Y., 1999, Bcl-2 family proteins regulate the release of apoptogenic cytochrome c by the mitochondrial channel VDAC. *Nature* **399**: 483-7.

Siegel, R. M., Frederiksen, J. K., Zacharias, D. A., Chan, F. K., Johnson, M., Lynch, D., Tsien, R. Y., and Lenardo, M. J., 2000, Fas preassociation required for apoptosis signaling and dominant inhibition by pathogenic mutations. *Science* **288**: 2354-2357.

Sprong, H., Kruithof, B., Leijendekker, R., Slot, J. W., van Meer, G., and van der Sluijs, P., 1998, UDP-galactose: ceramide galactosyltransferase is a class I integral membrane protein of the endoplasmic reticulum. *J Biol Chem* **273**: 25880-25888.

Stennicke, H. R., and Salvesen, G. S., 2000, Caspases - controlling intracellular signals by protease zymogen activation. *Biochim Biophys Acta* **1477**: 299-306.

Susin, S. A., Lorenzo, H. K., Zamzami, N., Marzo, I., Snow, B. E., Brothers, G. M., Mangion, J., Jacotot, E., Costantini, P., Loeffler, M., Larochette, N., Goodlett, D. R., Aebersold, R., Siderovski, D. P., Penninger, J. M., and Kroemer, G., 1999, Molecular characterization of mitochondrial apoptosis-inducing factor. *Nature* **397**: 441-446.

Tepper, C. G., Jayadev, S., Liu, B., Bielawska, A., Wolff, R., Yonehara, S., Hannun, Y. A., and Seldin, M., 1995, Role for ceramide as an endogenous mediator of Fas-induced cytotoxicity. *Proc. Natl. Acad. Sci. USA* **92**: 8443-8447.

Testi, R., 1996, Sphingomyelin breakdown and cell fate. *Trends Biochem. Sci.* 21: 468-471.

Tomiuk, S., Zumbansen, M., and Stoffel, W., 2000, Characterization and subcellular localization of murine and human magnesium-dependent neutral sphingomyelinase. *J Biol Chem* **275**: 5710-5717.

Tschopp, J., Irmler, M., and Thome, M., 1998, Inhibition of fas death signals by FLIPs. *Curr Opin Immunol* **10**: 552-558.

Varfolomeev, E. E., Schuchmann, M., Luria, V., Chiannilkulchai, N., Beckmann, J. S., Mett, I. L., Rebrikov, D., Brodianski, V. M., Kemper, O. C., Kollet, O., Lapidot, T., Soffer, D., Sobe, T., Avraham, K. B., Goncharov, T., Holtmann, H., Lonai, P., and Wallach, D., 1998, Targeted disruption of the mouse Caspase 8 gene ablates cell death induction by the TNF receptors, Fas/Apo1, and DR3 and is lethal prenatally. *Immunity* **9**: 267-276.

Wallach, D., Varfolomeev, E. E., Malinin, N. L., Goltsev, Y. V., Kovalenko, A. V., and Boldin, M. P., 1999, Tumor necrosis factor receptor and Fas signaling mechanisms. *Annu Rev Immunol* **17**: 331-367.

Wang, J., Zheng, L., Lobito, A., Chan, F. K., Dale, J., Sneller, M., Yao, X., Puck, J. M., Straus, S. E., and Lenardo, M. J., 1999, Inherited human Caspase 10 mutations underlie defective lymphocyte and dendritic cell apoptosis in autoimmune lymphoproliferative syndrome type II. *Cell* **98**: 47-58.

Yang, X., Khosravi-Far, R., Chang, H., and Baltimore, D., 1997, Daxx, a novel Fas-binding protein that activates JNK and apoptosis. *Cell* **89**(7): 1067-1076.

Yeh, W. C., Itie, A., Elia, A. J., Ng, M., Shu, H. B., Wakeham, A., Mirtsos, C., Suzuki, N., Bonnard, M., Goeddel, D. V., and Mak, T. W., 2000, Requirement for Casper (c-FLIP) in regulation of death receptor-induced apoptosis and embryonic development. *Immunity* 12: 633-642.

Yu, Z. F., Nikolova-Karakashian, M., Zhou, D., Cheng, G., Schuchman, E. H., and Mattson, M. p., 2000, Pivotal role for acidic sphingomyelinase in cerebral ischemia-induced ceramide and cytokine production, and neuronal apoptosis. *J Mol Neuroscience* **15**: 85-97.

Zhang, J., Cado, D., Chen, A., Kabra, N., and Winoto, A., 1998, Fas-mediated apoptosis and activation-induced T-cell proliferation are defective in mice lacking FADD/Mort1. *Nature* **392**: 296–99.

Zhang, Y., Yao, B., Delikat, S., Bayoumy, S., Lin, X. H., Basu, S., McGinley, M., Chan-Hui, P. Y., Lichenstein, H., and Kolesnick, R., 1997, Kinase suppressor of Ras is ceramide-activated protein kinase. *Cell* **89**: 63-72.

Zhong, S., Salomoni, P., Ronchetti, S., Guo, A., Ruggero, D., and Pandolfi, P. P., 2000, Promyelocytic leukemia protein (PML) and Daxx participate in a novel nuclear pathway for apoptosis. *J Exp Med* **191**: 631-640.

Chapter 16

Phosphatidylinositol 3-kinase, Phosphoinositides and Apoptosis

Polyphosphoinositol phosphatase and apoptosis

Gabriella Sarmay
Lorand Eotvos University, Budapest, Hungary

1. INTRODUCTION

Phosphoinositides (PtdIns) are the target molecules of several intracellular enzymes activated by a variety of extracellular stimuli, such as growth factors, antigens and cytokines. The cells respond to these stimuli by transcribing genes encoding for proteins that control cell activation, proliferation, and differentiation or - by terminating these processes,- induce programmed cell death (Liscovitch and Cantley 1994, Divecha and Irvine, 1995). PtdIns phosphorylated at various positions act as membrane embedded second messengers mediating signals from the cell membrane to the nucleus. Phosphorylation of and phosphate removal from the inositol moiety of PtdIns performed by various kinases and phosphatases, respectively, as well as the release of the inositol group by phospholipases are crucial events controlling inositol lipid metabolism and signaling (Erneux, *et al.*, 1998, Zhang and Majerus, 1998) (Figure 1). The membrane bound metabolism of PtdIns is linked to the metabolism of soluble inositol polyphosphates by the action of two enzymes, phospholipase C and phosphatidyl inositol synthase. Phospholipase C activated by receptor- or receptor associated tyrosine kinases, hydrolyses the phosphodiester bond between the inositol polyphosphate and the glycerol backbone of the lipid in phosphatidylinositol bisphosphate (PtdIns(4,5)P_2), releasing inositol trisphosphate (IP_3) (Cockcroft and Thomas, 1992). Phosphatidylinositol synthase is responsible for the *de novo* synthesis of PtdIns, which is then phosphorylated by various kinases. Perhaps the most significant group, the phosphoinositide 3-kinases (PI3-K) phosphorylate the inositol ring at the third position, resulting PtdIns(3)*P*, PtdIns(3,4)P_2, or PtdIns(3,4,5)P_3 (Vanhaesebroeck *et al.*, 1997). The membrane bound D-3 PtdIns recruit various cytosolic signaling molecules having a pleckstrin homology (PH) domain to the inner layer of the cell membrane (Lemmon *et al.*, 1997). Thereby the products of PI3-K control the subcellular localization and the activity of these proteins involved in diverse physiological functions, such as vesicle trafficking, cell adhesion, actin rearrangement, cell growth and cell survival (Carpenter and Cantley, 1996). Three major classes of signaling molecules are regulated by D-3 PtdIns–PH domain interaction: guanine nucleotide exchange proteins for Rho family GTPases, TEC family protein tyrosine kinases, such as Btk and Itk, and serine/threonin kinases, like the PtdIns-dependent kinase 1 (PDK-1) and Akt/PKB (Cantrell, 2001). The latter kinases are implicated in signaling pathways leading to cell survival.

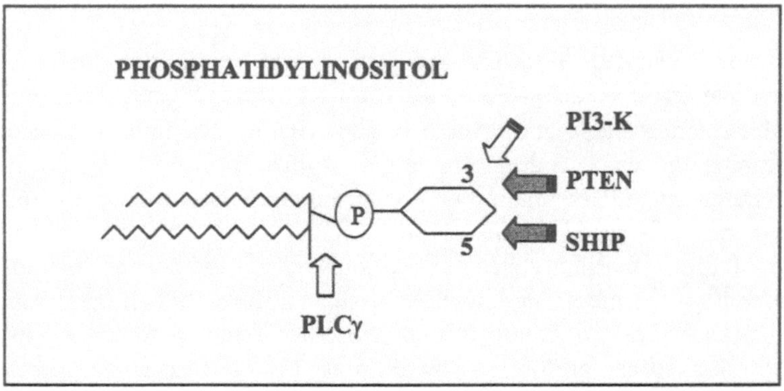

Figure 1. Phosphorylation, dephosphorylation and cleavage of PtdIns. PI3-K phosphorylates the inositol ring at D3, while PTEN dephosphorylates it at the same position. SHIP dephosphorylates 5-phosphorylated PtdIns, PLCγ hydrolyses the phosphodiester bond between the glycerol backbone and the inositol headgroup of PtdIns (3,4)P_2, releasing IP3.

Since various PH domains have different affinities for PtdIns phosphorylated at different positions, the integrated activity of PtdIns kinases and PtdIns phosphatases has a vital role in triggering the biological function (Rameh and Cantley, 1999). Akt/PKB, mediating an anti-apoptotic signal via the phosphorylation of various molecules, binds -although controversial- preferentially to PtdIns(3,4)P_2 (Franke *et al.*, 1997), while PH domain of PDK1, a kinase that regulates Akt by phosphorylating the activation loop of Akt on Thr^{308} binds to PtdIns(3,4,5)P_3 with high affinity. Btk, the tyrosine kinase responsible for the phosphorylation and full activation of PLCγ as well as the Gab1 docking/scaffolding protein bind to PtdIns(3,4,5)P_3. Phosphoinositol phosphatases controlling the level of the central player of inositol phospholipid metabolism, PtdIns(3,4,5)P_3, play a decisive role in signaling for survival or programmed cell death in response to a given stimuli. These phosphatases are the SH2 containing PtdIns 5-phosphatases SHIP1 and SHIP2, and the PtdIns 3-phosphatase, PTEN (phosphatase and tensin homology deleted on chromosome ten) (Figure 2).

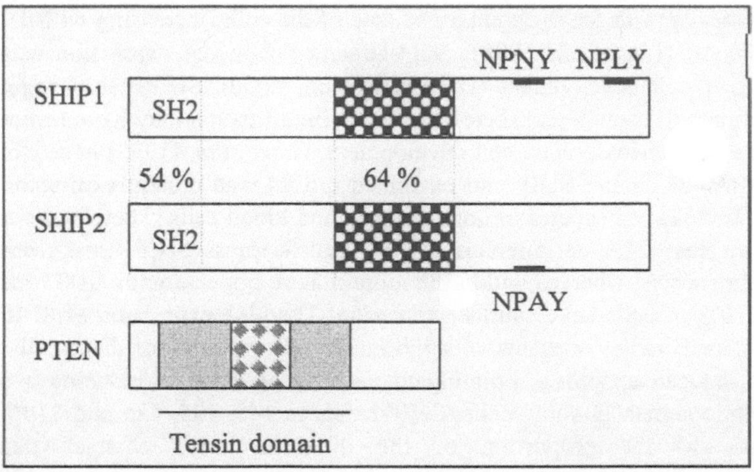

Figure 2. Structure of phosphoinositid phosphatases

2. SHIP 1

2.1 Structure and expression

SHIP1 (hereafter SHIP) was identified in the early 1990s as a 145 kDa intracellular protein that is tyrosine phosphorylated upon the stimulation of hemeatopoietic cells via B and T-cell receptor, and by multiple cytokines (Damen *et al.*, 1996, Lioubin *et al.* 1996). Molecular cloning of cDNA for SHIP revealed that the molecule consists of an N terminal SH2 domain, a highly conserved, centrally located motif having PtdIns 5-phosphatase activity, two NPXY sequences, which are phosphorylated on Tyr upon various stimuli and a proline-rich C terminus (Figure 2). SHIP hydrolyses PtdIns(3,4,5)P_3, and inositol 1,3,4,5-tetrakisphosphate both *in vitro* and *in vivo*, thus belongs to the type II 5-phosphatases, characterized also by their relative tolerance towards phosphate groups at the 3- or 4 position of the inositol. However, type II 5-phosphatases require the presence of 3-phosphate group for substrate recognition. The activity of SHIP does not change with cytokine stimulation or after cross-linking BCR *in vivo*, suggesting that it is regulated by the translocation of the molecule to the cell membrane (Damen *et al.*, 1996, Huber *et al.*, 1999). The SH2 domain as well as the phosphorylated NPXY motifs may play a role in this process.

SHIP's expression is restricted to hemeatopoietic cells and it changes dramatically with the differentiation state of the cells, increasing with T cell maturation (Liu *et al.*, 1998), and showing a bimodal expression pattern during B cell development (Helgason *et al.*, 2000). SHIP has a negative effect on cell growth and therefore loss or modification may have profound effects on hemeatopoietic cell development. Geier *et al.* (1997) have cloned a cDNA for human SHIP and examined mRNA and protein expression of SHIP and related species in bone marrow and blood cells. They have found that at least 74% of immature $CD34^+$ cells express SHIP cross-reacting protein species, whereas within the more mature population of $CD33^+$ cells, only 10% of cells have similar expression. The high expression of SHIP in immature B cells correlates with the increased sensitivity of these cells for BCR induced apoptosis, pointing to the role of SHIP in inducing a pro-apoptotic signal. In some cells, SHIP exists in 145, 135, 125 and 110 kDa forms and the proportion of the different forms changes during hemeatopoiesis (Geier *et al.*, 1997) and leukemogenesis (Odai *et al.*, 1997). Immunoblotting detected up to seven different cross-reacting SHIP species, with peripheral blood mononuclear cells exhibiting primarily a 100-kD protein and a $CD34^+$ acute myeloblastic leukemia expressing mainly 130-kD and 145-kD forms of SHIP (Geier *et al.* 1997). The lower molecular weight forms appeared to be produced by the C terminal truncation of the molecule. The truncated forms, although phosphorylated on tyrosine, did not co-precipitate with Shc, thus might carry out distinct functions (Damen *et al.*, 1998).

2.2 Binding molecules

Since the activity of SHIP does not change upon cell activation, it appears to mediate its effects by translocating to its substrate. SHIP may bind to phosphorylated proteins via its SH2 domain, and to phosphotyrosine binding (PTB) domains via its two phosphorylated motifs, in addition, its C terminal Pro rich region might modify interactions of SHIP. Molecules binding to any of the motifs mentioned above might regulate SHIP's phosphatase activity. SHIP has a negative influence on cell growth. This was first demonstrated in myeloid cells (Lioubin *et al.*, 1996) and later in B cells (Ono *et al.*, 1996,

Chacko *et al.*, 1996). Cell growth is arrested when the type II Fcγ receptor b1 (FcγRIIb1) associates with the B cell receptor (BCR) by the interaction of their extracellular domains with immune complexes. Clustering FcγRIIb1 with BCR induces the phosphorylation of the immunoreceptor tyrosine-based inhibitory motif (ITIM) within the FcγRIIb intracellular tail, which then allows SHIP to bind through its SH2 domain. Recruitment of SHIP to the cell membrane by FcγRIIb p-ITIM is believed to block extracellular Ca^{2+} uptake and cell growth via the inositol 5-phosphatase activity of SHIP.

SHIP SH2 domain binds to the phosphorylated ITIM (p-ITIM) of other inhibitory co-receptors (gp49B1) (Kuroiwa *et al.*, 1998) and to the p-ITIM-like motif of the adhesion receptor, PECAM-1 (Pumphrey *et al.*, 1999), furthermore, *in vitro* it binds to the phosphorylated immunoreceptor tyrosine based activation motifs (p-ITAM) within the β and γ subunits of the high affinity IgE receptor (Kimura *et al.* 1997)), and the zeta chain of T cell receptor (Osborne *et al.*, 1996). However, *in vivo* coprecipitation of these molecules was not observed.

SHIP binds to Shc (<u>S</u>rc <u>h</u>omology and <u>c</u>ollagen) adaptor protein and to SHP-2 protein tyrosine phosphatase via SHIP's SH2 domain (Liu *et al.*, 1997), and in some cells (B-cells), but not in others (myeloid cells) SHIP binds Grb2 via its proline rich region (Harmer and DeFranco, 1999, Liu *et al.*, 1994). The difference in the binding molecules in various cell types could be due to the activation of different PTKs resulting in diverse phosphorylation pattern of SHIP and its binding partners. The phosphorylated NPXY motifs were shown to interact with the PTB domains of Shc adaptor protein, while the tyrosine-phosphorylated motif of Shc (Tyr^{317}) binds to the SH2 domain of SHIP. It was also shown that the SH2 domain is not only required for the association of SHIP with Shc, but it is essential for tyrosine phosphorylation of SHIP and its induction of apoptosis in response to IL-3 stimuli (Liu *et al.*, 1997). In antigen receptor stimulated B cells SHIP forms a ternary complex with Shc and Grb2, and these molecules may localize SHIP to the cell membrane where its substrates reside (Harmer and De-Franco, 1999). SHIP induced loss of viability could occur through the ability of SHIP to compete with Grb2 for Shc, thus reducing ras activation (Tridandapani *et al.* 1997). Alternatively, via hydrolyzing the product of PI3-K, PtdIns(3,4,5)P_3, SHIP might reduce viability by inhibiting Akt and/or PDK binding to the cell membrane. These Ser/Thr kinases are implicated in anti-apoptotic signaling (Coggeshall, 1998).

SHIP binds to SHP-2 upon stimulation with IL-3 (Liu *et al.*, 1997). Sattler *et al.* (1997) reported that interleukin-3 (IL-3) and erythropoietin (EPO) induce the formation of a complex in BaF3 cells, composed of two SH2-containing phosphatases, the tyrosine phosphatase SHP-2 and SHIP, both are known to be involved in growth factor mediated signal transduction and controlling cell viability. Both phosphatases in the complex were shown to be tyrosine phosphorylated, and the amount of SHIP coprecipitating with SHP-2 was inversely related to the amount of SHIP coprecipitating with SHC, suggesting a competitive interaction. Phosphorylation of both phosphatases suggests that an additional molecule with SH2 domains might be involved in the complex. Transformation by Bcr/ABL was associated with a reduced SHIP protein expression, which could result in the accumulation of various inositol polyphosphates and an enhanced survival/proliferation signal in these leukemic cells. The authors suggested that SHP-2 and SHIP functions are linked in normal cells, furthermore, that increased SHIP activity leads to decreased viability (Sattler *et al.,* 1997).

Gupta *et al.* (1999) provided evidence that coligation of FcγRIIb1 with BCR induces binding of PI3-K to SHIP. This interaction was mediated by the binding of the SH2 domains of the p85 subunit of PI3K to a tyrosine-based motif in the C-terminal region of SHIP. The authors suggested that association with SHIP may serve to activate PI3K and to regulate downstream events such as B cell activation-induced apoptosis, since *in vitro* SHIP catalyzes the conversion of the PI3-K product, PtdIns(3,4,5)P_3. Thus co-localization of SHIP and PI3K may assure a very precise control of the level of various PtdIns phosphates in the cell membrane. Since recently both SHIP and PI3K were shown to bind to the tyrosine phosphorylated Grb2 associated binder adaptor/scaffolding protein, Gab1 upon erythropoietin (Lecoq-Lafon *et al.,* 1999) and BCR (Ingham *et al.* 1998, Saxton *et al.,* 1994) induced signaling, it is likely that the association of SHIP and PI3-K is indirect, an may be mediated by Grb2 and Gab1. The same might be true for SHP-2 - SHIP interaction, since SHP-2 also binds to Gab1. Thus Gab1 scaffolding protein provides a platform where all these molecules may assemble.

Mikhalap *et al.* (1999) reported that BCR activation leads to the phosphorylation of an 80 kDa transmembrane protein. This molecule was identified as CDw150, also called signaling lymphocytic activation molecule (SLAM), which is highly expressed in B, T and dendritic cells. When tyrosine phosphorylated -due to cell activation- SLAM binds SHIP via the SH2 domain of the latter. Ligation of CDw150/SLAM induces a rapid dephosphorylation of SHIP and itself as well as the association of lyn and frg kinases with SHIP. Many of the SLAM-associated molecules are involved in apoptosis, and a defect in one of them (SLAM associated protein, SAP) leads to the X-linked B cell lymphoproliferative disease (Sayos, *et al.,* 1998). Moreover, analysis of SLAM expression revealed that it is invariably co-expressed with CD95/Fas. CD95/Fas mediated apoptosis was enhanced by signaling via CDw150. Moreover, the CDw150 mediated regulation of cell-death did not correlate with the serine phosphorylation state of Akt but correlated with the dephosphorylation of SHIP. The authors suggested that dephosphorylation might inactivate SHIP, leading to the enhanced Btk and PLCγ activation, consequently, the sustained increase of intracellular [Ca^{2+}], which is required for apoptosis (Oshimi and Miyazaki, 1995). Thus CDw150 may function to regulate the survival of activated B cells via SHIP.

Upon FcγRIIb–BCR co-engagement tyrosine phosphorylated SHIP also interacts with the p62dok via the PTB domain of the latter and translocates to the cell membrane (Tamir *et al.*, 2000). In the vicinity of BCR, p62dok becomes hyperphosphorylated by the BCR activated tyrosine kinases, and in turn associates with the ras regulator protein, rasGAP, potentiating GTP hydrolysis and blocking ras activity. Consequently, a downstream effector from ras, ERK activity is inhibited (Yamanashi *et al.*, 2000). Thus SHIP, translocating p62dok to the cell membrane, inhibits ras activity suggesting one possible explanation for the previously observed decrease of ras activity upon BCR-FcγRIIb co-clustering (Sármay *et al.*, 1996).

2.3 Tyrosine phosphorylation

SHIP has two potential tyrosine phosphorylation sites near to its C terminus, both containing the consensus sequences interacting with PTB domains (NPXY). These sites are tyrosine phosphorylated upon BCR clustering, which is enhanced when BCR and FcγRIIb are co-clustered (Saxton *et al.*, 1994; Smit *et al.*, 1994). The kinase responsible for the phosphorylation of SHIP appears to be lyn (Wang *et al.*, 1996; Sármay *et al.*, 1999). Several growth factors and cytokines will also induce the tyrosine phosphorylation of SHIP in cultured murine hemeatopoietic cell lines. In appropriate cells, tyrosine phosphorylation of the 140- to 150-kD SHIP has been detected after treatment with erythropoietin (Liu *et al.*, 1994), interleukin-2 (IL-2), IL-3, granulocyte-macrophage colony-stimulating factor (GM-CSF) (Matsuguchi, 1994), Kit ligand, M-CSF, thrombopoietin (Drachman *et al.*, 1995), and upon T-cell activation (Ravichandran *et al.*, 1993).

The tyrosine phosphorylated motifs and its N terminal SH2 domain might enable SHIP to link various proteins as an adaptor. As mentioned earlier, tyrosine phosphorylation of SHIP induces its interaction with Grb2, Shc and p62dok. SHIP may form a ternary complex with Shc and Grb2 in antigen receptor-stimulated B cells, namely, the PTB domain of Shc binds to NPXp-Y in SHIP, then Shc-SHIP interaction is stabilized by Grb2, whose SH3 domain binds to SHIP, and its SH2 domain binds to p-YVNV motif in Shc (Harmer and DeFranco, 1999). SHIP has been shown to associate with the Gab1 scaffolding protein, probably via Shc-Grb2 complexes (Lecoq-Lafon *et al.,* 1999). Tyrosine phosphorylation of SHIP plays an important role in its localization at the cell membrane and its PtdIns(3,4,5)P_3 hydrolysis. Thus dephosphorylation of the molecule mediated via the interaction with SLAM, or by SHP-2 might weaken the interaction with Shc/Grb2, resulting in the dissociation of the complex and the inactivation of SHIP (Mikhalap *et al.,* 1999; Koncz, *et al.,* 1999).

2.4 Function

Inositol 5'-phosphatase activity of SHIP regulates the level of various PtdIns-phosphates in the cell membrane, thus modulates the ability of PH domain containing proteins (Btk, PKB/Akt, PDK1, Vav, PLCγ1) to target the plasma membrane and to be activated (Rameh and Cantley, 1999, Huber *et al.,* 1999). SHIP has the potential to regulate many, if not all, PI3-K induced events including proliferation, differentiation, cell activation, cell movement and adhesion as well as apoptosis. Most of the molecules binding to PI3-K products on the cell membrane have a regulatory potential on cell survival: Btk phosphorylates PLCγ, thus regulates the intracellular [Ca^{2+}], while PDK activates Akt, and Akt may deliver anti-apoptotic signals by phosphorylating Forkhead (an inducer of Fas ligand), and Bad, a member of the Bcl 2 family (Anderson *et al.*, 1998, Kennedy, *et al.*, 1997). Akt also can prevent cytochrome c dependent proteolytic activation of caspase-9 (Cardone *et al.,* 1998). Removing 5-phosphate groups from PtdIns polyphosphates by SHIP reduces survival and induces apoptosis.

The *in vivo* function of SHIP was studied in cells overexpressing SHIP and in SHIP knockout mice. Transfecting DA-ER cells with wild type SHIP resulted at not more than twice endogenous SHIP levels, perhaps reflecting a reduced ability of these cells to survive. Even at this level of over-expression SHIP increased the rate of apoptosis in DA-ER cells at confluence. This might be explained by the SHIP mediated inhibition of ras activity, which is consistent with other results suggesting that ras pathway plays an important role in preventing apoptosis in IL-3 stimulated cells (Kinoshita *et al.,* 1995). Alternatively, SHIP might reduce cell viability directly by hydrolyzing PtdIns(3,4,5)P_3 (Liu *et al.,* 1997).

The SHIP knockout mice overproduce granulocytes and macrophages, due to the hyperresponsiveness of granulocyte and macrophage progenitors, indicating that SHIP is a negative regulator of hemeatopoietic cell proliferation/survival (Helgason *et al.,* 1998; Liu *et al.,* 1998). Liu *et al.,* (1999) have shown that SHIP-deficient mice exhibit dramatic chronic hyperplasia of myeloid cells resulting in splenomegaly, lymphadenopathy, and myeloid infiltration of vital organs. It was shown that neutrophils and bone marrow-derived mast cells (BMMC) from SHIP-/- mice are less susceptible to programmed cell death induced by various apoptotic stimuli or by growth factor withdrawal. Engagement of IL3-R and GM-CSF-R in SHIP-/- cells leads to increased and prolonged PI3-K-dependent PtdIns(3,4,5)P_3 accumulation and Akt/PKB activation. These data indicate that SHIP is a negative regulator of growth factor-mediated PKB activation and myeloid cell survival.

SHIP plays a critical role in regulating both B cell development and responsiveness to antigen stimulation (Brauweiler *et al.*, 2000; Helgason *et al.*, 2000). SHIP(-/-) mice have a reduced numbers of precursor and immature B cells in the bone marrow and manifest elevated serum immunoglobulin levels and an exaggerated IgG response to T cell-independent type 2 antigens. *In vitro*, purified SHIP (-/-) B cells exhibit enhanced proliferation in response to BCR stimulation in both the presence and absence of FcγRIIb1 coligation. This is associated with increased phosphorylation of both mitogen-activated protein kinase (MAPK) and Akt, as well as with increased survival and cell cycling. An accelerated development of transitional B cells and immature BMMC were also characteristics for SHIP -/- mice. SHIP-/- B cells exhibit an elevated PtdIns(3,4,5)P_3 level upon BCR stimulation compared with SHIP+/+ cells (Figure 3A). Since a small but significant conversion of PtdIns(3,4,5)P_3 to PtdIns (3,4)P_2 was observed in SHIP-/- cells (Figure 3B), these data suggested the presence of a second 5'inositol phosphatase, which may be identical with the recently described SHIP2 (Muraille *et al.*, 1999). However, the majority of the degradation of PtdIns(3,4,5)P_3 was mediated through SHIP, suggesting that SHIP plays a significant role in regulation of PtdIns(3,4,5)P_3 dependent BCR signaling pathways (Brauweiler *et al.*, 2000).

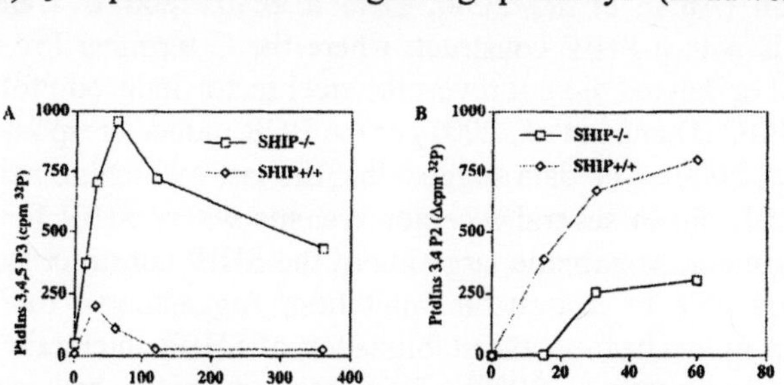

Figure 3. SHIP terminates signals generated upon BCR aggregation. Reproduced with permission from Brauweiler *et al.*, 2000, *J. Exp. Med.* **191**. © Rockefeller Univ. Press.

At high concentration of antigen and at an immature developmental stage of B cells BCR cross-linking results in apoptosis (Norvell *et al.*, 1995). In many cells SHIP mediated degradation of PtdIns polyphosphates promotes programmed cell death in response to BCR ligation. Recent data suggest that SHIP-/- mice are more resistant to BCR induced cell death, compared with the SHIP+/+ ones (Brauweiler *et al.*, 2000) (Figure 4).

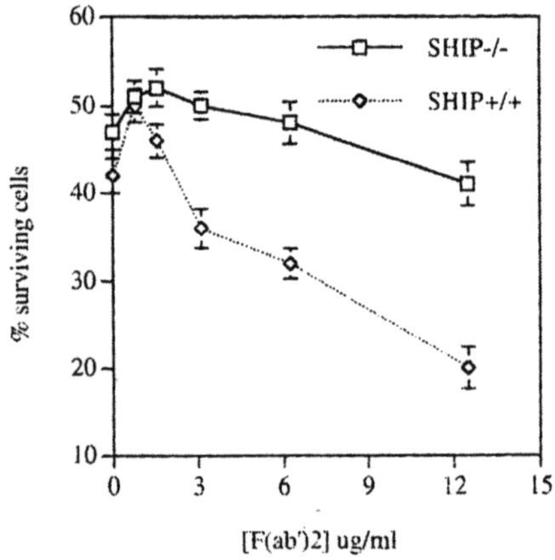

Figure 4. SHIP-/- splenic B cells are resistant to BCR mediated induction of cell death. B cells were treated with anti-Ig(Fab)$_2$ and surviving cells were detected after 12 h by propidium iodide staining. Reproduced with permission from Brauweiler *et al.*, 2000, *J. Exp. Med.* **191**. © Rockefeller Univ. Press.

Interestingly enough, recent data revealed that the catalytic domain alone is not sufficient, and SHIP's C terminus is essential for the hydrolysis of PtdIns(3,4,5)P_3 and for the inhibition of both B cell activation and mast cell degranulation (Aman *et al.,* 2000; Damen *et al.,* 2001). Transfection of SHIP-/- cells with a SHIP construct, where the C terminal Pro rich region was mutated or deleted did not revert the steel factor-induced PtdIns(3,4,5)P_3 level in BMMC (Damen *et al.,* 2001) or the BCR induced response in B cells (Aman *et al.,* 2000). The data suggest that the last 190 amino acids of SHIP play a critical role in several receptor systems where SHIP functions as a negative regulator. Membrane targeting of the SHIP construct lacking the C terminus was able to restore the inhibition, suggesting a role for the C terminal part in localization and stabilization of SHIPs interaction in the cell membrane (Aman *et al.,* 2000). Translocation to the cell membrane is essential for SHIP activity, but the binding structure is not identified yet. According to recent data SHIP may translocate to lipid rafts on the membrane (Petrie *et al.,* 2000), alternatively it might bind to scaffolding proteins, like Gab1/Gab2. In any case, the proline rich C terminal region may regulate the protein-protein interaction.

SHIP recruitment by the BCR cross-linked FcγRIIb was suggested to attenuate a pro-apoptotic signal initiated by the latter (Ono *et al.,* 1996).

FcγRIIb1 may also signal independently of BCR to directly mediate an apoptotic response, by the association of SHIP. Failure to recruit SHIP, either by deletion of SHIP or mutation of FcγRIIb1, results in enhanced FcγRIIb1-triggered apoptosis. Based on these data it was suggested that in the germinal center, where immune complexes are retained by follicular dendritic cells, FcγRIIb1 might actively participate in the negative selection of B cells having low affinity BCR (Ono, *et al.,* 1997). Selection of B cells is regulated by opposing signals generated by the interaction of immune complexes with the BCR and FcγRIIb1 through pathways modulated by SHIP. The SHIP effector in this case is Btk, which together with Syk, phosphorylates PLCγ leading to the rise of intracellular Ca^{2+}.

Aman *et al.* (1998) reported that SHIP inhibits Akt/PKB activation in B cells. Akt is activated by stimulation through BCR in a PI3-K–dependent manner, and co-clustering BCR and FcγRIIb1 inhibits this activation. Using mutants of FcγRIIb1 and SHIP-deficient B cells their data demonstrate that the p-ITIM of FcγRIIb by binding SHIP mediates the inhibition of Akt. The SHIP dependent inhibition of Akt activation suggests that PtdIns(3,4,5)P_3 plays a greater role in Akt activation than PtdIns(3,4)P_2 *in vivo*. Carver *et al.* (2000) have shown that SHIP inhibits Akt activation in B cells through regulation of Akt membrane localization. Akt activity is regulated by PDK, and PH domain of PDK binds to PtdIns(3,4,5)P_3 at the cell membrane. The results revealed that PDK is not a target of SHIP mediated inhibition and that SHIP inhibits Akt primarily through regulation of its membrane localization. They also observed that BCR-FcγRIIb1 cross-linking had no effect on $p70^{s6k}$ activation. Thus blocking membrane localization of Akt by SHIP upon co-clustering FcγRIIb1 with BCR will inhibit survival, and probably in collaboration with other signals, induce apoptosis.

3. SHIP2

A second SH2-containing inositol 5' phosphatase was cloned in 1997 (Pesesse *et al.,* 1997). SHIP2 is slightly larger than SHIP1 (160-150 kDa) and its central part shows a 64% amino acid identity with SHIP1. The overall structure is similar, SHIP2 possesses an N terminal SH2 domain, a central PtdIns phosphatase domain, only one NPXY motif and a C terminal proline rich region (Figure 2). Its substrate specificity is controversial, it hidrolyzes PtdIns(3,4,5)P_3 (Wisniewski *et al.,* 1999), but Pesesse *et al.* (1998) reported that it also hydrolyzes Ins(1,3,4,5)P_4. SHIP2 expression is not restricted to hemeopoetic cells, although it is co-expressed with SHIP1 in B and T cells, in spleen and thymus, however, exclusively SHIP2 is expressed in other tissues, such as heart and brain. SHIP2 contains a sterile alpha motif (SAM) which is an evolutionary conserved protein-binding domain of 60-70 residues. SAM has been found in diverse organisms, such as fungi, protozoans and animals. The possible function of SAM containing proteins is the regulation of the developmental process. Like SHIP1, SHIP2 is tyrosine phosphorylated and associates with Shc in response to growth factors, cytokines and BCR or TCR stimulation. Its tyrosine phosphorylation is the highest when BCR and FcγRIIb are cross-linked, and it may participate in down-regulation of PtdIns(3,4,5)P_3 level similarly to SHIP1 (Muraille *et al.,* 1999). SHIP2, just as SHIP1 could play a central role in determining PtdIns(3,4,5)P_3 and PtdIns(3,4)P_2 levels in many cell types. Regarding the binding molecules, SHIP2 binds selectively to SH3 domain of Abl, while does not react with SH3 of Grb2. This and other observations suggest that SHIP1 and SHIP2 may have a different hierarchy of binding SH3 containing proteins and therefore may modulate different signaling pathways and/or localize to different cellular compartments. SHIP1 and SHIP2 may be substrates for phosphorylation by different tyrosine kinases. The finding that both SHIP1 and SHIP2 are constitutively tyrosine phosphorylated in CML primary hemeatopoietic progenitor cells may have important implications in p210(bcr/abl)-mediated myeloid expansion (Odai *et al.,* 1997, Wisniewsky *et al.,* 1999). The different C terminal proline rich region of the two molecules might explain the differences between SHIP1 and SHIP2, and suggests that the function of the two molecules is not totally redundant. SHIP2 may dephosphorylate PtdIns(3,4,5)P_3, thus inhibit Akt/PKB, however SHIP2 cannot compensate the effect of SHIP1 in SHIP1 deficient B cells. Ishihara *et al.* (1999) reported that stably over-expression of SHIP2 in Rat1 fibroblasts inhibits insulin induced Akt/PKB activation and proliferation, suggesting that SHIP2 has a negative role in insulin induced mitogenesis/survival.

4. PTEN

4.1 Structure, expression and binding molecules

PTEN was cloned in 1997 by three independent groups (Li *et al.,* 1997; Steck *et al.,* 1997; Li and Sun, 1997). The gene is located in chromosome 10q23, a region frequently deleted in various tumors (rev. in Cantley and Neel, 1999). Mutation of PTEN is observed in human cancer with a similar rate as that of p53 tumor suppressor gene, indicating that it is involved in controling cell growth.

The gene encodes for a 43 kDa protein, containing a phosphatase domain (Figure 2). It shows an extensive homology with tensin, a cytoskeletal protein that interacts with actin filaments at focal adhesion, and auxilin, involved in synaptic vesicle transport. Overexpression of PTEN showed that it is primarily a cytoplasmic molecule, but some PTEN is capable to directly interact with focal adhesion kinase (FAK). PTEN also may dephosphorylate FAK, and thus it is involved in formation of focal adhesion, cell migration and integrin mediated spreading (Li *et al.*, 1997). The cDNA sequence of PTEN suggested that it is a member of the protein tyrosine phosphatase family, since it contained the "signature motif" (HCXXGXXRS/T) of these molecules and it exerted a tyrosine phosphatase activity. The sequence also suggested that PTEN is a dual specificity phosphatase, and its catalytic activity is essential for its biological function. Finally, the data of Maehama and Dixon (1998) revealed that PTEN is a relatively poor tyrosine phosphatase and its primary function is dephosphorylation of D3 phospholipids. Furthermore, the lipid phosphatase activity is required for the tumor suppressor effect of PTEN and for its role in normal development (Myers *et al.*, 1998).

Besides its catalytic domain PTEN contains a PDZ domain-binding site at its C terminus, thus interaction with PDZ proteins might potentially regulate PTEN activity. PDZ proteins have been shown to direct the assembly of multiprotein complexes at membrane/cytoskeletal interfaces, such as synapsis (Craven and Bredt, 1998). Recent findings of Wu *et al.* (2000) have shown that PTEN binds to a PDZ domain of a novel inverted membrane-associated guanylate kinase (MAGI3) that localizes to epithelial cell tight junctions. Importantly, the data of this group revealed that MAGI3 and PTEN cooperate to modulate the kinase activity of AKT/PKB. Based on these data the authors suggested that MAGI3 allows for the juxtaposition of PTEN to phospholipid signaling pathways involved with cell survival.

4.2 Function

Analysis of the PTEN gene revealed that the homozygotic inactivation of PTEN occurs in a large fractions of glioblastomas, furthermore, PTEN mutations are extremely common in melanoma cell lines, breast and prostate cancers, and endometrial carcinomas (Li *et al.*, 1997; Steck *et al.*, 1997). Germ-line mutations in PTEN cause rare autosomal dominant inherited cancer syndromes, such as Cowden disease (rev. in: Cantley and Neel, 1999) characterized by benign tumors in which differentiation is normal, but the cells are highly disorganized.

Immortalized embryonic fibroblasts from PTEN-/- mice were shown to have elevated PtdIns(3,4,5)P_3 level, constitutively elevated PKB/AKT activity and a decreased sensitivity to cell death in response to apoptotic stimuli (Stambolic *et al.*, 1998). Homozygotic mutation of PTEN induced early embryonic lethality in mice, demonstrating that PTEN has an essential role in normal development (Di Cristofano *et al.* 1998; Suzuki *et al.*, 1998; Podsypanina *et al.*, 1999). Heterozygotes showed an increased tumor incidence consistent to the definition of PTEN as tumor suppressor gene. However, different mutations resulted in different phenotypes (DiCristofano *et al.*, 1998). Together these data suggest that dephosphorylation of PtdIns(3,4,5)P_3 to PtdIns(4,5)P_2 by PTEN keeps PKB/Akt activity under control and this has a critical role in PTEN's function both as a tumor suppressor and as a regulator of normal development.

Lu *et al.* (1999) demonstrated that loss of PTEN function in breast cancer cells resulted in an increased basal phosphorylation of multiple components of the PI3-K signaling cascade as well as an increased duration of ligand-induced signaling through the PI3-K cascade. These alterations were reversed by wild-type but not by phosphatase inactive PTEN. Interestingly, high concentrations of serum enforced expression of PTEN inducing a predominant G1 arrest, while low serum concentration resulted in a marked increase in cellular apoptosis. This finding is consistent with the capacity of PTEN to alter the phosphorylation, and presumably function, of the apoptosis regulators: AKT, BAD, p70S6 kinase and GSK3.

Constitutive activation of the PI3-K - Akt/PKB "survival signaling" pathway is a likely mechanism by which many cancers become refractory to cytotoxic therapy. In LNCaP prostate cancer cells, the PTEN is inactivated, leading to constitutive activation of Akt/PKB and resistance to apoptosis. However, apoptosis and inactivation of Akt/PKB can be induced in these cells by treatment with PI3-K inhibitors. Surprisingly, androgen, epidermal growth factor, or serum can protect these cells from apoptosis, even in the presence of PI3-K inhibitors and without activation of Akt/PKB, indicating the activity of a novel, Akt/PKB-independent survival pathway. This pathway blocks apoptosis at a level prior to caspase 3 activation and release of cytochrome c from mitochondria (Carson *et al.*, 1999)

Hemizygous deletions of PTEN were found in approximately 40% of hemeatological malignancies, while a smaller number of samples, in particular non-Hodgkin's lymphomas, carried PTEN mutations. PTEN expression and the phosphorylation level of AKT/PKB were inversely correlated in the large majority of samples supporting the hypothesis that PTEN regulates PtdIns(3,4,5)P_3 level and suggests a role for PTEN in apoptosis (Dahia *et al.*, 1999). Biochemical abnormalities associated with the development of multiple myeloma have been difficult to define. Ge and Rudikoff (2000) have identified such a defect associated with lack of expression of PTEN. They have found that in myeloma cells, loss of PTEN results in a failure to dephosphorylate PtdIns(3,4,5)P_3, and a corresponding increase in Akt phosphorylation, a key event leading to inhibition of apoptosis. OPM-2 cells lacking PTEN expression had the highest level of Akt phosphorylation of eight lines examined. Ectopic expression of wild type PTEN inhibited Akt phosphorylation, which was correlated with an increase in apoptosis. The *in vivo* relevance of PTEN expression was demonstrated by injecting OPM-2 cells lacking of PTEN and transfected with wild type PTEN, respectively, into SCID mice. Tumors arose at an incidence of 100% in controls without PTEN, but only 50% (and of smaller size and longer latency) in low PTEN expressing clones. These results demonstrate that PTEN deletion/mutation may play an important role in tumor development in a subset of multiple myeloma patients.

The functions of integrins can be negatively regulated by PTEN. PTEN inhibits cell migration and invasion by directly dephosphorylating two key tyrosine-phosphorylated proteins, thereby antagonizing interactions of integrins with the extracellular matrix and integrin-triggered signaling pathways. Persad *et al.* (2000) demonstrated that the activity of integrin linked kinase (ILK) is constitutively elevated in PTEN-mutant prostate cancer cells, and transfection of wild-type (WT) PTEN into these cells inhibited ILK activity. Transfection of a kinase-deficient, dominant-negative form of ILK suppressed the constitutive phosphorylation of Akt/PKB on Ser-473, but not on Thr-308, a phosphorylation site in the activation loop. Transfection of dominant-negative ILK and WT PTEN also resulted in the inhibition of PKB/Akt kinase activity, furthermore, it induced G(1) phase cycle arrest and enhanced apoptosis (Figure 5). These results demonstrate that PTEN is a negative regulator of PI3-K mediated signaling from external stimuli to cell proliferation and survival through inactivating ILK and Akt/PKB.

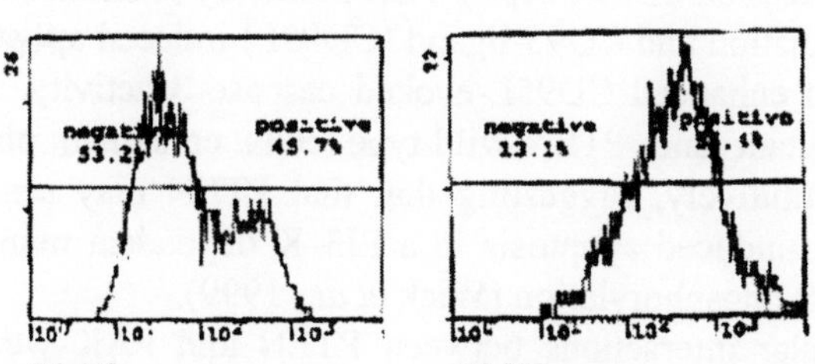

CONTROL PTEN

Figure 5. Induction of apoptosis in PC3 cells by PTEN. Cells were transplanted with empty vector (control) or PTEN, serum starved for 48h, and stained with annexin-FITC and 50µg/ml propidium iodide. Histogram shown is annexin-FITC staining of propidium iodide-negative cells. Reproduced with permission from Persad *et al.*, 2000, *Proc. Natl. Acad. Sci. U S A*, **97**. © National Academy of Sciences, USA.

PTEN, when ectopically expressed in human breast carcinoma cells, exhibited an inhibition of phosphorylation of both activating residues of PKB/AKT at Ser-473 and Thr-308 without any significant alteration of AKT expression, and transcriptionally repressed both exogenous and endogenous c-myc expressions in human breast carcinoma cells (Ghosh *et al.,* 1999). Furthermore, introduction of PTEN into human breast carcinoma cells induced apoptotic cell death and inhibited cell growth and tumor formation in nude mice. These data demonstrate that PTEN acts as a transcriptional repressor, inhibits the AKT-mediated cell survival signaling pathway, and negatively regulates human breast carcinoma cell growth.

In contrast to the findings obtained in breast, prostate and endometrium cancer and in hemeopoietic cell malignancies, PTEN expression level does not seem to be related in general to neuronal apoptosis (Kyrylenko *et al.,* 1999). However, the expression of PTEN seems to regulate certain apoptotic signals affecting PI3-K function. Transient expression study of PTEN in glioma cells indicated that PTEN plays an important role in cellular proliferation, tumorigenicity, cell migration, and focal adhesions. Cells stably expressing wild-type PTEN have suppressed proliferation and also showed higher expression of glial fibrillary acidic protein and changed morphologically from spindle-shaped to elongated cell bodies with multiple slender processes, suggesting that these cells have undergone differentiation. Additionally, telomerase activity decreased more than 10-fold in PTEN-expressing cells when compared with control cells. Apoptosis was detected in about 5% of PTEN-expressing cells, representing a 17-fold increase over the control cells. These results suggest that PTEN plays an important role in regulation of cell homeostasis by maintaining a balance between proliferation, differentiation, and apoptosis (Tian *et al.,* 1999).

Ectopic expression of wild type PTEN markedly sensitizes PTEN mutant gliomas to irradiation and CD95-ligand (CD95L) induced apoptosis. This is associated with enhanced CD95L-evoked caspase 3 activity. Interestingly, both PTEN-mutant and PTEN-wild-type cells contained phosphorylated Akt/PKB constitutively, suggesting that that PTEN may sensitise glioma cells to CD95L-induced apoptosis in a PI3–K dependent manner that may not require PKB phosphorylation (Wick *et al.*, 1999).

The molecular interactions between PTEN and FAK were studied in glioblastoma and breast cancer cells lacking PTEN. The PTEN trapping mutant bound wild-type FAK, requiring FAK autophosphorylation site Tyr397. In PTEN-mutated cancer cells, FAK phosphorylation was retained even in suspension after detachment from extracellular matrix, accompanied by enhanced PI3-K association with FAK and sustained PI3-K activity. PTEN-mutated cells were resistant to apoptosis in suspension, but most of the cells entered apoptosis after expression of exogenous PTEN or wortmannin treatment. These data indicate that FAK Tyr^{397} is important in PTEN interactions with FAK, and that PTEN regulates FAK phosphorylation and molecular associations after detachment from matrix, suggesting that PTEN negatively regulates the extracellular matrix-dependent PI3-K/Akt cell survival pathway in a process that can include FAK (Tamura *et al.*, 1999).

Programmed cell death plays an important role during lymphocyte development, by eliminating autoreactive cells, as well as in the effector phase of the immune response, when antigen induced cell death halt cell activation. Several groups have been studied the function of PTEN in the immune response. PTEN heterozygous (PTEN+/-) mutants develop a lethal polyclonal autoimmune disorder with features reminiscent of those observed in Fas-deficient mutants. Fas-mediated apoptosis was impaired in Pten+/- mice, and T lymphocytes from these mice show reduced activation-induced cell death and increased proliferation upon activation. PI3-K inhibitors restored Fas responsiveness in PTEN+/- cells. These results indicate that PTEN is an essential mediator of the Fas response and a repressor of autoimmunity, thus implicate the PI3-Kinase/Akt pathway in Fas-mediated apoptosis (DiCristofano *et al.*, 1999)

The chicken B cell line, DT40 cells undergo apoptosis in response to BCR cross-linking, while cells overexpressing Akt show a greatly diminished apoptotic response. By contrast, limiting the activation of Akt, either by inhibiting phosphatidylinositol 3-kinase or by ectopic expression of PTEN, results in a significant increase in the percentage of apoptotic cells after BCR cross-linking. Taken together, the data demonstrate that Akt plays an important role in B cell survival and that Akt is activated in a Syk-dependent pathway (Poque *et al.*, 2000).

Active (but not mutant) PTEN also decreased T cell receptor (TCR) induced activation of the mitogen-activated protein kinase ERK2 (extracellular signal-related kinase 2), as seen after inhibition of PI3-K (Wang *et al.*, 2000). PTEN may play a role in the regulation of T cell survival and TCR signaling by directly opposing PI3-K. Transgenic mice expressing in T lymphocytes an active form of PI3K developed an infiltrating lymphoproliferative disorder and autoimmune renal disease with an increased number of T lymphocytes exhibiting a memory phenotype and reduced apoptosis. This pathology was strikingly similar to that described in mice exhibiting heterozygous loss of the tumor suppressor PTEN. Overexpression of PTEN selectively blocks p65(PI3K)-induced 3T3 fibroblast transformation, while constitutive activation of PI3K extends T cell survival *in vivo*, affects T cell homeostasis, and contributes to tumor generation (Borlando *et al.*, 2000).

4.3 Therapeutic applications

Adenovirus-mediated gene transfer was carried out using endometrial cancer cell lines with completely inactivated PTEN, together with control cells expressing wild-type PTEN. The PTEN transgene significantly suppressed cell growth *in vitro* through induction of apoptosis in cells lacking wild-type PTEN. Furthermore, the ex vivo tumor formation was completely inhibited by the introduction of wild-type PTEN. However, neither regression nor progression was observed in inoculated tumors of either cell line by *in vivo* introduction of the PTEN gene. These results suggest that PTEN may be a good candidate for gene therapy in patients with endometrial carcinoma (Sakurada *et al.,* 1999).

PtdIns phosphatases regulate the level of membrane embedded second messenger phosphoinositids that are responsible for binding and activating PH domain containing signaling molecules involved in apotosis. SHIP activity determines the level of PtdIns(3,4)P_2 and PtdIns(3,4,5)P_3 in the cell membrane, and controls both proliferating and survival signals in hemeatopoietic cells (Figure 6). The commonly expressed PTEN, by dephosphorylating both key signal transduction lipids, PtdIns(3,4)P_2 and PtdIns(3,4,5)P_3 has a more profound and general effect on cell survival in many, although not all organs. In the absence of PTEN, the lipid signal transduction pathway activating Akt/PKB can protect tumor cells from apoptosis. Thus PTEN and the signaling pathways it regulates may provide novel targets for potential therapy against cancer.

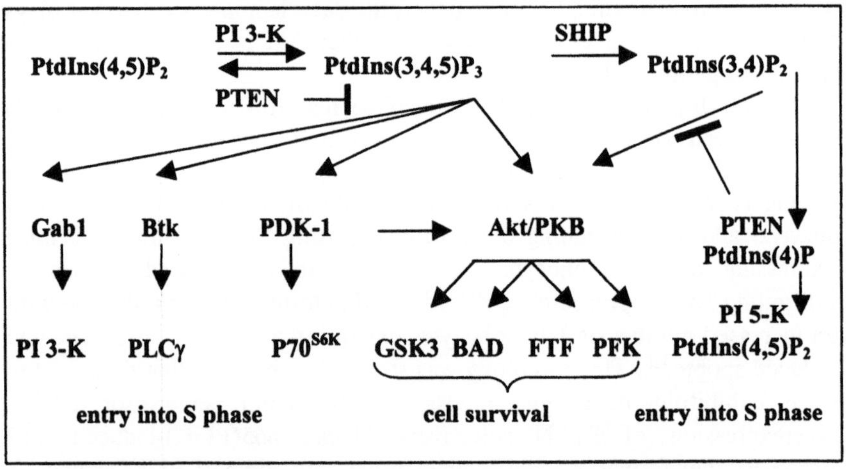

Figure 6. Regulation of cell cycling and survival by phosphoinositids.

REFERENCES

Aman, M.J., Lamkin, T.D., Okada, H., Kurosaki, T., and Ravichandran, K.S., 1998, The inositol phosphatase SHIP inhibits Akt/PKB activation in B cells. *J. Biol. Chem.* **273**: 33922-33928.

Aman, M.J., Walk, S.F., March, M.E., Su, H-P. Carver, D.J., and Ravichandran, K.S., 2000, Essential role for the C-terminal noncatalytic region of SHIP in FcRIIb1 mediated inhibitory signaling. *Mol. Cell. Biol.* **20**: 3576-3589.

Anderson, K.E., Coadwell, J., Stephens, L.R., and Hawkins, P.T., 1998, Translocation of PDK1 to the plasma membrane is important in allowing PDK-1 to activate protein kinase B. *Curr. Biol.*, **8**: 684-691.

Borlando, L.R., Redondo, C., Alvarez, B., Jimenez, C., Criado, L.M., Flores, J., Marcos, M.A. Martinez, A.C., Balomenos D., and Carrera, A.C., Increased phosphoinositide 3-kinase activity induces a lymphoproliferative disorder and contributes to tumor generation *in vivo*. (2000) *FASEB J.* **14**: 895-903.

Brauweiler, A., Tamir, I., Dal Porto, J., Benschop, R.J., Helgason, C.D., Humphries, R.K., Freed, J.H. and Cambier, J.C., 2000, Differential regulation of B cell development, activation, and death by the Src homology 2 domain-containing 5'inositol phosphatase (SHIP). *J. Exp. Med.* **191**: 1545-1554.

Cantley, L.C., and Neel, B., 1999, New insights into tumor suppression: PTEN suppresses tumor formation by restraining the phosphoinositide 3-kinase/Akt pathway. *Proc. Natl. Acad. Sci.* **96**: 4240-4245.

Cantrell, D.A., 2001, Phosphoinositide 3-kinase signaling pathways. *J. Cell Science,* **114**: 1439-1445.

Cardone, M.H., Roy N., Stennicke, H.R., Salvesen, G.S., Franke, T.F., Stanbridge, E., Frisch, S., and Reed, J.C., 1998, Regulation of cell death protease caspase-9 by phosphorylation. *Science*, **282**: 1318-1321

Carpenter, C. L., and Cantley, L. C., 1996, Phosphoinositide 3-kinase and the regulation of cell growth. *Biochim. Biophys. Acta*, **1288**: M11-16.

Carson, J.P., Kulik, G., Weber, M.J. 1999, Antiapoptotic signaling in LNCaP prostate cancer cells: a survival signaling pathway independent of phosphatidylinositol 3'-kinase and Akt/protein kinase B. Cancer Res Apr 59: 1449-1453

Carver, D.J., Aman, M.J., and Ravichandran, K.S., 2000, SHIP inhibits Akt activation in B cells through regulation of Akt membrane localization. *Blood,* **96**: 1449-1456.

Clement S, Krause U, Desmedt F, Tanti J.F, Behrends J., Pesesse X., Sasaki T., Penninger J., Doherty M., Malaisse W., Dumont J.E., Le Marchand-Brustel Y., Erneux C., Hue L., and Schurmans S., 2001 The lipid phosphatase SHIP2 controls insulin sensitivity. *Nature,* **409**: 92-97.

Chacko GW, Tridandapani S, Damen, J.E., Liu L., Krystal G., and Coggeshall, K.M., 1996, Negative signaling in B lymphocytes induces tyrosine phosphorylation of the 145-kDa isositol polyphosphate 5-phosphatase, SHIP. *J Immunol* **157**: 2234-2238.

Cockroft, S., and Thomas, G.M.H., 1992, Inositol lipid specific phospholipase C isoenzymes and their differential regulation by receptors. *Biochem. J.* **288**: 1-14.

Coggeshall, K.M., 1998, Inhibitory signaling by B cell FcγRIIb. *Curr. Op. Immunol.* **10**: 306-312

Craven, S.E., and Bredt, D.S., 1998, PDZ proteins organize synaptic signaling pathways. *Cell,* **93**: 495-498.

Dahia, P.L., Aquiar, R.C., Alberta, J., Kum, J.B., Caron, S., Sill, H., Marsh, D.J., Ritz, J., Freedman, A., Stiles, C., and Eng, C., 1999, PTEN is inversely correlated with the cell survival factor Akt/PKB and is inactivated via multiple mechanismsin hemeatological malignancies. *Hum. Mol. Genet.* **8**: 185-193

Damen, J.E., Liu, L., Rosten, P., Humphries, R.K., Jefferson, A.B., Majerus, P., and Krystal, G., 1996, The 145 kDa protein induce to associate with Shc by multiple cytokines is an inositol tetraphosohate and phosphatidylinositol 3,4,5-trisphosphate 5-phosphatase. *Proc. Natl. Acad. Sci. USA* 93: 1689-1693.

Damen, J.E., Liu, L., Ware, M.D., Ermolaeva, M., Majerus, P.W., and Krystal, G., 1998, Multiple forms of the SH2-containing inositol phosphatase, SHIP, are generated by C-terminal truncation. *Blood,* **92**: 1199-1205.

Damen, J.E. Ware, M.D., Kalesnikoff, J., Hughes, M.R. and Krystal, G., 2001, SHIP's C terminus is essential for ist hydrolysis of PIP3 and inhibition of mast cell-degranulation. *Blood,* **97**: 1343-1351.

Di Cristofano, A., Pesce, B., Cordon-Cardo, C., Pandolfi, P.P., 1998, Pten is essential for embryonic development and tumor suppression. *Nat. Genet.,* **19**: 348-355.

Divecha, N., and Irvine, R.F., 1995, Phospholipid signaling. *Cell,* **80**: 269-278.

Drachman, J.G., Griffin, J.D., and Kaushansky, K., 1995, The c-Mpl ligand (thrombopoetin) stimulates tyrosine phosphorylation of Jak2, Shc, and c-Mpl. *J. Biol. Chem.* **270**: 4979-4982.

Erneux, C., Goaverts, C., Communi, D., and Pesesse, X., 1998, The diversity and possible functions of the inositolpolyphosphate 5-phosphatase. *Biochim. Biophys. Acta* **1436**: 185-199.

Franke, T.F., Kaplan, D.R., and Cantley, L.C., 1997, PI3K: Downstream action blocks apoptosis. *Cell,* **88**: 435-437.

Furnari, F.B., Huang, H.J., Cavenee, W.K., 1998, The phosphatidylinositol phosphatase activity of PTEN mediates a serum-sensititve G1 growth arrest in glioma cells. *Cancer Res.* **58**: 5002-5008.

Ge, N.L., and Rudikoff, S., 2000, Expression of PTEN in PTEN-deficient multiple myeloma cells abolishes tumor growth *in vivo. Oncogene* **19**: 4091-4095.

Geier, S.J., Algate, P.A., Carlberg, K., Flowers, D., Friedman, C., Trask, B., and Rohrschneider, L.R. 1997, The human SHIP gene is differentially expressed in cell lineages of the bone marrow and blood. *Blood,* 89: 1876-1885.

Gold, M.R., Ingham, R.J., McLeod, S.J., Christian, S.L., Scheid, M.P., Duronio, V., Santos, L., and Matsuuchi, L., 2000, Targets of B cell antigen receptor signaling: the phosphatidylinositol 3-kinase/Akt/glycogen synthase kinase-3 signaling pathway and the Rap1 GTPase. *J. Exp. Med.* **176**: 47-68.

Ghosh, A.K., Grigorieva, I., Steele, R., Hoover, R.G., Ray, R.B., 1999, PTEN transcriptionally modulates c-myc gene expression in human breast carcinoma cells and is involved in cell growth regulation. *Gene,* **235**: 85-91.

Gupta, N., Scharenberg, A.M., Fruman, D.A., Cantley, L.C., Kinet, J-P., and Long, E.O., 1999, The SH2 domain containing inositol 5'-phosphatase (SHIP) recruits the p85 subunit of phosphoinositide 3-kinase during Fc□RIIb1-mediated inhibition of B cell receptor signaling. *J. Biol. Chem.* **274**: 7489-7494.

Harmer, S.L., and DeFranco, A.L., 1999, The Src homology domain 2-containing inositol phosphatase SHIP forms a ternary complex with Shc and Grb2 in antigen receptor stimulated B lymphocytes. *J. Biol. Chem.* **274**: 12183-12191.

Helgason, C.D., Damen, J.E., Rosten, P.,. Grewal, R., Sorensen, P., Chappel, S.M., Borowsky, A., Jirik, F., Krystal, G. and Humphries, R.K., 1998, Targeted disruption of SHIP leads to hemopoietic perturbations, lung pathology and a shortened lifespan. *Genes Dev.* **12**: 1610-1620

Helgason, C.D., Kalberer, C.P., Damen, J.E., Chappel, S.M., Pineault, N., Krystal, G., and Humphries, R.K., 2000, A dual role for Src homology 2 domain-containing inositol-5-phosphatase (SHIP) in immunity: aberrant development and enhanced function of B lymphocytes in ship -/- mice. *J Exp Med.* **191**: 5 781-5794.

Huber, M., Helgason, CD, Damen, J.E., Scheid, M., Duronio, V., Liu, L., Ware, M.D., Humphries R.K., and Krystal, G., 1999, The role of SHIP in growth factor induced signaling. *Progress Biophys. Mol. Biol.* (Pawson A.J. ed.) pp. 423-434. Elsevier Science, The Netherlands.

Ingham R.J., Holgado-Madruga, M., Siu, C., Wong, A.J., and Gold, M.R., 1998, The Gab1 protein is a docking site for multiple proteins involved in signaling by the B cell antigen receptor. *J. Biol. Chem.* **273**: 30630-30637.

Ishihara, H., Sasaoka, T., Hori, H., Wada, T., Hirai, H., Haruta, T., Langlois, W.J., and Kobayashi, M., 1999, Molecular cloning of rat SH2-containing inositol phosphatase 2 (SHIP2) and its role in the regulation of insulin siganling. *Biochim. Biophys. Res. Commun.* **260**: 265-272.

Kennedy, S.G., Wagner, A.J., Conzen, S.D., Jordán, J., Bellacosa, A., Tsichlis, P.N. and Hay, N., 1997, The PI3-kinase/Akt signaling pathway delivers an anti-apoptotic signal. *Genes Develop.* **11**: 701-713.

Kimura T., Sakamoto, H., Appella, E., and Siraganian, R.P. 1997. The negative signaling molecule SH2 domain containing inositol polyphosphate 5-phosphatase (SHIP) binds to the tyrosine phosphorylated β subunit of the high affinity IgE receptor. J. Bio.Chem. 272: 13991-13996.

Kinoshita, T., Yokota, T., Arai, K., and Miyajima, A., 1995, Suppression of apoptotic death in hematopoietic cells by signaling through the IL-3/GM-CSF receptors. *EMBO J.* **14**: 266-275.

Koncz, G., Pecht, I., Gergely, J., and Sármay, G., 1999, Fcγ receptor mediated inhibition of human B cell activation: the role of SHP-2 phosphatase. *Eur. J. Immunol.* **29**: 1980-1989.

Kuroiwa, A., Yamashita, Y., Inui, M., Yuasa, T., Ono, M., Nagabukuro, A., Matsuda, Y., and Takai, T., 1998, Association of tyrosine phosphatases SHP-1 and SHP-2, inositol 5-phosphatase SHIP with gp49B1 and chromosomal assignment of the gene. *J. Biol. Chem.* **273**: 1070-1074.

Kyrylenko, S., Roschier, M., Korhonen, P., Salminen, A. 1999, Regulation of PTEN expression in neuronal apotosis. *Brain Res. Mol. Brain Res.* **73**: 198-202.

Lecoq-Lafon, C., Verdier, F., Fichelson, S., Chretien, S., Gisselbrecht, S., Lacombe, C., Mayeux, P., 1999, Erythropoietin induces the tyrosine phosphorylation of Gab1 and its association with SHC, SHP2 SHIP ans phosphatidylinositol 3-kinase. *Blood,* **93**: 2578-2585

Lemmon, M.A., Falsca, M., Ferguson, K.M., and Schlessinger, J., 1997, Regulatory recruitment of signaling molecules to the cell membrane by pleckstrin homology domains. *Trends Cell. Biol.* **7**: 237-242.

Li, J., Yen, C., Liaw, D., Podsypanina, K., Bose, S., Wang, S.I., Puc, J., Miliaresis, C., Rodgers, L., McCombie, R., Bigner, S.H., Giovanella, B.C., Ittmann, M., Tycko, B., Hibshoosh, H., Wigler, M.H. and Parsons, R., 1997, PTEN, a putative protein tyrosine phosphatase gene mutated in human brain, breast and prostate cancer. *Science,* **275**: 1943-1947.

Li, D.M., and Sun H., 1997, TEP1, encoded by a candidate tumor suppressor locus, is a novel protein tyrosine phosphatase regulated by transforming growth factor. *Cancer Res.* **57**: 2124-2129.

Lioubin, M.N., Algate, P.A., Tsai, S., Carlberg, K., Aebersold, R., and Rohrschneider, L.R., 1996, $p150^{Ship}$, a signal transduction molecule with inositol polyphosphate-5-phosphatase activity. *Genes Dev.* **10**: 1084-1095.

Liscovitch, M. and Cantley, L.C., 1994, Lipid second messengers. *Cell,* **77**: 329-334.

Liu L, Damen JE, Cutler RL, and Krystal G., 1994, Multiple cytokines stimulate the binding of a common 145-kilodalton protein to Shc at the Grb2 recognition site of Shc. *Mol Cell Biol.* **14**: 6926-6935.

Liu, L., Damen, J.E., Hughes, M.R., Babic, I., Jirik, F.R., and Krystal, G., 1997. The Src homology 2 (SH2) domain of SH2-containing inositol phophatase (SHIP) is essential for tyrosine phosphorylatio of SHIP, its association with Shc and its induction of apoptosis. *J. Biol. Chem.* **272**: 8983-8988.

Liu, Q., Shalaby, F., Jones, J., Bouchard, D., and Dumont, D.J., 1998, The SH2-containing inositol poliphosphate 5-phosphatase, Ship, is expressed during hematopoiesis and spermatogenesis. *Blood,* **91**: 2753-2759.

Liu, Q., Sasaki, T., Kozieradzki, I., Wakeham, A., Itie, A., Dumont, D.J., and Penninger, J.M., 1999, SHIP is a negative regulator of growth factor receptor-mediated PKB/Akt activation and myeloid cell survival. *Genes Dev* **13**: 786-791.

Lu, Y., Lin, Y.Z., LaPushin, R., Cuevas, B., Fang, X., Yu, S.X., Davies, M.A., Khan, H., Furui, T., Mao, M., Zinner, R., Hung, M.C., Steck, P., Siminovitch, K., and Mills, G.B., 1999, The PTEN/MMAC1/TEP tumor suppressor gene decreases cell growth and induces apoptosis and anoikis in breast cancer cells. *Oncogene,* **18**: 7034-7045.

Maehama, T., and Dixon, J.E., 1998, The tumor suppressor, PTEN/MMAC1, dephosphorylates the lipid second messenger, phosphatidylinositol 3,4,5-trisphosphate. *J. Biol. Chem.* **273**: 13375-13378.

Matsuguchi T., Salgia R, Hallek M, Eder M, Druker B, Ernst TJ, and Griffin JD., 1994, Shc phosphorylation in myeloid cells is regulated by granulocyte macrophage colony-stimulating factor, interleukin-3, and Steel factor is constitutively increased by $p210^{BCR/ABL}$. *J. Biol. Chem.* **269**: 5016-5021.

Mikhalap, S.W., Shlapatska L.M., Berdova, A.G., Law, C.-L., Clark, E.A, and Sidorenko, S.P., 1999, CDw150associates with Src-homology 2-containinng inositol phosphatase and modulates CD95-mediated apoptosis. *J. Immunol.* **162**: 5719-5727.

Muraille, E., Pesesse, X., Kuntz, C., and Erneux, C., 1999, Distribution of the Src homology 2 domain-containing inositol 5-phophatase, SHIP-2 in both non-hemeatopoietic and hemeopoietic cells and possible invovlvement of SHIP-2 in negative signaling of B cells. *Biochem. J.,* **342**: 697-705.

Myers, M.P., Pass, I.,Batty, I.H., Van der Kaay, J., Stolarov, J.P., Hemmings, B.A., Wigler, M.H., Downes, C.P., and Tonks, N.K., 1998, The lipid phosphatase activity of PTEN is critical for its tumor supressor function.*Proc. Natl. Acad. Sci. USA*, **95**: 13513-13518.

Norvell, A., Mandik, L., and Monroe, J.G., 1995, Engagement of the antigen receptor on immature murine B lymphocytesresults in death by apoptosis. *J. Immunol.* **154**: 4404-4413.

Odai, H., Sasaki, K., Iwamatsu, A., Nakamoto, T., Ueno, H., Yamagata, T., Mitani, K., Yazaki, Y., and Hirai, H., 1997, Purification and molecular cloning of SH2 and SH3 containing inositol polyphosphat 5'phosphatase, which is involved in the signaling pathway of granulocyte macrophage coclony stimulating factor, erythropoietin, and Bcr-Abl. *Blood,* **89**: 2745-2756.

Ono, M., Bolland, S., Tempst, P., and Ravetch, J.V., 1996, Role of the inositol phosphatase SHIP in negative regulation of the immune system by the receptor FcγRIIB. *Nature,* **383**: 263-266.

Ono, M., Okada, H., Bolland, S., Yanagi, S., Kurosaki, T., and Ravetch, J.V., 1997, Deletion of SHIP or SHP-1 reveals two distinct pathways for inhibitory signaling. *Cell,* **90**: 293-301.

Osborn, M.A., Zenner,G., Lubinus M., Zhang X., Songyang, Z., Cantley, L.C., Majerus, P., Burns, P., and Kochan, J.P., 1996, The inositol 5-phosphatase SHIP binds to the mmunoreceptor signaling motifs and responds to high affinity IgE receptor aggregation. *J. Biol. Chem.* **271**: 29271-29278.

Oshimi, Y., and Miyazaki, S., 1995, Fas antigen mediated DNA fragmentation and apoptotic morphological changes are regulated by elevated cytosolic Ca^{2+} level. *J. Immunol.* **154**: 599-609

Pearse, R.N., Kawabe, T., Bolland, S., Guinamard, R., Kurosaki, T., and Ravetch, J.V., 1999, SHIP recruitment attenuates Fc gamma RIIB-induced B cell apoptosis. *Immunity,* **10**: 6753-6760.

Persad, S., Attwell, S. Gray, V., Delcomenne, M., Troussard, A., Sanghera, J., and Dedhar, S., 2000, Inhibition of integrin-linked kinase (ILK) suppresses activation of protein kinase B/Akt and induces cell cycle arrest and apoptosis of PTEN-mutant prostate cancer cells. *Proc. Natl. Acad. Sci. U S A,* **97**: 3207-3212.

Pesesse, X., Deleu, S., De Smedt, F., Drayer, L., Erneux, C., 1997, Identification of a second SH2-domain containing protein closely related to the phosphatidyl inositol polyphosphate 5-phosphatase SHIP. *Biochem. Biophys. Res. Commun.* **239**: 697-700.

Pesesse, X., Moreau, C., Drayer, A.L., Woscholski, R, Parker, P., and Erneux, C. 1998, The SH2 domain containing inositol 5-phosphatase SHIP2 displays phosphatidylinositol 3,4,5 trisphosphate and inositol 1,3,4,5 tetrakisphosphate 5-phosphatase activity. *FEBS Lett.* **437**: 301-303.

Petrie, R.J., Schnetkamp, P.P., Patel, K.D., Awashtikalia M., and Deans, J.P., 2000, Transient translocation of the B cell receptor and src homology 2 domain –containing inositol 5 phosphatase to lipid rafts evidence toward a role in calcium regulation. *J. Immunol.* **165**: 1220-1227.

Podsypanina, K., Ellenson, L.H., Nemes, A., Gu, J., Tamura, M., Yamada, K.M., Cordon-Cardo, C., Catoretti, G., Fisher, P.E., and Parsons, R., 1999, Mutation of PTEN/MMAC1 in mice causes neoplasia in multiple organ sytems. *Proc. Natl. Acad. Sci. USA* **96**: 1563-1568.

Poque, S.L., Kurosaki, T., Bolen, J., Herbst, R., 2000, B cell antigen receptor-induced activation of Akt promotes B cell survival and is dependent on Syk kinase. *J. Immunol.* **165**: 1300-1306.

Pumphrey N.J., Taylor, V.; Freeman, S.; Douglas, MR.; Bradfield, PF.; Young, SP.; Lord, JM.;Wakelam, MJ.; Bird, IN.; Salmon, M.; Buckley, CD., 1999, Differential association of cytoplasmic signaling molecules SHP-1, SHP-2 SHIP and phospholipase Cγ1 with PECAM1/CD31. *FEBS Lett.* **450**: 77-83.

Rameh, L.E., and Cantley, L.C., 1999, The role of phosphoinositide 3-kinase lipid products in cell function. *J. Biol. Chem.* **274**: 8347-83550.

Ravichandran, K.S., Lee, K.K., Songyang, Z., Cantley, L.C., Burn, P., and Burakoff, S.J., 1993, Interaction of Shc with the zeta chain of the T cell receptor upon T cell activation. *Science,* **262**: 902-905

Sakurada, A., Hamada, H., Fukushige, S., Yokoyama, T., Yoshinaga, K., Furukawa, T., Sato, S., Yaiima, A., Sato, M., Fujimura, S., and Horii, A., 1999, Adenovirus-mediated delivery of the PTEN gene inhibits cell growth by induction of apoptosis in endometrial cancer. *Int. J. Oncol.* **15**: 1069-1074

Sármay, G., Koncz, G. and Gergely, J., 1996, Human type II Fcγ receptors inhibit B cell activation by interacting with the $p21^{ras}$-dependent pathway. *J. Biol. Chem.* **271**: 30499-30504.

Sármay, G., Koncz, G., Pecht, I., and Gergely, J., 1999, Cooperation between SHP-2 phosphatidyl inositol 3-kinase and phosphoinositol5-phosphatase in the FcγRIIb mediated B cell regulation. *Imm. Lett.* **68**: 25-34.

Sattler, M., Salgia, R., Shrikhande, G., Verma, S., Choi, J.L., Rohrschneider, L.R., and Griffin, J.D., 1997, The phosphatidylinositol polyphosphate 5-phosphatase SHIP and the protein tyrosine phosphatase SHP-2 form a complex in hematopoietic cells which can be regulated by BCR/ABL and growth factors. *Oncogene,* **15**: 2379-2384.

Saxton, T.M., Oostween, I., Bowtell, D., Aebersold R., and Gold, M.R., 1994, B cell antigen receptor crosslinking induces phosphorylation of the ras activators SHC and mSOS1 as well as the assembly of complexes containing SHC, Grb2 and SOS1 and a 145 kDa tyrosine phosphorylated protein. *J. Immunol.* **153**: 623-636.

Sayos, J., Wu, C., Morra, M., Wang, N., Zhang, X., Allen, D., van Schaik, S., Notarangelo, L., Geha, R., Roncarolo, M.G., Oettgen, H., De Vries, J.E. and Terhorst, C., 1998, The X-linked lymphoproliferative disease gene product, SAP regulates signals induced through the co-receptor, SLAM. *Nature,* **395**: 462-464.

Smit, L., de Vries-Smits, A.M.M., Bos, J.L., and Borst, J., 1994, B cell antigen receptor stimulation induces formation of a Shc-Grb2 complex containing multiple tyrosine-phosphorylated proteins. *J. Biol. Chem.* **269**: 20209-20212

Stambolic, V., Suzuki, A., de laPompa, J.L., Brothers, G.M., Mirtsos, C., Sasaki, T., Ruland, J., Penninger, J.M., Siderovski, D.P., and Mak, T.W., 1998, Negative regulation of PKB/Akt–dependent cell survival by the tumor suppressor PTEN. *Cell,* **95**: 29-39.

Steck, P.A., Pershouse, M.A., Jasser, S.A., Yung, W.K., Lin, H., Ligon, A.H., Langford, L.A., Baumgard, M.L., Hattier, T., Davies, T., 1997, Identification of a candidate tumor suppressor gene, MMAC1, at chromosome 10q23.3 that is mutated in multiple advanced cancers. *Nat. Genet.* **15**: 356-362.

Suzuki, A., de la Pompa, J.L., Stambolic, V., Elia, A.J., Sasaki, T., del Barco Barrantes, I., Ho, A., Wakeham, A., Itie, A., Khoo, W., Fukumoto, M., and Mak, T.W., 1998, High cancer susceptibility and embryonic lethality associated with mutation of the PTEN tumor suppressor gene in mice. *Curr. Biol.* **8**: 1169-1178.

Tamir, I., Stlopa, J.C., Helgason, C.D., Nakamura, K., Bruhns, P., Daeron, M., and Cambier, J.C., 2000, The RasGAP-binding protein p62 dok is a mediator of inhibitory FcγRIIb signals in B cells. *Immunity,* **12**: 347-358.

Tamura, M., Gu, J., Danen, E.H., Takino, T., Miyamoto, S., Yamada, KM., 1999, PTEN interactions with focal adhesion kinase and suppression of the extracellular matrix-dependent phosphatidylinositol 3-kinase/Akt cell survival pathway. *J Biol Chem* **274**: 20693-20703.

Tian, X.X., Pang, J.C., To, S.S., Ng, H.K., 1999, Restoration of wild type PTEN expression leads to apaotosis, induces differentiation, and reduces telomerase activity in human glyoma cells. *J. Neuropathol. Exp. Neurol.* **58**: 472-479.

Tridandapani, S., Kelley, T., Cooney, D., Pradhan, M., and Coggeshall, K.M., 1997, Negative signaling in B cells: SHIP, Grbs, Shc. *Immunol. Today,* **18**: 424-427.

Yamanashi Y., Tamura, T., Kanamori, T., Yamane, H., Nariuchi, H., Yamamoto, T., and Baltimore, D., 2000, Role of the rasGAP-associated docking protein p62(dok) in negative regulation of B cell receptor mediated signaling. *Genes and Dev.* **14**: 11-16.

Vanhaesebroeck, B., Leevers, S.J., Panayotou, G., and Waterfield, M.D., 1997, Phosphoinositide 3-kinases: A conserved family of signal transducers, *Trends Biochem. Sci.* **22**: 267-272.

Wada T, Sasaoka T, Funaki M, Hori H, Murakami S, Ishiki M, Haruta T, Asano T, Ogawa W, Ishihara H, Kobayashi M. 2001, Overexpression of SH2-containing inositol phosphatase 2 results in negative regulation of insulin-*induced metabolic actions in 3T3-L1 adipocytes via its 5'-phosphatase catalytic activity. Mol. Cell. Biol.* **21**: 1633-1646.

Wang, J., Koizumi, T., and Watanabe, T., 1996, Altered antigen receptor signaling and impaired Fas-mediated apoptosis of B cells in Lyn-deficient mice. *J.Exp. Med.* **184**: 831-838.

Wang, X., Gjörloff-Wingren, A., Saxena, M., Pathan, M., Reed, J.C., and Mustelin, T. 2000. The tumor suppressor PTEN regulates T cell survival and antigen receptor signaling by acting as a phosphatidylinositol 3-phosphatase. *J Immunol.* **164**: 1934-1939.

Wick, W., Furnar, F.B., Naumann. U., Cavenee, W.K., Weller, M. 1999, PTEN gene transfer in human malignant glioma: sensitization to irradiation and CD95L-induced apoptosis. *Oncogene* **18**: 3936-3943.

Wisniewski D, Strife A, Swendeman S, Erdjument-Bromage H, Geromanos S, Kavanaugh WM, Tempst P, Clarkson B., 1999, A novel SH2-containing phosphatidylinositol 3,4,5-trisphosphate 5-phosphatase (SHIP2) is constitutively tyrosine phosphorylated and associated with src homologous and collagen gene (SHC) in chronic myelogenous leukemia progenitor cells. *Blood*, **93**: 2707-2720.

Wu, Y., Dowbenko, D., Spencer, S., Laura, R., Lee, J., Gu Q., Lasky L.A. Interaction of the tumor suppressor PTEN/MMAC with a PDZ domain of MAGI3, a novel membrane-associated guanylate kinase. 2000, *J. Biol. Chem.* **275**: 21477-21485

Zhang, X., and Majerus, P.W., 1998, Phosphatidylinositol signaling reactions. *Semin. Cell. Dev. Biol.* **9**: 153-160.

Index

MIX
Papier aus verantwortungsvollen Quellen
Paper from responsible sources
FSC® C105338

If you have any concerns about our products,
you can contact us on
ProductSafety@springernature.com

In case Publisher is established outside the EU,
the EU authorized representative is:
Springer Nature Customer Service Center GmbH
Europaplatz 3, 69115 Heidelberg, Germany

Printed by Libri Plureos GmbH
in Hamburg, Germany